The Power Supply Industry

Panos Konstantin · Margarete Konstantin

The Power Supply Industry

Best Practice Manual for Power Generation and Transport, Economics and Trade

Springer

Panos Konstantin
Burgstetten, Baden-Württemberg
Germany

Margarete Konstantin
Burgstetten, Baden-Württemberg
Germany

ISBN 978-3-319-72304-4 ISBN 978-3-319-72305-1 (eBook)
https://doi.org/10.1007/978-3-319-72305-1

Library of Congress Control Number: 2017964251

Printed on acid-free paper

This Springer imprint is published by the registered company Springer International Publishing AG part of Springer Nature
The registered company address is: Gewerbestrasse 11, 6330 Cham, Switzerland

Preface

My first book of the series "Best Practice Manual" with the title "**Power & Energy Systems Engineering Economics**" was dedicated to provide a concise yet very comprehensive coverage of engineering economics required for techno-economic evaluation of investments in the energy supply business.

This book of the same series with the title "**The Power Supply Industry**" mainly deals with technologies used for power and energy supply and evaluates their economics by applying concepts and knowledge built in the first book.

Core aim of the books is to transfer know-how of power and energy systems in a practical way rather than pure theoretical knowledge, thereby avoiding the detail of voluminous reference texts as needed by experts in specific fields. This is also demonstrated in numerous application examples and case studies derived from experience of real world projects. The examples and case studies are also available as softcopies on my website to assist readers practicing the books' contents.

The books are neither scientific papers nor literature research. In writing these books, I have drawn on my cumulative knowledge gained from more than 35 years of experience as a consultant in engineering and power economics for energy business projects worldwide and also from delivering numerous training courses to junior utilities' staff in several countries. It is my ambition after retirement to make my knowledge and experience available through practically oriented books applicable to real world scenarios.

Target audience of the books are primarily international consultants, staff members of engineering companies, utility personnel, energy economists and lawyers, as well as employees of government agencies entrusted with regulating the energy and utility sector and, finally, students in related fields of engineering and economics.

Although being a non-native English speaker, I have written these books in English because this is the most suitable language in this field among others as most techno-economic terms are available in English only. I ask native English readers for their understanding for any linguistic shortcomings.

Comments and recommendations for improvements from readers are highly appreciated and will be thankfully considered in forthcoming editions.

Burgstetten, Germany, October 2017 Panos Konstantin

Other practical books of the author:

Power and Energy Systems – **Engineering Economics**
SpringerVieweg, Germany, 2018

The book provides practical knowhow for appraisal and technical-economic evaluation of investments in the power and energy sector.

The book comprises eight chapters: Financial Mathematics, Inflation/Interest and Cost of Capital, Investment Appraisal, Financial and Economic Analysis, Introduction of Cost Allocation to Cogeneration Projects, Project Analysis under Uncertainties, Overview of Energy Markets and Price Mechanisms and finally case studies. The text part is supported by about 36 tables, 105 figures, 53 application examples and 13 case studies.

For German readers: **Praxisbuch Energiewirtschaft**[1], 4th Edition 2017, SpringerVieweg, Germany, ISBN 978-3-642-37264-4

Der Inhalt: Der Primärenergiemarkt, Beschaffung leitungsgebundener Energien, energierechtliche Rahmenbedingungen, Investitionsrechnung in der Energiewirtschaft, Physikalisch-technisches Grundwissen, Energieumwandlung und Emissionen, Kraftwerke – Technik & Kosten, Kraft-Wärme-Kopplung – Technik, Kostenaufteilung, Energietransport- und -verteilung, Abwicklung von Energieprojekten.

Der Textteil wird möglichst knapp gehalten und durch ca. 140 Tabellen und rund 160 Abbildungen ergänzt. Zum besseren Verständnis enthält das Buch auch ca. 80 praxisbezogene Beispiele.

For German readers: **Praxisbuch Fernwärmeversorgung.** In Bearbeitung, geplante Veröffentlichung Anfang 2018, SpringerVieweg.

Der Inhalt: Das Buch beginnt mit einem kurzen historischen Überblick und umfasst den Fernwärmenetzaufbau sowohl mit Heißwasser als auch mit Dampf als Wärmeträger, Leitungs-Verlegeverfahren, Fernwärme-Erzeugung durch Kraft-Wärme-Kopplung (KWK) in Kombination mit Spitzenlastkesseln, eine eingehende Behandlung der Kosten- und Aufwandaufteilung bei KWK, bis hin zur Fernwärme-Preisgestaltung für den Endkunden. Die Anwendungsbeispiele und Fallstudien im Buch stehen auch auf der Website des Autors als Softkopien in MS-Excel® zum Download zur Verfügung.

English Version „District Heating“ forthcoming in 2018.

---

[1]) In English: “Practice Oriented Book on Energy Economy”

Acknowledgments

First and foremost I am particularly thankful **to Fichtner GmbH & Co KG** in Stuttgart, Germany for their support and the opportunity to have access to their technical and human resources during my employment and beyond. The book mainly reflects the cumulative knowledge I have acquired and further developed from over 35 years' experience working for the Company as a consultant and trainer for energy business projects worldwide.

Many thanks are also due to the colleagues of **HelpDesk Görlitz GmbH**, Germany for their help in properly formatting the book.

I am grateful to Markus Groissböck, who has developed and maintains my Website and is always available for support.

I also acknowledge the support of many of my Fichtner colleagues, friends and clients for their advice and contribution to the development of this and previous books.

In particular, I wish to mention the following persons for reviewing chapters of the book: Adriana Mejia Gomez (Application examples and case studies), Dr. Ursula Haller (Modelling and development of KPRO®), Christian Mayr (Project development), Johannes Kretschmann and Christoph Scherer (Concentrated Solar Power), Ursula Mayr (Photovoltaics), Markus Schüller (Wind power), Dr. Liliana Oprea (Transmission & Distribution of Power), Till Aldinger (Electricity trading) and Arcady Greeshpoon (Nuclear power).

Many thanks also to Amy Gooderum and Maggie Konstantin for proofreading and linguistic revision of the book's text.

All my professional life as a consultant, I wrote hundreds of reports for projects and attained a certain routine in writing. I have furthermore greatly benefited from the experience in writing my first book "Praxisbuch Energiewirtschaft",[2] first published by Springer in 2006, and now available in its 4th edition in 2017 published by SpringerVieweg.

Last but not least I wish to thank Maggie Konstantin, my wife, for her support in editorial design and proofreading, and for her understanding for the long hours and evenings we have been spending in front of the computer.

[2] In English: "Practice Oriented Book on Energy Industry"

Downloads

Readers of the book can access the author's website under the addresses
www.PK-Energy-Practical-Knowhow.com or
www.PK-Energie-Praxiswissen.com
and download the following items and software tools[1]:

Softcopies in Excel® of all **Application Examples** and **Case Studies** included in the book

Software tool **FluidEXL,** for calculations of water/steam properties.
The developer, University for Applied Sciences – Zittau/Goerlitz/Germany, Department of Thermodynamics, Prof. Hans-Joachim Kretzschmar and his co-worker Matthias Kunick, make available the software tool exclusively for readers of this book, free of charge.

You will find a link for download on the example page on the author's website, along with the installation instruction and read me file. A license code is automatically sent by email after registration.

Use of the software for purposes other than for the book or commercial use requires a special license from the developer.

Software tool **KPRO®**, for modelling and performance simulation of power generation thermodynamic cycles and power & steam supply systems. The German Consulting Company Fichtner, Stuttgart, announced that they will make available the software tool for registered readers of the book for a period of six months, upon direct request. Note, however, that this is a highly professional tool and requires a strong background on thermodynamics of cycle calculations and in information technology. Again, commercial use requires a special license from Fichtner.

> **Note:** Purchasers of the book are highly advised to register in the author's website in order to be kept informed about updates and changes in the software

1) Brief instructions of the tools are available in the toolbox chapter of the book. Detailed instructions are available for download on the author's website.

Important notes on the chapters

Examples: All chapters contain numerous practical application examples. The examples as well as the case studies are intended to practice the contents of the book only and are not applicable for commercial use.

Download examples: Website www.pk-energy-practical-knowhow.com

Almost all examples and case studies are developed in MS-Excel® spreadsheets and inserted into the text as pictures. We tried to keep them relatively simple; nevertheless, it is not always easy to retrace the calculation steps because they often include complex calculation formulas. However, it is not possible to include these in the examples depicted in the chapters due to limited space. Readers have the opportunity to download softcopies of the examples from my website above.

Currencies: The book is written for an international audience in countries with different currencies. In formulas which are generally applicable, the term "CU" (currency unit) is used. In application examples, which are mainly derived from projects, either € (Euro) or US$ are used, depending on the origin of the projects. The real origin of the projects, however, is not disclosed.

Unit system: Throughout the book, the Standard International Unit System is used (based on MKS system: meter, kilogram, second). This system is based on physics, includes only a few basic units, and all the other units are derived from the basic units. The units are easy to handle in calculations without the need for conversions. In the European Union, its use is obligatory for public projects and in most countries it is the standard unit system.

Heating values: For energy balances, price references etc. the lower heating values LHV are used (also referred to in literature as net calorific values NCV or inferior heating value Hi). Worth mentioning is that natural gas is commonly traded based on its HHV and is to be converted in LHV.

Table of Main Chapters and Contents

Table of Contents

List of Tables

List of Figures

List of Examples

1 Basics of Technical Thermodynamics

1.1 Key Thermodynamic Concepts and Definitions

1.1.1 Energy, work and power

Technical thermodynamics deals with the processes of interconversion of energy forms into each other. It includes the behavior of related properties of substances involved in the process flow, e.g., pressure and temperature.

Energy E is defined as the ability to produce work; *Work W* is performed by moving a body with a force F at a distance l in the direction of the force ($W = F \times l$). *Power P* is work divided by the time the work has been performed ($P=W/t$).

Based on these definitions several forms of energy are distinguished:

- Mechanical energy may occur as kinetic energy (movement energy) or potential energy (energy of the height)
- Thermal energy is contained in energy carriers as steam, hot water, thermooil etc.
- Chemical energy is contained, e.g., in fossil fuels
- Nuclear energy is contained in the nucleus of atoms
- Radiation energy such as solar irradiation

A different classification is done according to the state of energy conversion and usage chain. There are following forms:

- Primary Energy
- Secondary energy including the sub forms
  - Final energy
  - Useful energy

Primary energy is extracted from stocks of natural resources through mining or exploration such as coal, uranium, crude oil and natural gas, or captured from natural energy flows such as solar radiation or wind. Primary energy has not undergone any conversion other than separation and cleaning.

P. Konstantin and M. Konstantin, *The Power Supply Industry*,
https://doi.org/10.1007/978-3-319-72305-1_1

Final energy is produced from primary energy through a conversion process. The conversion process may take place in a refinery, power generation plant or a different type of energy converter. Examples include oil products as light and heavy fuel oil or gasoline, natural gas, electricity. Other forms of final energy are blast furnace gas, converter gas, district heat or chilled water.

Final energy is converted in end-use appliances into *useful energy* such as electrical light, space heating or cooling, movement or rotation of tools.

1.1.2 Thermodynamic systems

A thermodynamic system is an imaginary confined volume of matter that is separated from the surrounding by its *system boundaries.* The boundaries may be fixed or movable. The system may have some exchange of mass or energy or both with its surroundings or neither of them – Figure 1-1.

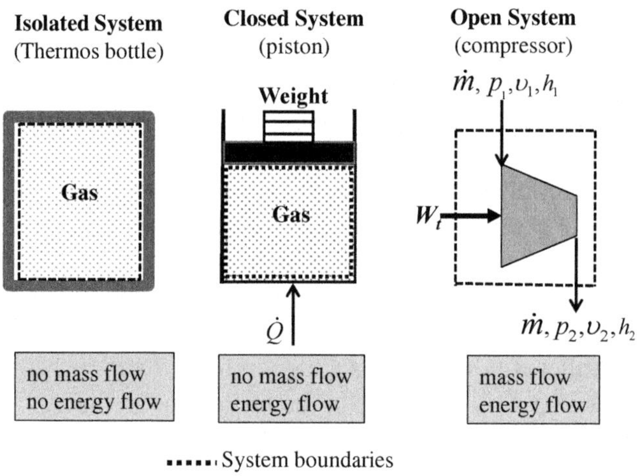

Figure 1-1: Examples of thermodynamic systems

The boundaries of thermodynamic systems are fixed in such a way that an analysis of mass and energy-exchange with the surroundings is feasible. In an open system, the space within the system boundaries is the control volume.

1.1.3 The Standard International Unit System

The Standard International (SI) Unit System comprises seven quantities and their *base units* are shown in Table 1-1. All the other units are derived from these base units by applying laws and principles of the physics – Table 1-2.

Table 1-1: SI base units

Quantity		Base Unit	
Name	Symbol (*in italics*)	Name	Unit (in standard)
Length	l	Meter	m
Mass	m	Kilogram	kg
Time	t	Second	s
Thermodynamic temperature	T	Kelvin	K
Amount of substance	n	Mole	mol
Electric current	I	Ampere	A
Luminous intensity	L	Candela	Cd

Table 1-2: SI Selected derived units used in thermodynamics

Quantity	Symbol Unit	Definition
Speed Velocity	υ, w, c m/s	Displacement of a body in a distance l m in a time t in seconds. $\upsilon = l/t$
Acceleration	$\boldsymbol{a}$ m/s^2	The change of velocity in m/s per s $\boldsymbol{a} = \upsilon / t$
Force	$\boldsymbol{F}$ N - Newton ($1 N = 1 kg{\cdot}m/s^2$)	1 N is the force which, when applied to a body having a mass of 1 kg, gives it an acceleration s of 1 m/s^2 $\boldsymbol{F}=\boldsymbol{m} \cdot \boldsymbol{a}$
Work, Energy	$\boldsymbol{W}$ J - Joule J=N·m=W· s ($1 J = 1 kg \cdot m^2/s^2$)	Work is force multiplied by the displacement in the direction of the force. $\boldsymbol{W}=\boldsymbol{F} \cdot \boldsymbol{l}$
Power	$\boldsymbol{P}$ W - Watt W = J/s ($1 W= 1 kg \cdot m^2/s^3$)	Power is the rate of work divided by the time it is done. $\boldsymbol{P} = \boldsymbol{W} / \boldsymbol{t}$
Pressure	$\boldsymbol{p}$ Pa - Pascal $Pa = N/m^2$ ($1 Pa = kg \cdot / (m s^2)$)	Force divided by the area of its application: $\boldsymbol{p} = \boldsymbol{F/A}$ Gravity pressure: $\boldsymbol{p}= \boldsymbol{\rho} \cdot \boldsymbol{g} \cdot \boldsymbol{h}$

Note: A more detailed list of SI derived units, their prefixes, decimals and multiples as well as conversion tables are shown in **Annex 1** and **Annex 2**.

1.1.4 Definitions and rules

Rules for writing quantities and units (see Table 1-1 and Table 1-2):

- Symbols of quantities are written in italics, e.g.: *l, m, T, F, P, υ*
- Units are written in standard characters, e.g.: m, kg, K, N, kW, m/s
- Quantities are denoted as a product of their symbol, equal sign, value and unit with a space in between e.g.: *P*=100 kW, *p*=200 Pa
- Compound units are denoted as a product or quotient, e.g.: N⋅m or Nm, m/s, kg⋅m/s^2
- Subscripts might be used to distinguish quantities which have the same unit, e.g.: electrical power P_e, thermal power P_t

Mass, molar mass and amount of substance: Mass and amount of substance are related quantities. The symbol for mass is "*m*" and its base unit "kg". The symbol of amount of substance is "*n*" and its base unit "mol" where $n = m/M$ [mol]. *M* is the symbol for *molar mass* with the unit kg/kmol. This is a specific physical property of substance referred to chemical elements or compounds **Annex 4.** The molar mass of compounds is calculated from that of their elements (the unit kmol instead of mol is common):

M kg/kmol: Carbon C=12; Oxygen O=16; Hydrogen H=1

M kg/kmol: Oxygen O_2= 32; Carbon dioxide CO_2=44; Water H_2O=18

Temperature: SI base unit for *thermodynamic temperature* and for temperature difference is Kelvin "K" denoted with the symbol "*T*" and "*ΔT*" respectively. In practice temperature is metered in degrees Celsius °C, the symbol is the small *t* (or *θ*). Where:

$$T = t\ °C + 273.15\ K;\quad 0\ °C = 237.15\ K;\quad \text{or}\quad 0\ K = -\ 237.5\ °C$$

The temperature $T = 0$ K (respectively -237.5 °C) is the lowest feasible temperature and is characterized in physics as the absolute zero temperature.

In the USA and some other countries the customary unit degrees Fahrenheit °*F* is used for temperature (non-SI unit).

$$°F = 9/5 \times °C + 32;\ °C = 5/9 \times (°F - 32)$$

Pressure: SI base unit for *pressure p* is Pascal [N/m^2]. In practice the unit [bar] is used.

$$1\ \text{bar} = 10^5\ \text{Pa} = 100\ \text{kPa} = 0.1\ \text{MPa}$$

Pressure is always measured as *over pressure* or *gauge pressure* "p_o, *or* p_g" on top of the atmospheric or barometric pressure p_b. For thermodynamic calculations the *absolute pressure* is used: $p = p_b + p_o$.

In the USA and some other countries the customary unit "psi" is used (pound per square inch – psia, psig). This is a non-SI unit.

$$1\,\text{bar} \approx 14.5\ \text{psi};\ 1\ \text{psi} \approx 0.069\ \text{bar}$$

Conversion tables for SI-units to imperial units are listed in **Annex 3**.

1.1.5 Properties and quantities, definitions, symbols and units

A *pure substance* is a matter[4] that has a defined composition, and its physical and chemical properties are uniform and constant. *Mixtures* can consist of two or more pure substances. In thermodynamics, pure substances or mixtures are characterized by their thermal and energetic properties. They are constituents of thermodynamic systems.

Properties are intrinsic characteristics for the actual state of the substance or a system and can be observed, metered and quantified. *Quantities* express the magnitude of properties and have a number and a unit.

Thermal properties are: temperature T [K], pressure p [Pa, bar], mass m [kg], Volume V [m^3] and density ρ [kg/m^3] or specific volume υ [m^3/kg]. For solid matter and for liquids usually the density ρ in kg/m^3 is used, while for gases the specific volume υ in m^3/kg is more common. Thermal properties are independent from the mass; they are also known as *intensive properties*.

Energetic properties are: internal energy, enthalpy, entropy and exergy. In absolute terms their symbols are specified with capitals, while mass related symbols are specified with small characters: internal energy U [kJ] or u [J/kg], enthalpy H [kJ] or h [kJ/kg], entropy S [kJ/K] *or* s [kJ/kg K] and exergy E [kJ] or e [J/kg]. The value of energetic properties depends on the mass included in a system. They are also known as *extensive properties*.

Process properties are heat Q [kJ] and mechanical work W [kW]

Time rated properties (flows) are denoted with a dot on top of the symbol:

$$\dot{m}\left[\frac{kg}{s}\right],\quad \dot{V}\left[\frac{m^3}{s}\right],\quad \dot{Q}\left[\frac{\text{kJ}}{\text{s}}\right],\quad \dot{H}\left[\frac{\text{kJ}}{\text{s}}\right]$$

Equations describing the correlations between system properties are called *equations of state*. Thermal and energetic properties of different substances are shown in **Annex 4.**

---

[4] Matter is everything that has a mass and occupies space. It can occur in solid, liquid and gaseous state.

1.1.6 Mass, weight and acceleration of gravity

Mass and weight are different properties. The *mass* of matter is equal at any place on earth (or even on the moon), its unit is kg: In contrary, *weight* is the force of gravity; it is defined as mass multiplied with the acceleration of gravity; it has the unit of a force 1 N.

$$\text{Weight: } F_G = m[\text{kg}] \times g\left[\text{m/s}^2\right] \ [N]$$

The *acceleration of gravity* depends on the latitude and altitude of a certain place. At the equator at sea level (altitude 0) it is 9.78 m/s^2, while in the poles it is 9.932 m/s^2. For technical calculations the value of 9.81 m/s^2 is used. This corresponds to the acceleration of gravity at latitude of 45° and sea level.

1.1.7 Normal reference conditions

Ambient temperature, atmospheric pressure and humidity are different in different places on earth and change also during the time of the day and elevation. Therefore it is necessary to define standard reference conditions in order to enable testing and comparison of the performance of machinery and to define amounts of matter and of commodities (e.g. natural gas). There are different definitions of normal conditions. The most common is the definition according DIN 1343 commonly used in Continental Europe and most other countries:

Table 1-3: Reference normal conditions, DIN 1343

Quantity	Symbol	State
Temperature	T_n	273.15 K (or 0 °C)
Pressure	p_n	1.01325 bar (1 atm)
Relative humidity	φ_n	0%
Normal volume *)	V_{mn}	22.41383 m^3/kmol

*) referred to ideal gas

In *Imperial and US customary* system, normal conditions are defined:
Temperature: 60 °F (15.6°C), pressure 14.969 psia (1atm). The reference temperature is different compared to the DIN definition. Hence, also the normal conditions are not identical.

ISO conditions of gas turbines: The performance parameters of gas turbines are defined according to ISO for the following reference conditions:

- Ambient temperature: 15°C
- Ambient pressure: 1.013 bar
- Relative humidity: 60%

1.2 The Principal Laws of Thermodynamics

1.2.1 The First Law of thermodynamics

The First Law of thermodynamics states that energy can neither be created nor be destroyed, but one form of energy can be converted into another form. The total energy associated with an energy conversion remains constant. The First Law is also known as the law of energy conservation. It is applied differently for closed and open systems.

The energy balance for a **closed system** that undergoes a change of state from its initial state 1 to the final state 2 is according to First Law:

$$Q_{12} + W_{12} = U_2 - U_1 \qquad (1.1)$$

The heat Q_{12} and work W_{12}, transferred into a system, are utilized to increase its internal energy U_2-U_1. *Internal energy* U [kJ] *or* u [kJ/kg] *is* the total energy contained in a closed system after the state of equilibrium is reached. Thereby the following conventions apply:

- Heat transfer into the system (Q> 0, +), out of the system (Q< 0, -)
- Work transfer into the system (W> 0, +), out of the system (W< 0, -)

The work is defined $\boldsymbol{W_{12} = -p \cdot (V_1 - V_2) = W_{V12}}$ as the *volume expansion work* (note: subscript v) also known as *pressure-volume (PV)* work; it is utilized in a closed system with movable boundaries to increase its volume (see Figure 1-1). The equation (1.1) transformed becomes:

$$Q_{V12} = U_2 - U_1 + p \cdot (V_1 - V_2) \quad \text{or} \quad dQ_V = dU + p \cdot dV \qquad (1.2)$$

For V constant is dV=0, and the above equation becomes:

$$Q_{V12} = U_2 - U_1 = m \cdot c_\upsilon \cdot (T_2 - T_1) \qquad (1.3)$$

Where: $\boldsymbol{c_\upsilon}$ [J/(kg K)] is *the specific heat capacity* at a constant volume. This is the amount of heat required to raise the temperature of one kg of a substance by 1 K (values of c_υ see **Annex 5**).

For **open systems** that also include mass flow, the enthalpy is preferred instead of internal energy. *Enthalpy* denoted with H or h is defined:

$$H = U + p \cdot V \ [\text{kJ}] \text{ or in specific form: } h = u + p \cdot \upsilon \left[\frac{\text{kJ}}{\text{kg}}\right] \qquad (1.4)$$

The energy balance for an open system that undergoes a change of state from its initial state 1 to the final state 2 is according to First Law:

$$Q_{12} + W_{S12} = H_2 - H_1 \,[\mathrm{kJ}] \quad \text{or} \quad q_{12} + w_{S12} = h_2 - h_1 \left[\frac{\mathrm{kJ}}{\mathrm{kg}}\right] \tag{1.5}$$

Where $W_{S12} = \int_1^2 V \cdot dp$ is *the shaft work.* This is the work transferred into or out of a system by a shaft. It is also known as *technical work.* Thereby equation (1.5) becomes:

$$dH = dQ + V \cdot dp \tag{1.6}$$

For constant pressure is $dp = 0$, $W_{S12} = 0$ and equation (1.5) becomes:

$$Q_{12} = H_2 - H_1 = m \cdot c_p \cdot (T_2 - T_1) \tag{1.7}$$

where c_p is the *specific heat capacity at constant pressure.* The following conventions and relations exist:

- Specific capacity for solids and liquids is practically constant $c_p = c_v = c$
- Specific capacities of gases depend on the temperature
- Relation $\boldsymbol{c_p = c_\upsilon + R}$

Specific heat capacities for different substances see **Annex** 5.
For an open system mostly *time rates* are used. Equation (1.5) expressed in time rates, becomes:

$$\dot{Q}_{12}\left[\frac{\mathrm{kJ}}{\mathrm{s}}\right] + P\,[\mathrm{kW}] = \dot{m} \cdot \left(h_2 - h_1\right)\left[\frac{\mathrm{kJ}}{\mathrm{s}}\right] \tag{1.8}$$

Note: kJ/s = kW the former unit is used for heat, the latter for power. Instead, it can also be written $\mathrm{kW_m}$ ($_\mathrm{m}$ for mechanical) and $\mathrm{kW_t}$ ($_\mathrm{t}$ for thermal). Time rates of quantities are denoted with a dot on the symbol, e.g.:

$$\dot{Q}[\mathrm{kJ/s}], \quad \dot{H}[\mathrm{kJ/s}], \quad \dot{m}[\mathrm{kg/s}], \quad \dot{V}\left[\mathrm{m^3/s}\right]$$

1.2.2 The Second Law of thermodynamics

The First Law of thermodynamics is known as the law of the conservation of energy; it states that all forms of energy are equivalent and can be mutually converted one into another as long as energy is not created or destroyed.
We know from experience, however, that mutual conversion of different energies is limited and that real processes follow a distinct direction, namely:

- While mechanical energy can be fully converted to thermal energy, heat can never be fully converted to mechanical energy
- Heat cannot be transferred by itself from a lower (heat sink) to a higher temperature level (heat source)
- Mass cannot be transferred by itself from a lower to a higher pressure level
- Work cannot be created solely from internal energy of a system

The Second Law of thermodynamics, in contrary to the First Law, restricts the convertibility of energy forms by introducing the *privileged process direction* and the irreversibility of processes.

The Second Law distinguishes between reversible and irreversible processes. For *reversible* processes it is deemed that the initial state of the system can be fully restored solely by changing the direction without any energy expense. In contrary, *irreversible* processes occur in one direction only, and the initial state can only be restored by expense of energy. The Second Law is also known as the *law of the preferential process direction.*

The Second Law further states that there is a property of state known as *entropy*, denoted with the symbol S [kJ/K] or s [kJ/kg K], that in a closed system undergoing a change of state never decreases. All natural processes are irreversible; they are characterized by entropy increase. Entropy is mathematically defined as the ratio of heat transfer to the thermodynamic temperature of a system.

$$dS = \frac{dQ}{T} \geq 0 \quad \left[\frac{\mathrm{kJ}}{\mathrm{K}}\right] \tag{1.9}$$

The following conventions apply for thermodynamic systems undergoing a process:

- $dS <$ this is a technically not feasible process
- $dS =$ this is an ideal, reversible process
- $dS> =$ this is a technically feasible, real, irreversible process

Under consideration of equation (1.9) the main equations of thermodynamics (1.2) and (1.6) become:

$$dU = T \cdot dS - p \cdot dV \quad \text{and} \quad dH = T \cdot dS + V \cdot dp \tag{1.10}$$

The entropy is an essential parameter for cycle calculations with Mollier h-s diagrams.

1.3 Gas Thermodynamics

1.3.1 The equation of state of the ideal gas

The state of a gas is defined by the properties of state pressure *p*, volume *V* and thermodynamic temperature *T*. The equation of state of the ideal gas is:

Specific form: $$p \cdot \upsilon = R_i \cdot T \quad \text{or} \quad p \cdot V = m \cdot R_i \cdot T \tag{1.11}$$

Molar form: $$p \cdot V = n \cdot \bar{R} \cdot T \quad \text{or} \quad p \cdot \bar{\upsilon} = \bar{R} \cdot T \tag{1.12}$$

Where:
P: Pressure [N/m²]
V: υ[m³/kg]: Gas volume, specific volume [m³]
T: Thermodynamic temperature [K]
R_i: Specific gas constant of the gas *i* [kJ/kg· K]:
$\bar{R}$: Universal gas constant [KJ/kmol· K]
m: Mass of the gas *i* [kg]
$n=m/M_i$ Amount of substance [kmol], M_i Molar mass of the gas *i* [kg/kmol]
$\bar{\upsilon}$ Molar volume [m³/kmol]

Ideal gases are gases which satisfy the above equations (1.11) and (1.12) !whereas r*eal gases* satisfy the above equations only approximately. For technical applications, however, with atmospheric air or flue gases and some other technical gases the margin of accuracy is sufficient.

Universal gas constant $\bar{R} = M \cdot R = 8.31451 \left[\frac{\text{kJ}}{\text{kmol} \times \text{K}} \right]$

Molar volume: $$\bar{\upsilon} = \frac{\bar{R} \cdot T}{p} = \frac{M \cdot R \cdot T}{p} = M \cdot \upsilon \quad \left[\frac{\text{m}^3}{\text{kmol}} \right] \tag{1.13}$$

The specific molar volume of all ideal gases at the same temperature and pressure is the same.

The molar volume $\bar{\upsilon}_n$ of all ideal gases for normal conditions (273 K, 1.01325 bar)[5] is 22.4141 m³/kmol:

$$\bar{\upsilon}_n = \frac{\bar{R} \cdot T_n}{p_n} = \frac{8.314 \cdot 10^3 \cdot 273.15}{1.01325 \cdot 10^5} \left[\frac{\text{kg} \cdot \text{m}^2 \cdot \text{K} \cdot \text{m} \cdot \text{s}^2}{\text{s}^2 \cdot \text{kmol} \cdot \text{K} \cdot \text{kg}} \right] = 22.4141 \left[\frac{\text{m}^3}{\text{kmol}} \right] \tag{1.14}$$

The specific volume can be obtained from equations (1.11) or (1.14) and is:

[5] see Table 1-2 for unit control

$$\upsilon_i = \frac{V}{m} = \frac{R_i.T}{p_i} = \frac{\bar{\upsilon}}{M_i}\left[\frac{\text{m}^3}{\text{kg}}\right] \quad (1.15)$$

Important Note: Thermodynamic calculations include parameters with complex compound units. It is therefore highly recommended and necessary to conduct unit control as it is done in equation (1.14).

Table 1-4: Molar mass and gas constant of selected technical gases

Gas	Exact Value		Technical Applications	
	Molar mass	Gas constant	Molar mass	Gas constant
	kg / kmol	Nm / kg K	kg / kmol	Nm / kg K
Atmospheric air	28.964	287.1	29.0	287.0
Hydrogen H_2	2.016	4,124.4	2.0	4,124.4
Nitrogen N_2	28.013	296.8	28.0	297.0
Oxygen O_2	31.999	259.8	32.0	260.0
Carbon dioxide CO_2	44.010	188.9	44.0	189.0
Water vapor (super heated)	18.015	461.5	18.0	461.5

Example 1-1: Specific volume ideal gases

Find the molar volume of ideal gases at a pressure of 5 bar and 120 °C (393 K) and the specific volume of Oxygen (32 kg/kmol).

$$\bar{\upsilon} = \frac{\bar{R}\cdot T}{p} = \frac{8,314\left[\frac{\text{N m}}{\text{kmol}\times\text{K}}\right]\cdot 393.15\,[K]}{5\cdot 10^5\left[\frac{\text{N}}{\text{m}^2}\right]} = 6.54\left[\frac{\text{m}^3}{\text{kmol}}\right]$$

$$\upsilon = \frac{\bar{\upsilon}}{M} = \frac{6.54\left[\frac{\text{m}^3}{\text{kmol}}\right]}{32\left[\frac{\text{kg}}{\text{kmol}}\right]} = 0.204\left[\frac{\text{m}^3}{\text{kg}}\right]$$

Example 1-2: Mass content of a gas tank

Two tanks of 3 m^3 each contain hydrogen and oxygen at a pressure of 25 bar_a. Find the mass content in the tanks. Conduct also unit control[6].

[6] see Table 1-2 for unit control

Item		Unit	Hydrogen	Oxygen
Given:				
Tank content		m^3	3.0	3.0
Pressure	25 bar	N/m^2	2.50E+06	2.50E+06
Temperature	20 °C	°C	293.15	293.15
Gas constant		Nm/ (kg K)	4124.4	259.83
Sought				
Mass $m = p \cdot V/R \cdot T$	Tn=273.15	kg	6.20	98.47

$$m = \frac{p\left[\mathrm{N/m^2}\right] \cdot V\left[\mathrm{m^3}\right]}{R\left[\mathrm{Nm/kg\,K}\right] \cdot T\left[\mathrm{K}\right]} \quad [\mathrm{kg}]$$

Example 1-3: Density of ideal gases at actual and at normal conditions

Item		Unit	Hydrogen	Oxygen	Air
Given:					
Temperature t	150 °C	K		423 °C	
Pressure p	25 bar	bar		2.5.E+06	
Molar mass M		kg / kmol	2.016	31.999	28.964
Gas constant R		Nm / (kg K)	4124.4	259.83	287.06
Molar volume of ideal gas		m^3 / kmol		22.41	
Results					
Density at normal conditions		kg / m^3	0.09	1.43	1.29
Density at actual conditions		kg / m^3	1.43	22.74	20.58

1.3.2 Mixtures of ideal gases

In technical thermodynamics usually working fluids are mixtures of gases such as combustion air or flue gas.

According to *Dalton's law* each gas of a mixture occupies the whole available volume as if there were no other gases. Each gas of the mixture exerts a *partial pressure*. The sum of the partial pressures is the total pressure. The following equations apply (subscripts: $_i$ for gas, $_\mathrm{m}$ for mixture):

Volume fraction: $$\upsilon_i = \frac{V_i}{V_m} = \frac{p_i}{p_m} = \frac{M_i}{M_m} \qquad \sum \upsilon_i = 1 \tag{1.16}$$

Mass fraction: $\mu_i = \frac{m_i}{m_m} = \upsilon_i \cdot \frac{M_i}{M_m} \quad \sum \mu_i = 1$ (1.17)

Molar mass $M_m = \sum \upsilon_i \cdot M_i$ (1.18)

Gas constant $R_m = \sum \mu_i \cdot R_i$ (1.19)

Example 1-4: Properties of combustion air

Combustion air is assumed to be composed of 79% nitrogen (N_2) and 21% oxygen (O_2). Where:

N_2: M=28 kg/kmol; R=297 J/(kg K)

O_2: M=32 kg/kmol; R=260 J/(kg K)

Properties of mixture:

Molar mass:

$M_m = \Sigma\, \upsilon_i \cdot M_i = 0.79 \cdot 28 + 0.21 \cdot 32 = 28.84$ kg/kmol

Mass fractions:

$\mu_i = \upsilon_i \cdot M_i / M_m$: $\mu_{N2} = 0.79 \cdot 28/28.84 = 0.767$; $\mu_{O2} = 0.21 x 32/28,84 = 0.233$

Gas constant: $R_m = \Sigma \mu_i \cdot R_i$: $= 0.767 \cdot 297 + 0.233 \cdot 260 = 288.4$ J/(kg K)

1.3.3 Thermodynamic processes

A *thermodynamic process* is an operation in which a thermodynamic system changes its properties of state and passes from one equilibrium to another. A thermodynamic system is in equilibrium when all its properties of state remain constant. In the course of a thermodynamic process conversion of thermal energy and work takes place.

Thermodynamic processes are distinguished according to the property of the system that remains constant during the process (denoted with prefix "iso") namely: isochoric (constant volume), isobaric (constant pressure), isothermal (constant temperature), and isentropic (constant entropy). Based on the equation of state of the ideal gas the following equations apply [1] [2] [3]:

Isochoric process, V constant: $\frac{p_1}{T_1} = \frac{p_2}{T_2}$ (1.20)

$$W_{V12} = 0; \quad W_{S12} = V \cdot (p_2 - p_1) = m \cdot R \cdot T \cdot (T_2 - T_1) \qquad (1.21)$$

$$Q_{12} = m \cdot c_\upsilon \cdot (T_2 - T_1) \qquad (1.22)$$

Technical application: e.g. heat storage

Isobaric process, p constant: $\frac{V_1}{T_1} = \frac{V_2}{T_2}$ (1.23)

$$W_{V12} = -p \cdot (V_2 - V_1) = -m \cdot R \cdot (T_2 - T_1); \quad -W_{S12} = 0 \tag{1.24}$$

$$Q_{12} = m \cdot c_p \cdot (T_2 - T_1) \tag{1.25}$$

Technical application: Heat exchanger (friction loss)

Isothermal process, T constant: $p_1 \cdot V_1 = p_2 \cdot V_2$ (1.26)

$$W_{V12} = m \cdot R \cdot T \cdot \left(\frac{p_2}{p_1}\right) = -R \cdot T \cdot \ln\frac{V_2}{V_1} = W_{S12} = -Q \tag{1.27}$$

Technical application: ideally cooled compressor

Adiabatic process, heat Q=0: $p_1 \cdot V_1^{\kappa} = p_2 \cdot V_2^{\kappa}$ (1.28)

Approximately is: $\frac{T_2}{T_1} = \left(\frac{p_2}{p_1}\right)^{\frac{\kappa-1}{\kappa}} = \left(\frac{V_1}{V_2}\right)^{k-1}$ and $\frac{p_2}{p_1} = \left(\frac{T_2}{T_1}\right)^{\frac{\kappa}{\kappa-1}}$ (1.29)

$$W_{V12} = m \cdot \frac{R}{\kappa - 1} \cdot (T_2 - T_1) = m \cdot c_\upsilon \cdot (T_2 - T_1); \quad W_{S12} = m \cdot c_p \cdot (T_2 - T_1) \tag{1.30}$$

Technical application: e.g. adiabatic process (q=0), pumps, compressors, turbines;

Isentropic process is a reversible adiabatic process at constant entropy (Q=0; S=constant).

Table 1-5: Isentropic exponent

Type of gas	Gas examples	Isentropic exponent
Monoatomic	Inert gases, He, Ar, Ne	κ=1.67
Diatomic	N_2, O_2, CO, atm., air	κ=1.4
Triatomic	CO_2, H_20, SO_2	κ=1.33

Relation $\kappa = c_p/c_\upsilon$

Polytropic process, p, V, T changing: $p_1 \cdot V_1^n = p_2 \cdot V_2^n$ (1.31)

$$\frac{T_2}{T_1} = \left(\frac{p_2}{p_1}\right)^{\frac{n-1}{n}} = \left(\frac{V_1}{V_2}\right)^{n-1} \quad \text{and} \quad \frac{p_2}{p_1} = \left(\frac{T_2}{T_1}\right)^{\frac{n}{n-1}} \tag{1.32}$$

$$W_{V12} = m \cdot \frac{R}{n-1} \cdot (T_2 - T_1) = m \cdot \frac{R \cdot T_1}{n-1} \cdot \left[\left(\frac{p_2}{p_1} \right)^{\frac{n-1}{n}} - 1 \right]; \quad W_{S12} = n \cdot W_{V12} \tag{1.33}$$

$$Q_{12} = c_\upsilon \cdot \frac{n - \kappa}{n-1} \cdot (T_2 - T_1) \tag{1.34}$$

Polytropic exponent: $1<n<\kappa$

Technical application: e.g., cooled compressor

Isenthalpic process: $H_2 = H_1 \quad W_{V12} = W_{S12} = 0$ (1.35)

Pressure reduction without generation of work, virtually destruction of work

Ideal gas: $H_2 - H_1 = m \cdot c_p \cdot (T_2 - T_1) = \text{constant} \quad \Rightarrow T_2 = T_1$ (1.36)

Technical application: Throttling of fluids to reduce pressure

Example 1-5: Isochoric process, tanks containing O_2 and H_2

Two tanks with the same volume contain oxygen and hydrogen respectively. Due to direct exposure to solar radiation their temperature increases. Calculate mass content, pressure, final pressure and the heat transfer. **Note:** Properties to be taken from **Annex 4**.

Item	Symbol	Unit	Values	
Given				
Substance	-	-	Oxygen	Hydrogen
Volume	V	m^3	0.25	0.25
Initial pressure	p_1	bar	5.0	5.0
Initial temperature 25 °C	T_1	K	298.15	298.15
Final temperature 60 °C	T_2	K	333.15	333.15
Properties				
Density (0°C)	ρ	kg /m^3	1.43	0.09
Gas constant	R	J/kg K	259.80	4,124.80
Heat capacity	c_p	kJ/kg K	0.92	14.30
Heat capacity	$c_\upsilon=c_p-R$	kJ/kg K	0.66	10.18
Results				
Mass	$m=pV/RT$	kg	1.61	0.10
Pressure, isochoric process	$p_2=T_2 \cdot p_1/p_2$	bar	5.59	5.59
Heat transfer	$Q=m \cdot c_\upsilon \cdot (T_2-T_1)$	kJ	37.3	36.2

Comments: Although the properties of the two gases are completely different, the two gases absorb almost the same amount of heat.

Example 1-6: Isobaric process, preheating of combustion air

The combustion air of a small industrial boiler is preheated with some waste heat from an industrial process. Calculate the heat flow for the given temperature increase.

Item		Symbol	Unit	Values
Given				
Boiler capacity		Q	kJ/s	20,000
Air	20,000 m3/h	V	m^3 /s	5.56
Pressure, constant	1 bar	p	bar	1.00E+05
Temperature	20 °C	T_1	K	293 °C
Temperature	80 °C	T_2	K	353 °C
Air Properties				
Gas constant		R	J/kg K	287.1
specific heat		c_p	kJ/kg K	1.00
Results				
Mass flow of air		$m=pV/RT$	kg/s	6.60
Heat flow		$Q=m.c_p.(T_2-T_1)$	kJ/s	396

Note: Temperature dependency of c_p is neglected

Example 1-7: Expansion process of a fluid

Atmospheric air contained in a vessel of 20 liter and at a pressure of 12 bar expands to environmental pressure of 1 bar. Calculate the final volume, temperature, volume expansion work and heat transfer if the expansion is isothermal, adiabatic or polytropic.

Item		Symbol	Unit	Isothermal	Adiabatic	Polytropic
Given						
Volume, air	20 liter	V_1	m^3		0.02	
Initial pressure	12	p_1	N/m^2		1.2.E+06	
Final pressure	1	p_2			1.E+05	
Temperature	25 °C	T_1	K		298.15	
Properties						
Gas constant		R	kJ/kg K		0.287	
Specific heat	cp= 1.00	$c_\upsilon=c_p-R$	kJ/kg K	-	0.713	0.71
Exponent		κ, n	-	1.00	1.40	1.30
Results						
Mass		$m=p\cdot V/R\cdot T$	kg		0.280	
Final volume		V_2	m^3	0.240	0.122	0.142
Final temperature in K		T_2	K	298.2	144.4	165.6
Final temperature in °C		t_2	°C	25.0	-128.7	-107.6
Volume expansion work		W_{V12}	kJ	-59.6	-30.9	-35.6
Heat transfer		Q_{12}	kJ	59.6	0.0	8.8

Note: Formulas are different in the columns

Notes: Properties to be taken from **Annex 4**, Equations for process calculations are available in section 1.3.3. Note: Formulas are different in the columns! Work leaving the system is negative, heat transferred into the system is positive. Carrying out unit control of all calculations is indispensable.

1.4 Thermodynamic Cycles

1.4.1 Definitions

A thermodynamic cycle is a cyclic succession of thermodynamic processes that involve changes of temperature, pressure, specific volume and entropy of the involved substance (called working medium). At the restart of each cycle the system returns to its initial state. The cycle's result is generation or utilization of mechanical work and transfer of heat.

Cycles are visualized in different forms of diagrams; the most simplified forms are energy flow schematics as shown in Figure 1-2.

A device performing a thermodynamic cycle, generating work and exchanging heat from a heat reservoir to a heat sink is known as a *heat engine* (e.g. Rankine cycle in steam power plants or Brayton cycle in gas turbines). A *work-to-heat* device is a *driven engine* it consumes work and lifts heat from a lower to a higher temperature level. The process flow is right-moving in the former and left-moving in the latter.

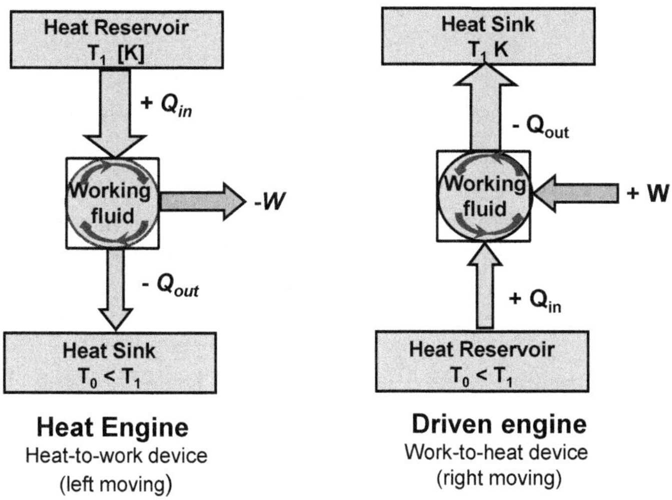

Figure 1-2: Simplified cycle schematics with energy flow

The performance of heat engines is given by the term *thermal efficiency* denoted as η (eta) which is the ratio of generated work to the amount of heat consumed by the engine.

$$\eta = \frac{W}{Q_{in} - Q_{out}} \tag{1.37}$$

It is usually given in percent. The thermal efficiency of modern gas turbines is, e.g., between 30 % and 40%.

The performance of a heat-to-work device is given by the *coefficient of performance COP*. This is the ratio of the useful thermal energy obtained to the work spent.

For a compression type refrigerator the *COP* is expressed by the ratio of heat extracted from the refrigerator (Q_{in}) to the work spent.

$$COP = \frac{Q_{in}}{W} = \frac{Q_{in}}{Q_{out} - Q_{in}} \tag{1.38}$$

For a heat pump the *COP* is the ratio of the useful heat supplied *(Q_{out})* to the work spent.

$$COP = \frac{Q_{out}}{W} = \frac{Q_{out}}{Q_{out} - Q_{in}} \tag{1.39}$$

COP is usually given in absolute terms in both cases.

1.4.2 The Carnot Cycle

The *Carnot cycle* is an idealized cycle consisting of four reversible processes, two isothermal and two isentropic – Figure 1-3. The work done is the maximum that can be produced between the upper temperature of the heat reservoir and the lower temperature of the heat sink. Its thermal efficiency only depends on these two temperatures and is the maximum that a cycle can theoretically achieve.

$$\eta_C = \frac{W}{Q_{in}} = \frac{T_1 \cdot \Delta S - T_0 \cdot \Delta S}{T_1 \cdot \Delta S} = \frac{T_1 - T_0}{T_1} = 1 - \frac{T_0}{T_1} \tag{1.40}$$

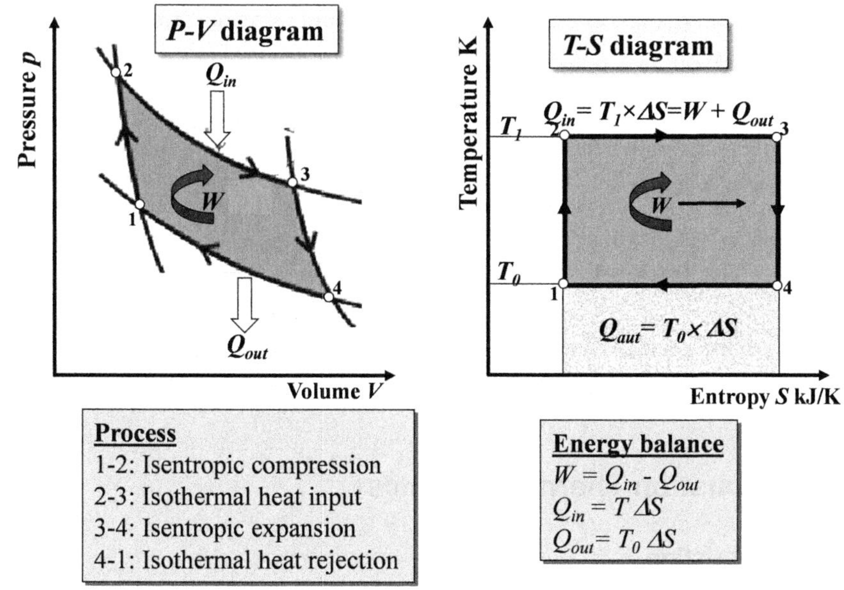

Figure 1-3: The Carnot cycle in p-V and in T-s diagram

The efficiency of the Carnot cycle mainly serves as a benchmark for evaluation of the performance of real cycles. It provides an indication of the potential for performance improvement of cycles.

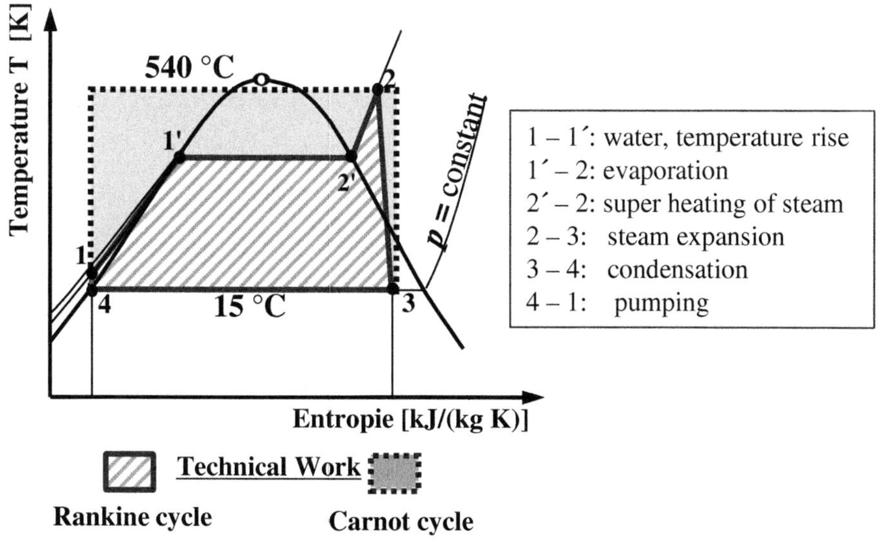

Figure 1-4: Carnot cycle vs. Rankine cycle in T-s diagram

Thermodynamic cycles of power generation plants are described in the related chapter 3, 4 and 5.

Example 1-8: Electrical efficiency of gas turbine vs. Carnot cycle

The gas temperature at the entry of a modern gas turbine is 1200°C, the air inlet temperature to the compressor is 15°C. The maximum plant efficiency in a state-of-the-art simple cycle gas turbine cycle (SCGT) is 44%, in a combined gas-steam turbine cycle (CCGT) it is 60%. What is the maximum efficiency of the Carnot cycle between the two temperatures? What is the theoretical improvement potential?

Carnot efficiency: $\eta_c = 1 - \frac{273+15}{273+1200} = 0.8$ or 80%

The theoretical improvement potential is 20 percentage points.

1.5 Combustion Thermodynamics

1.5.1 Combustion

Combustion is the chemical process in which an oxidant reacts with a fuel to release the chemically stored energy in the form of high temperature flue gases. For technical application the oxidant for combustion is oxygen (O_2). Conventional fuels contain mainly carbon and hydrogen, and their combustion products are carbon dioxide (CO_2) and water vapour (H_2O). The key characteristics of combustion are:

- The heating value of the fuel (*HHV or LHV*)
- The amount of oxygen for a stoichiometric combustion (O_{2_min})
- The minimum combustion air (V_{A_min})
- The air-to-fuel ratio (λ)
- The minimum flue gas volume ($V_{FG\text{-}min}$)
- The maximum CO_2 content in the flue gases

The key characteristics of combustion for the main fuels are shown in **Annex 10** and are described in the following sections.

1.5.2 The heating values

The *heating value* is the measurement for determination of the chemical energy stored in fuels. There are two heating values for each type of fuel. The terms used in American English are: "lower heating value (LHV)" and "higher heating value (HHV)"; whereas in British English the terms are "net calorific value (NCV)" and "gross calorific value (GCV)". In German literature the terms H_u and H_o[7] are common. In scientific papers the terms "inferior heating value (H_i)" and "superior heating value (H_s)" have been introduced but they are rarely used in practice.

> The *higher heating value HHV* is defined as the amount of heat released by complete combustion of one unit of fuel. All combustion products are cooled down to the temperature before the combustion and the water vapor formed during the combustion is condensed into water and its condensation heat is included in the HHV.
>
> In contrary, the *lower heating value* does not include the condensation heat of the water vapor formed during the combustion.

The heating values for pure fuels and for those, for which the chemical composition is known, can be calculated – see **Annex 6** and **Annex 7**. They are given in kJ, MJ or kWh_t, MWh_t per unit of fuel (kg, metric ton, and normal cubic meters nm^3). The notation in kWh_t or MWh_t is practical for energy balances and is preferably applied in this book.
It is worth mentioning that for combustion calculations in British and US literature the higher heating values are commonly used, while in Continental Europe the lower heating values are preferably used.

> In power engineering and in this book the lower heating values are exclusively used for energy balances and for financial calculations.

Some guide values of HHV to LHV ratio for selected fuels are given below:

Fuel	LHV/ HHV
Natural gas	0.903
Heating oil	0.940
Hard coal	0.958

Coal and HFO are internationally traded in US$ per metric ton. For the conversion in thermal units based on MWh_t or GJ the commercial price is divided with the lower heating value of the fuel.

$$c_t = \frac{c_c\,[\text{US\$/t}]}{LHV\,[\text{MWh}_t/\text{t}]}\left[\frac{\text{US\$}}{\text{MWh}_t}\right] \quad \text{or} \quad c_t = \frac{c_c\,[\text{US\$/t}]}{LHV\,[\text{GJ/t}]}\left[\frac{\text{US\$}}{\text{GJ}}\right] \tag{1.41}$$

The wholesale prices of natural gas in international energy purchase contracts are referred to 1000 normal cubic meters (nm^3):

[7]) H_u: Unterer Heizwert, H_o **O**berer Heizwert (corresponding to: LHV, HHV)

$$c_t = \frac{c_c\left[\text{US\$/1000 nm}^3\right]}{LHV\left[\text{MWh}_\text{t}/1000\text{ nm}^3\right]}\left[\frac{\text{US\$}}{\text{MWh}_\text{t}}\right] \text{ or } c_t = \frac{c_c\left[\text{US\$/1000 nm}^3\right]}{LHV\left[\text{GJ/1000 nm}^3\right]}\left[\frac{\text{US\$}}{\text{GJ}}\right] \quad (1.42)$$

In general, gas utilities sell gas based on the HHV, for energy balances it must be converted in LHV.

1.5.3 Combustion air and flue gas volumes

The minimum quantity of air *(V_{A_min})* required for a complete combustion contains exactly the oxygen required for a stoichiometric *combustion* of all the combustible constituents of a fuel. However, for technical reasons, some excess air must be supplied to most practical combustion systems to ensure a complete combustion. This is expressed with the air-to-fuel (AF) ratio which is denoted with the Greek character lambda λ.

$$\lambda = \frac{V_A}{V_{A\_\min}} \quad \text{and} \quad V_A = \lambda \cdot V_{A\_\min} \quad \left[\frac{\text{m}_n^3}{\text{kg}}\right] \text{ or } \left[\frac{\text{m}_n^3}{\text{m}_n^3}\right] \quad (1.43)$$

The units used are: normal cubic meters flue gas per kg fuel for solid and liquid fuels and normal cubic meter flue gas per normal cubic meter for gaseous fuel.

A high amount of excess air results to higher fuel losses and must be limited to the actually necessary. Best practice air-to-fuel ratios for fuels used in power generation are given in **Annex 9**.

Conventional fuels contain mainly carbon and hydrocarbons and their main combustion products are carbon dioxide (CO_2) and water (H_2O). The flue gases contain the combustion products, the Nitrogen (N_2) of the combustion air and the excess air. A distinction is made between "dry flue gas" and "wet flue gas". The actual flue gas volumes are calculated with the formulas.

$$\text{Dry flue gas: } V_{FGD} = V_{FGD\_\min} + (\lambda - 1) \cdot V_{A\_\min} \quad \left[\frac{\text{m}_\text{n}^3}{\text{kg}}\right] \text{or} \left[\frac{\text{m}_\text{n}^3}{\text{m}_\text{n}^3}\right] \quad (1.44)$$

$$\text{Wet flue gas: } V_{FGW} = V_{FGW\_\min} + (\lambda - 1) \cdot V_{A\_\min} \quad \left[\frac{\text{m}_\text{n}^3}{\text{kg}}\right] \text{or} \left[\frac{\text{m}_\text{n}^3}{\text{m}_\text{n}^3}\right] \quad (1.45)$$

The minimum flue gas volumes of fuels used for heat and power generation are given in **Annex 10** (Fuel properties). The volumes of combustion air and flue gases are important parameters for designing and sizing the components of boilers and flue gas cleaning facilities. An important indicator is also the ratio of minimum combustion air to minimum flue gas volume ($V_{air\text{-}min}/V_{FG\_min}$). For most fuels it is approximately 1. For certain gaseous fuels there are, however, considerable deviations such as hydrogen 0.79 and blast furnace gas 1.42 – see **Annex 10** (Fuel properties).

1.5.4 Oxygen and carbon dioxide in the flue gas

There is a correlation between the air-to-fuel ratio of the combustion air and the O_2 content in flue gas; expressed as an equation it is:

$$O_2 \cdot V_{FG} = 21 \cdot (\lambda - 1) \cdot V_{A\_\min}$$

Where "21" percent is the O_2 content of the combustion air and "O_2" is the oxygen content in percent of the flue gas. For most fuels it can be assumed $V_{FG} \approx V_{air\text{-}min}$, and the ratio of minimum combustion air to minimum flue gas volume is nearly 1 ($V_{air\text{-}min}/V_{FG\_min} \approx 1$). The equation above becomes:

$$\lambda = 1 + \lambda \cdot \frac{O_2}{21 - O_2} \cdot \frac{V_{FG}}{V_A} \quad \text{or for } V_{FG} \approx V_A \Rightarrow \lambda = 1 + \frac{O_2}{21 - O_2} \qquad (1.46)$$

Emission standards specify emissions for the main pollutants such as sulphur dioxide (SO_2), nitrogen oxides (NO_x) and particulates referred to O_2-content in the flue gas. Reference values for O_2-content and lambda and the ratio $V_{air\text{-}min}/V_{FG\_min}$ are given in **Annex 9** and **Annex 10**.

1.5.5 Maximum CO_2 content and CO2 emission factors

The CO_2 content in the flue gas obtains its maximum value for stoichiometric combustion because all carbon of the fuel is converted into CO_2. However, due to the excess combustion air, its content is lower in practical applications. The CO_2 content serves therefore also an indicator for the quality of combustion.

$$CO_2 = \frac{CO_{2\_\max}}{\lambda} \qquad (1.47)$$

Burning fossil fuels is directly related to the increase of carbon dioxide emission in the atmosphere and is one reason for global warming. Different fuels emit different amounts of carbon dioxide in relation to the energy they produce when burned. An indicator of the carbon potential of fuels is the *CO_2 emission factor* referred to its lower heating value (kg_{co2}/GJ or kg_{co2}/MWh_t). These are stated in Annex 10.

Example 1-9: Combustion air and flue gas amounts of a power plant

Calculate the combustion air and flue gas volumes of a 600 MW coal fired steam power plant with the combustion parameters given in **Annex 10**.

Item		Symbol	Unit	Values
Given				
Type of plant		-	-	Steam coal
Electrical output		P_e	MW_e	600
Fuel, hard coal		LHV	MW_t/t	8.13
Capacity of boiler	η= 40%	P_Q	MW_t	1,500
Excess air		-	-	6%
Combustion characteristics				
Minimum combustion air		V_{A_min}	m_n^3/kWh_t	0.98
Minimum dry flue gas		V_{FGD_min}	m_n^3/kWh_t	0.95
Minimum wet flue gas		V_{FGW_min}	m_n^3/kWh_t	1.02
CO_2 Emissions			kg / MWh_t	342
Results				
Fuel consumption	t= 1 h	Q_F	MWh_t / h	1,500
mass flow		m_F	t / h	185
Air-to-fuel ratio		λ	-	1.06
Combustion air volume		V_A	m_n^3/h	1,558,200
Dry flue gas volume		V_{FGD}	m_n^3/h	1,510,500
Wet flue gas volume		V_{FGW}	m_n^3/h	1,621,800
CO_2 Emissions			t / h	513

1.6 Water and Steam Thermodynamics

1.6.1 States of water

Water is available in nature in three states, namely: solid (ice), liquid (water) and vapor (steam). It is a pure substance; its chemical composition is (H_2O), and its physical structure remains the same in all three states.

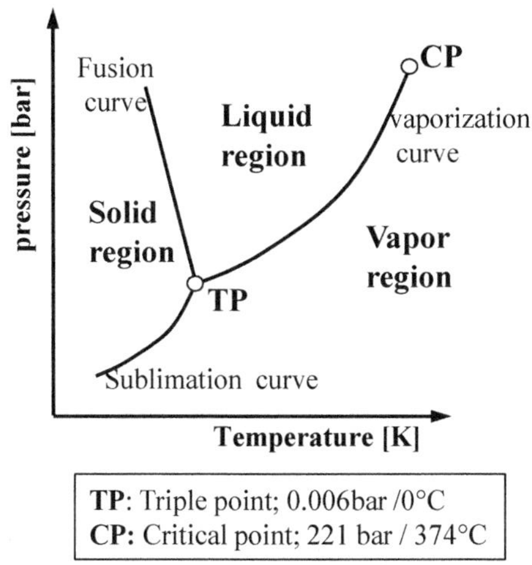

The three states of water are illustrated in a *p-T* diagram. At the triple point (TP) all three states are in equilibrium. Along the vaporization curve liquid and vapor states are in equilibrium. When heat is added and pressure is kept constant, the temperature increases and liquid water evaporates and becomes steam. This is the case up to the critical point (CP). At higher pressures and temperatures there is no more change of state. This state is known as the supercritical state. The fusion curve separates the state solid-liquid and the sublimation curve the states solid and vapor.

Figure 1-5: The pressure-temperature diagram of water

1.6.2 The steam generation process

Steam along with air is the most important working fluid for power generation cycles. Steam is vaporized water. The steam generation process takes place in three phases, namely: feed water preheating, evaporation and superheating. The process is isobaric; pressure remains constant during all the three phases. The steam generation process is depicted in Figure 1-6 in T-s and in h-s diagrams. Feed water (1) pumped to a high pressure is heated up to its boiling temperature (2´). At this point the water is saturated and if heat supply continues it starts to evaporate and gradually becomes vapor. During evaporation, pressure and temperature remain constant and there is a liquid-vapor mixture known as *wet steam.* At point 3´´ the entire liquid is completely evaporated and the steam is in *saturated* state and is *dry* (without any liquid content). Further heating of saturated steam rises its temperature and steam becomes superheated.

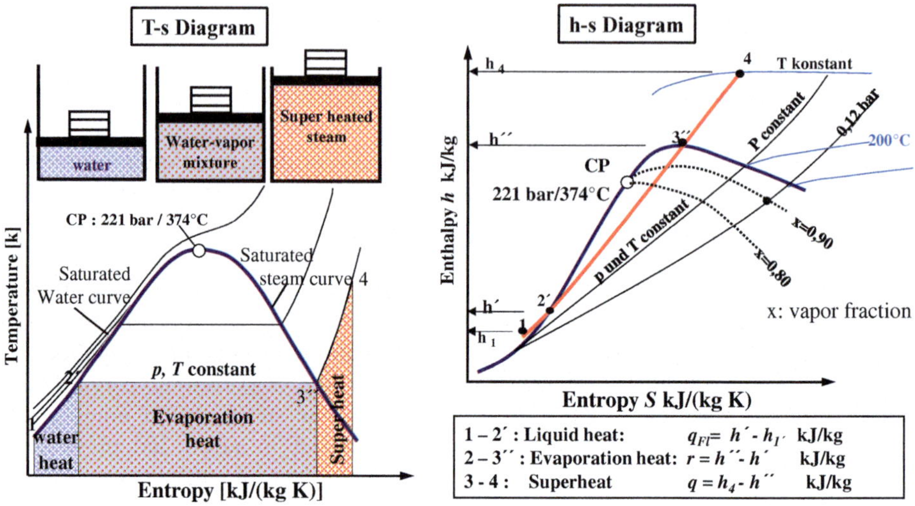

Figure 1-6: The steam generation process in T-s and h-s diagram

The bell-shaped curve in Figure 1-6 depicts the saturation characteristics of water. A distinct point of the curve is the *critical point (CP*). The part of the curve to the left of the CP is the saturated water curve; the part to the right of the CP is the saturated steam curve.

There is a fixed relationship between evaporation temperature and pressure. When feed water is pumped to a higher pressure its temperature increases to the corresponding evaporation temperature – Figure 1-5.

The clock-shaped curve of the water evaporation characteristic becomes increasingly narrow if evaporation temperature (and the corresponding pressure) increases. As a result, the required evaporation heat gradually decreases. For this reason high evaporation pressures are applied for power generation in steam power plants. At the critical point and above, change of phase from liquid to vapor takes place without addition of evaporation heat, evaporation heat becomes zero. The steam conditions above the critical point are known as *super critical.* Modern steam power plants are designed for super critical steam conditions.

1.6.3 Thermodynamic properties of steam

Table 1-6 shows the symbols and units used for key thermal properties of water and steam. Properties for saturated water are denoted by the symbol followed by an apostrophe (e.g. *h´*or *s´*) those of saturated steam by the symbol followed by double apostrophe (e.g. *h´´* or *s´´*).

Table 1-6: Symbols and Units of key properties of water and steam

Symbol	Unit	Properties
h' ; h''	kJ/kg	Enthalpy of saturated water, of saturated steam
s' ; s''	kJ/(kg·K)	Entropy of saturated water, of saturated steam
t_s	°C	Saturation, evaporation temperature
p_s	Bar	Saturation, evaporation pressure
υ' ; υ''	m³/kg	Spec. volume of water, of steam at saturation point
r	kJ/kg	Spec. evaporation enthalpy
x	g / kg	Steam content of the wet steam

Approximate calculations of the saturation temperature and pressure can be done with the formulas [4] stated below.

$$t_s = 100 \cdot \sqrt[4]{p_s} \quad [°C] \quad \text{and} \quad p_s = \left(\frac{1}{100} \cdot t_s\right)^4 \quad [\text{bar}] \qquad (1.48)$$

The thermodynamic properties for water and steam are given in steam tables and in Enthalpy-Entropy diagrams, first drawn by Mollier. The algorithms for their calculation are based on formulations provided by the International Association for the Properties of Water and Steam [5] (IAPWS), issued in so-called releases. The most recent is the IAPWS-IF97.

For thermodynamic cycle calculation there are currently software tools linked also to MS-Excel. In this book the software tool "FluidEXL^Graphics" is used [6] for cycle calculations.

Example 1-10: Heat demand for steam generation

Item	Symbol	Unit	Values	
Given				
Evaporation pressure	p_s	bar	80	180
Super heat temperature	t_4	°C	540	540
Intermediate Calculations *)				
Saturation temperature	t_s	°C	295	357
Enthalpy of saturated water	h'	kJ / kg	1,317	1,732
Enthalpy of saturated steam	h''	kJ / kg	2,759	2,510
Enthalpy of superheated steam	h_4	kJ / kg	3,497	3,390
Results				
Evaporation heat r	$h''-h'$	kJ / kg	1,442	778
Super heat enthalpy	$h_{sh}-h''$	kJ / kg	739	880
Total heat demand	$h_{sh}-h_1$	**kJ / kg**	**2,180**	**1,658**

*) Properties of steam are calculated with the Tool FluidEXL

1.6.4 Steam generation in boilers

Steam is produced in steam generators commonly known as boilers. A boiler is essentially a heat exchanger in which the thermal energy released by the combustion process is transferred from the combustion products to the working fluid. The increase in enthalpy of the working fluid represents the useful output. The parameter which indicates the process performance is the *thermal efficiency*. Two methods are applied to determine the boilers' efficiency during operation: the direct and the indirect method.

In the *direct method* the boiler's efficiency is defined as the quotient of the useful energy contained in the steam (kJ or kWh_t) divided by the heat input of the fuel and feed water (kJ or kWh_t). The symbols are shown in Figure 1-7.

$$\eta_B = \frac{\dot{m}_s \cdot (h_s - h_w)}{\dot{m}_F \cdot LHV + \dot{m}_F \cdot c_F \cdot t_F + \dot{m}_A \cdot c_A \cdot t_A} = \frac{\dot{m}_s \cdot (h_s - h_w)}{\sum Q_i} \qquad (1.49)$$

The term in the nominator is the mass flow of steam and the enthalpies of steam and feed water. The denominator includes the sensible heat of combustion air (index $_A$) and fuel (index $_F$) and the heating value of the fuel. All mass flows and temperatures are to be metered. The enthalpies of water and steam and the specific heat can be calculated with software tools (e.g. FluidEXL).

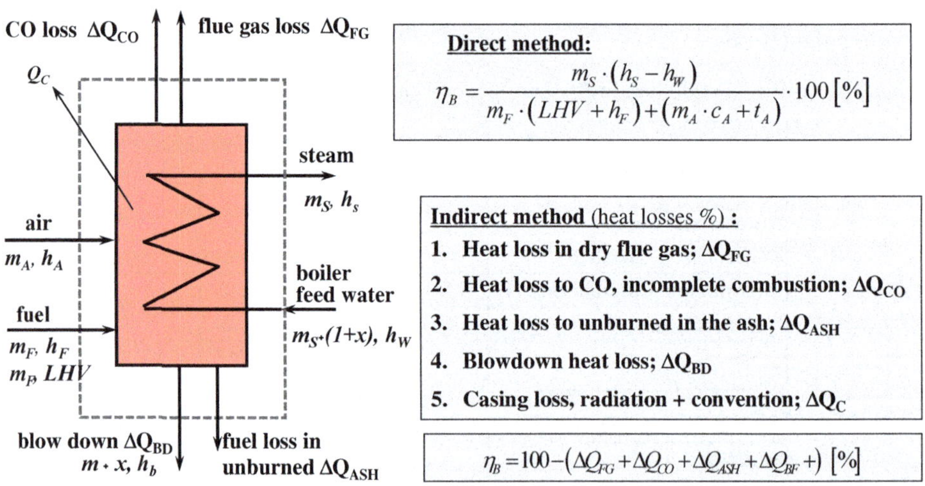

Figure 1-7: Boiler heat balance

The direct method works only for fuels which have a constant calorific value during operation and can be easily metered, such as natural gas or oil. The meth-

od is not applicable for coal and nuclear plants. Furthermore, to optimize the operation of boilers it is necessary to determine where and why energy losses occur. Therefore, the *indirect method* is applied in most cases.
For the calculation of the losses for the indirect method the following empirical formulas can be used:

Flue gas loss:

$$\Delta Q_{FG} = (t_{FG} - t_A) \cdot \left(\frac{A}{CO_2} - B \right) \; [\%] \tag{1.50}$$

Note: Flue gas t_{FG}, combustion air temperature t_A and the actual CO_2 (%) content in the flue gas are to be metered. The factors below are estimates for air-to-fuel ratio, specific heat capacities etc.

Factor	Fuel oil	Natural gas	Hard coal
A	0.5	0.37	0.7
B	0.007	0.009	0
f	50	30	60

CO Loss:

$$\Delta Q_{CO} = f \cdot \frac{CO}{CO_2 - CO} \; [\%] \tag{1.51}$$

CO can be formed if combustion air volume is insufficient. The CO content is zero if burners are properly set and controlled.

Loss to unburned occurs in coal fired boilers in the range of 1% to 2%.

Blowdown can be assumed to be about 1% of the steam mass flow.

Casing losses of modern boilers are in the range of 0.5%. However, it must be considered that they are constant during the entire operation of the boiler, while the other losses depend on the load.

Example 1-11: Boiler combustion losses

The maximum CO_2 content of a fuel (HFO) in flue gas is **15.90%**, the metered is **11.70%**.
Calculate: excess air ratio, O_2 content in the flue gas and flue gas losses for the following cases:
Case 1: base case as found
Case 2: lower excess air ratio higher CO_2
Case 3: further reduction of excess air, and higher CO_2
Case 4: as 2 and 20 °C lower flue gas temperature (frequent cleaning of boiler)
Case 5: as above and preheating of combustion air by waste heat

I t e m	Unit	Case				
		1	2	3	4	5
Given						
Maximum CO_2	%	15.9	15.9	15.9	15.9	15.9
Metered CO_2	%	11.70	**12.7**	**13.8**	12.70	12.70
Flue gas temperature	°C	230	230	230	***210***	210
Combustion air temperature	°C	25	25	25	25	***100***
Carbon monoxide CO	%	0	0	0.3	0	*0*
Results						
Excess air ratio λ	-	1.36	1.25	1.15	1.25	1.25
Oxygen content	%	5.55	4.23	2.77	4.23	4.23
Flue gas losses	%	10.20	9.51	8.86	8.58	5.10
CO losses	%	0.00	0.00	1.11	0.00	0.00
Efficiency of combustion	%	**89.80**	**90.49**	**90.03**	**91.42**	**94.90**
Plus percentage points	%	**0.00**	**0.69**	**0.22**	**1.62**	**5.09**

2 Basics Techno-economics of Power Systems

2.1 Types of Power Plants

Power generation plants are classified upon the type of thermodynamic cycle, the kind of the primary energy, and their operation mode. An overview of the types of power plants, which are described and evaluated in this book, is depicted in the figure below.

Table 2-1: Types of power plants

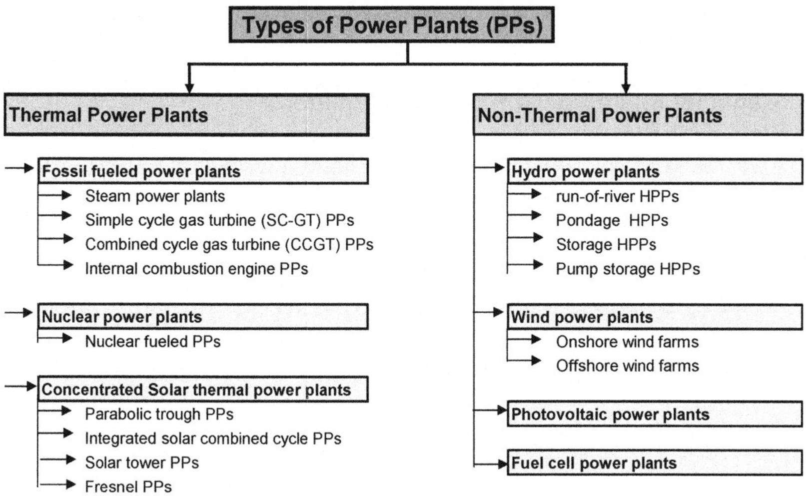

The first step for planning and techno-economic evaluation of new power plants is the development and modelling of technical concepts, which enable simulation and optimization of thermodynamic cycles. Based on the output of the thermodynamic models, the techno-economic evaluation can be conducted. In this context, understanding of the complex relationships, coaction and interplay between technology and economics is an indispensable prerequisite by dealing with energy economics. Professionals in this field may be either economists or engineers, but they must possess the required knowledge and background in either disciplines, in order to assess properly and evaluate energy and power system projects.

P. Konstantin and M. Konstantin, *The Power Supply Industry*,
https://doi.org/10.1007/978-3-319-72305-1_2

2.2 Key Performance Parameters of Power Generation

There is often some confusion in literature with regard to terms used in the techno-economic evaluation of power generation plants. Hence, we start with the definition of some key performance parameters. They are mainly referred to fossil fueled power plants; however, most of them are also applicable for all other types of power supply systems. Performance parameters, which are characteristic for other type of plants, are separately addressed in the respective sections. In general, we distinguish between *capacity related* and *energy related* performance parameters.

Note, however, that for calculation the Standard International Unit System (SI – Unit system) according ISO–1000 is applied in this book. A brief description is outlined in section 1.1.3 and in more detail in **Annex 1 and Annex 2**.

Essential points: *Symbols* in equations and formulas are written in italics, *units* in standard text characters (e.g. P_e = 500 MW; V=m^3/s; W_e = kWh/a).
In examples US Dollar (US$) or Euro (€) are used for currencies. In formulas and equations currency-units (CU) are used for neutral term for currency.

2.2.1 Capacity related performance parameters

The key performance parameters of power plants are calculated with thermodynamic cycle simulations' software tools. The author is familiar with the tool KPRO® [7].
The *output* of power generation plants is defined with the key parameters:

- Rated gross power output P_{gross} in [kW or MW]
- Rated thermal input[8] $\dot{Q}_{LHV}$ in kJ/s (=kW$_t$) or MJ/s (=MW$_t$) referred to the lower heating value ($_{LHV}$) of the fuel (in British English Net calorific value ($_{NCV}$).
- Power demand of the main auxiliaries ΔP_{aux}
- Net power output $P_{net} = P_{gross} - \Delta P_{aux}$ in [kW or MW]

Note: "Q" with upper dot denotes heat flow or heat power, e.g. MJ/s "$\dot{Q}$", without upper dot it denotes heat quantity or consumption, e.g. MWh$_t$/a.

The *performance* of power plants is defined with the *energy efficiency* or the *heat rate*:

---

[8] According to SI-unit system the unit for heat is J (Joule) and that for heat flow J/s. However 1J/s is numerically equal to 1 W. We prefer to use the term W, as it is more convenient for power balances. In practical applications the multiples are used (kJ; MJ, GJ, kW, MW, GW). The subscript "$_t$" denotes thermal "$_e$" electrical.

Electrical Efficiency: $\eta_{gross} = \dfrac{P_{gross}\,[\mathrm{kW_e}]}{\dot{Q}_{LHV}\,[\mathrm{kJ/s}]} \times 100 \quad [\%]$ (2.1)

Heat rate: $q = 3600 \times \dfrac{\dot{Q}_{LHV}}{P_{gross}} \quad \left[\dfrac{\mathrm{kJ}}{\mathrm{kWh_e}}\right]$ (2.2)

For the calculation of net electrical efficiency and heat *rate*, P_{gross} is to be replaced by P_{net} while Q_{lHV} is kept the same.

Reference site conditions (RSC): The *RSC* include the ambient temperature, relative humidity, atmospheric pressure or altitude (relevant for gas turbines) and cooling water temperature at the site. The RSC are representative for the time of the year when the annual on-peak load is expected in the grid.

The term *rated* means that the performance (kW, η) is determined based on *reference site conditions (RSC)*.

Table 2-2: Reference site conditions for selected regions

Site	Condenser Cooling	Ambient Temperature °C	Cooling water inlet °C
Europe, North Sea	Seawater once through	15	12
Central Europe	Cooling tower	15	25
Gulf	Seawater once through	46	36

2.2.2 Energy related performance parameters

The *annual electricity production* can be calculated either with the full capacity utilization hours (t_{FCH}) or with the capacity factor (*CF*).

Annual elec. production: $W_e = P_{gross} \cdot t_{FCH} \quad [\mathrm{MWh/a}]$ (2.3)

Annual elec. production: $W_e = CF \cdot P_{gross} \cdot 8760 \quad [\mathrm{MWh/a}]$ (2.4)

Note: Instead of equivalent full capacity hours (t_{FCH}), often the term equivalent full load hours (t_{FLH}) is used. The latter is referred to the peak load of the grid and is not correct for power generation plants. Nevertheless, it is customarily used for power generation, too.

The units kW and kWh are usually too small for utility size power plants, hence, the units MW and MWh or even GWh are used.

There is a distinction between the actual *operating hours* or running hours (t_{op}) and equivalent *full capacity hours* (t_{FCH}). The former denotes the time during which the plant is in operation and produces electricity also in part load. The term *full capacity (utilization) hours* (t_{FCH}) is the equivalent time period during which the power plant would have produced the same amount of electrical energy in continuous operation at *full power output*.

Full capacity hours: $$t_{FCH} = \frac{W_e\,[\mathrm{MWh/a}]}{P_{rated}\,[\mathrm{MW}]} \quad \left[\frac{\mathrm{h}}{\mathrm{a}}\right] \qquad (2.5)$$

The term *equivalent operating hours (EOH)*, is used in maintenance contracts, especially for gas turbines, and includes, in addition to the operation hours, also equivalent hours for start-ups, operation with different fuels etc. These cause wear and tear of machinery and additional maintenance expenses. *EOH* are therefore longer than operating hours. The prices for maintenance contracts are based on the *EOH*.

Capacity factor: $$CF = \frac{W_e\,[\mathrm{MWh}]}{P_{rated}\,[\mathrm{MW}] \times t_{op.}\,[\mathrm{h}]} \quad [-] \qquad (2.6)$$

The capacity factor CF is the ratio of the actually produced electrical energy to the electrical energy that could have been produced at continuous operation at full power output during the same period. It can also be expressed as the ratio of the annual average to the rated power output ($P_{average} / P_{max}$). The relation between full capacity hours (t_{FCH}) and CF is as follows:

$$t_{FCH} = CF \cdot t_{op.}\,[\mathrm{h/a}] \qquad (2.7)$$

Thermal energy consumption: $$Q = \frac{W_e\,[\mathrm{MWe}]}{\overline{\eta}_e\,[-]} \quad \left[\frac{\mathrm{MWh_t}}{\mathrm{a}}\right] \qquad (2.8)$$

Annual average efficiency: $$\overline{\eta}_e = \frac{W_e\,[\mathrm{MWh_e/a}]}{Q\,[\mathrm{MWh_t/a}]} \cdot 100 \quad [\%] \qquad (2.9)$$

Note: The subscript "*e*" is used for electrical (e.g. W_e *or* P_e) and the subscript "*t*" for thermal (e.g. the thermal efficiency of the boiler "η_t" or MW_t).

The annual average efficiency $\overline{\eta}_{e,net}$ is usually lower than the efficiency at rated conditions, due to part load operation, wear or degradation. It is estimated based on experience.

Availability: In general, availability is the state where a unit can provide energy within a reference period. The availability is reduced due to *planned outages* for maintenance and unplanned or *forced outages* due to failures during operation. There are different definitions of availability; most common is the time availability:

Time availability: $t_{av} = t_{ref} - t_{planned} - t_{forced} = f_{tav} \times t_{ref}$ h/a (2.10)

Availability factors are described in the Standard IEEE-762-2006 [8] as well as in [9]. The latter contains also statistics for thermal power plants.

The availability of power units and plants depends on the technology, age and several other factors. Two power units with the same net output (MW) but of different availability will produce different amounts of energy during the same reference operating period.

2.3 Technical-economic Evaluation of Projects

2.3.1 Overview of Investment appraisal methods

The methods for investment appraisal and cash flow analysis are thoroughly described in the author's book [10],

Power and Energy Systems – **Engineering Economics**

The book also includes numerous, real world application examples for a better understanding and practicing of the contents. In the following, an excerpt of the book's comprehensive contents is provided without any claim of completeness.

In general, *an investment* is a business activity during which capital is deployed to generate future returns. *Investment appraisal* is the process of assessing the financial viability of investment options. Methods applicable for long living energy sector projects are:

- The Net Present Value (NPV)
- The Internal Rate of Return (IRR)
- The Annuity, also known as the Annual Equivalent Amount

All three methods apply the discounting approach. In this book, the discount rate is based on the weighted average cost of capital (WACC); this is the minimum acceptable rate of return (hurdle rate). Guide values for power system projects for developed countries are shown in the following example.

Example 2-1: Weighted average cost of capital (WACC), guide values

Item	Conventional power projects		Renewable energy projects	
	Equity	Debt	Equity	Debt
Asset shares	**30%**	**70%**	**20%**	**80%**
Expected returns after tax				
Risk free rate of return / interest	5.0 %/a	5.0 %/a	5.0 %/a	5.0 %/a
Venture risks premium	6.0 %/a	1.0 %/a	5.0 %/a	0.0 %/a
Country risk premium *)	0.0 %/a	0.0 %/a	0.0 %/a	0.0 %/a
Cost of capital in nominal terms, after tax	**11.0 %/a**	**6.0 %/a**	**10.0 %/a**	**5.0 %/a**
Corporate tax **) 25%	3.7 %/a	0.0 %/a	3.3 %/a	0.0 %/a
Cost of capital in nominal terms, before tax	**14.7 %/a**	**6.0 %/a**	**13.3 %/a**	**5.0 %/a**
***WACC$_n$* in nominal terms, before tax**	**8.60 %/a**		**6.67 %/a**	
./. Inflation rate ***)	2.00 %/a		2.00 %/a	
***WACC$_r$* inflation adjusted**	**6.47 %/a**		**4.58 %/a**	

*) This is country specific, for most developed countries it is zero

**) This is a typical value, the rates depend on the country's tax legislation

**) The 2%/a inflation rate is the longterm target value in most economies

The weighted average costs of capital are used as discount rates: $WACC_n$ for calculations in nominal terms, and $WACC_r$ for calculations in real terms.

Important Note: Calculation of generation costs for renewable energy projects is conducted with a lower discount rate. This is because public grid operators are obliged to take-off electricity production from renewable energies and remunerate with a feed-in tariff set by the government. Hence, there is neither a volume risks nor a price risk for both, the investors and the creditors. This is considered in the WACC with one percentage point lower venture risk for both parties. It is further assumed that creditors will accept a lower equity share.

The **NPV of an investment** is calculated by discounting the time values of all payments during the lifetime of an investment project and adding the cumulative present value of the invested capital. Capital inflows are designated positive, capital outflows negative. This is mathematically expressed with the following equation (CU: Currency Units):

$$NPV = -CAPEX_0 + \sum_{t=1}^{t=n} \frac{(R_t - E_t)}{(1+i)^t} \quad [\text{CU}] \tag{2.11}$$

Where:

$CAPEX$: Capital expenditures present value

R_t: Time value of sale revenues of the year t

E_t: Time value of operating expenses at the year t (OPEX)

i : Discount rate in (% / a) in digital terms

n: Life time of the investment project in years

t_0: Reference year for discounting (start of commercial operation)

Profitability criterion: The Net Present Value of an investment option must be positive or at least zero NPV≥ 0.

Investment appraisal for energy sector projects is done in most of the cases, by applying a *least cost approach*. This means, the option with the lowest lifetime discounted costs is the most favorable investment. For this purpose the *Net Present Costs* (*NPC*) are appraised and only the cost part of the *NPV* equation is relevant. This is expressed mathematically with the following equation (as all terms are capital outflows, the negative sign is omitted):

$$NPC = CAPEX\,[\text{CU}] + \sum_{t=1}^{t=n} \frac{OPEX_t\,[\text{CU/a}]}{(1+i)^t\,[\text{1/a}]} \quad [\text{CU}] \tag{2.12}$$

Investment appraisal of power system projects requires the calculation of the specific, levelized electricity generation cost (*LEC*) over the project's lifetime. The formula is:

$$LEC = \frac{CAPEX + \sum_{t=1}^{t=n} \frac{OPEX_t}{(1+i)^t}}{\sum_{t=1}^{t=n} \frac{W_{e\_t}}{(1+i)^t}} \quad \left[\frac{\text{CU}}{\text{kWh}}\right] \tag{2.13}$$

This levelized electricity cost (*LEC*) is the net present costs *NPC* (numerator of the equation) divided by present value of the electricity generation (*We)* over the lifetime (denominator).

> In order to be comparable to different sources, LEC shall always be calculated in real terms, excluding inflation!

MS-Excel function: NPV (discount rate, range of values)

Own Add-In incl. escalation: BW_{esc}(escalation rate, discount rate, periods)

The Internal Rate of Return (IRR) method is a special form of the *NPV*. The *IRR* is the discount rate for which the *NPV* becomes zero. The *IRR* is the appraisal method that creditors and equity investors usually prefer. There are three different forms of the method:

- *IRR* on investment (*IRROI*)
- *IRR* on equity (*IRROE*) before tax
- *IRR* on equity (*IRROE*) after tax

IRROI is mathematically expressed with the following equation:

$$NPV = -I_0 + \sum_{t=1}^{t=n} \frac{(R_t - E_t)}{(1 + IRR)^t} = 0 \tag{2.14}$$

The sought value is the *IRR* stated in the denominator; however, the equation cannot be solved directly for this variable. Instead the *goal seek* function of MS-Excel is used first calculating with an assumed interest rate and (goal) seek the rate for which the *NPV* becomes zero.

Profitability criterion: the IRR must be higher than the WACC of capital. In particular: *IRROI≥WACC, IRROE≥* returns on equity before or after tax.

The equations for IRR on equity (*IRROE*) before and after tax are complex including many variables needing explanation that cannot be provided in this excerpt. It is therefore referred to the mentioned book Engineering Economics.

> MS-Excel provides two functions for calculation of the IRR (pls. refer to the help function of Excel):
>
> IRR(values, guess) and MIRR(values, finance rate, reinvest rate).
>
> IRR calculation must be done in nominal terms, including inflation!

The **Annuity or Annual Equivalent Amount** method requires that the discounted annual operating expenses (OPEX) during the lifetime of a project are added to the *annualized capital expenditures (CAPEX)* and multiplied with the *annuity factor (an)*. The result is the annual equivalent amount of the lifetime costs "*ANU*" (or annuity) of the investment.

$$ANU = an \cdot (CAPEX + \sum_{t=1}^{t=n} \frac{OPEX_t}{(1+i)^t}) = an \cdot NPV \quad \left[\frac{\text{CU}}{\text{a}}\right] \tag{2.15}$$

MS Excel function: PMT (*discount rate, nper, pv, type*)

Own Add-In w. escalation: AN_{esc}(*esc. rate, discount rate, nper, value$_0$*)

The formula for levelized electricity generation cost is:

$$LEC = \frac{PMT(CAPEX) + OPEX \left[\frac{\text{CU}}{\text{a}}\right]}{W_e \left[\frac{\text{MWh}}{\text{a}}\right]} \quad \left[\frac{\text{CU}}{\text{MWh}}\right] \tag{2.16}$$

The electricity generation during lifetime is assumed to be constant, discounting is not necessary.

Important Note: A distinct advantage of the annuity method is that investments with different lifetimes can be compared based on their annuities. In contrary, the NPV method requires that lifetimes must be equal. Furthermore, the calculations are much more transparent, compared to the NPV method.

2.3.2 Integrated model for levelized cost calculation

Investment appraisal of power system projects requires in most cases the calculation of annual costs and specific electricity generation cost. For cost calculations either the annuity method or the present value method can be applied. The overall structure of the integrated model for generation costs calculation applied in this book is presented in Figure 2-1.

The structure of the integrated model makes evident that investment appraisal of power system projects comprises both engineering and economic calculations, which are interlinked. Therefore it is called integrated techno-economic model. The model comprises several calculation modules which are explained in detail below:

Technical parameters: The starting point is the definition of the main technical parameters of the power plant project which are relevant for the technical-economic evaluation. In the development phase of such projects, several options are developed which are compared and evaluated from the technical and economic point of view. The objective is to define the most favorable option which will be analyzed in more detail in further steps of the project. For calculation of the electricity generation cost relatively few technical parameters are required as shown in the respective module of the model.

Energy & mass balance: The technical and operational parameters are the inputs for compilation of the *energy, mass and emission balance.* This is done on annual basis; it is commonly assumed that the electricity production remains constant throughout the lifetime.

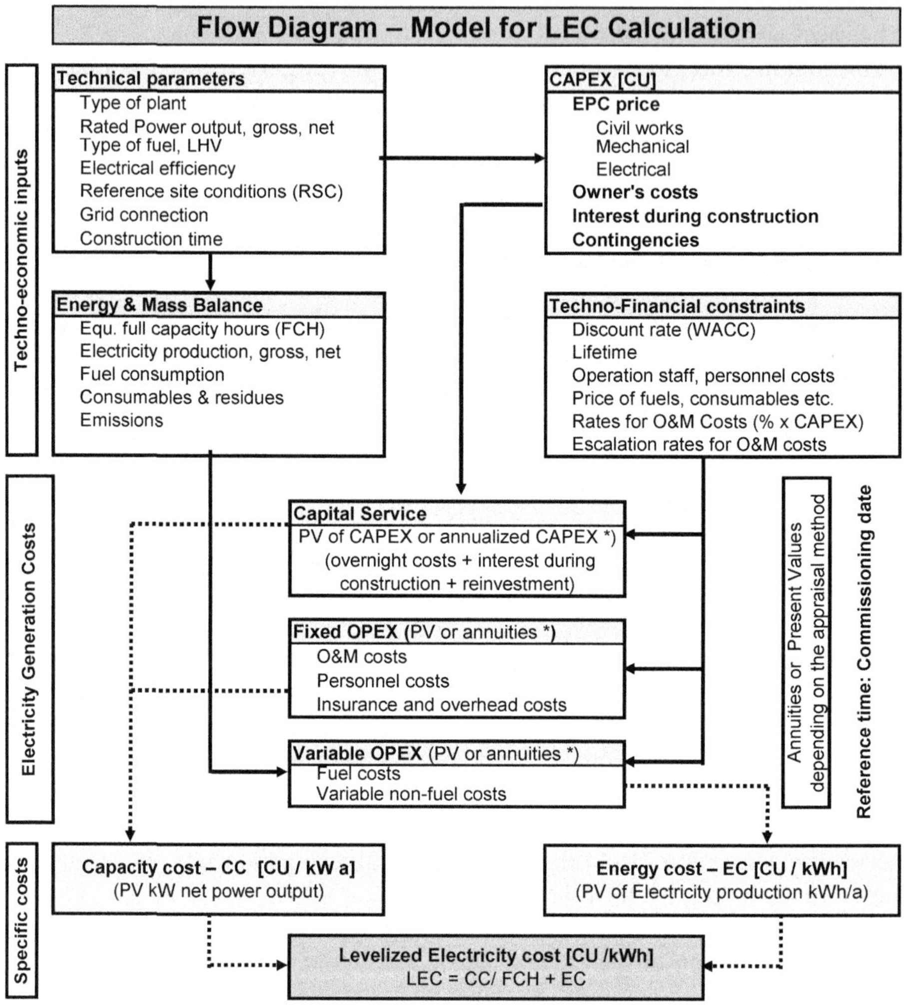

*) depending on the method used, PV or annuity **Source:** Author's own compilation

Figure 2-1: Integrated techno-economic model for electricity costs

CAPEX Estimates: A large number of power plant projects are implemented worldwide every year. So project developers and engineering companies in this business have the skills to estimate the capital expenditures, based on bids from previous projects – Example 2-2.

Overnight costs are the capital expenditures excluding costs of financing mainly *interest during construction*. Other important items are owner´s costs and contingencies for unforeseen items.

An indispensable note in CAPEX estimates is always *the reference year* of the costs, e.g. US$ 2015. The margin of the estimate depends on the stage of the project. For feasibility studies it is usually ±25%. The final CAPEX is negotiated during the tendering procedure and evaluation of bids and is reflected in the contract award.

Cost estimates for several types of power plant projects can be found in [11]. *Reinvestments* during the life time of the projects must also be considered as CAPEX. This may be done by discounting the respective amounts to the present value reference date (usually current year or commissioning date of the plant) see Example 2-2.

Example 2-2: CAPEX incl. interest during construction and reinvestment

Item	Unit	Steam PP	CCGT PP
Input data			
Construction time	years	5	2.5
Payments per year, equal	-	4	4
Overnight costs	mln. US$	1200	380
Reinvestment:			
Replacement of Gas turbine	mln. US$	0	95
Year after commissioning	a	0	15
Discount rate	-	6.5%	6.5%
Present values *)			
Overnight costs	mln. US$	1,200	380
Interest during construction	mln. US$	143	16
in percent of overnight costs	-	12%	4%
Present value of reinvestment	mln. US$	0	37
CAPEX, at commissioning time	**mln. US$**	**1,343**	**433**

*) PV Reference time, commissioning date

Techno-economic constraints: The definitions of the *techno-economic constraints* include the discount rate, the lifetime, prices for fuels and consumables, cost for residues' disposal as well as escalation rates for prices or costs. The discount rate is based on the weighted average costs of capital (WACC).

Lifetime for generation cost calculations and technical lifetime of plants are different. The former corresponds to the depreciation period and the time the investor expects the project to be amortized. The technical lifetime is usually longer. Lifetime depends on the type of power plant.

Fixed O&M costs are usually calculated based on rates referred to the CAPEX or the EPC price. Variable O&M costs are mainly fuel costs for power plant projects or energy costs for other types of projects. Variable non-fuel O&M costs are, e.g., costs for consumables and residues and costs of maintenance contracts which are charged based on equivalent operation hours (EOH).

Calculation of electricity generation Costs – LEC: The generation costs are composed of cost for capital service, fixed operating expenses, and variable operating expenses. LEC may also include opportunity costs such as use of own land for the site, cost as working capital; however, they are not depreciable and cannot be considered e.g. in Profit & Loss account for taxation purposes.
LEC can be calculated either by applying the annuity method or the present value method. The key characteristics of the methods are briefly shown in the table below.

Table 2-3: Calculation methods of electricity generation costs

Item	Annuity Method	Present Value Method
Capital service	Annualized CAPEX [CU/a]	CAPEX present value [CU]
OPEX	Levelized annual OPEX [CU/a]	OPEX present values over lifetime [CU]
Total	**Annual generation costs [CU/a]** Lifetime of options can be different	**Total lifetime costs present values [CU]** Lifetime of options must be equal
Electricity production	**Annual average production [MWh / a]**	**Present value over lifetime [MWh]**
Specific levelized electricity generation costs LEC	**Annual generation costs / annual electricity production [CU / MWh]**	**Present value of total lifetime cost / present value of electricity production [CU/MWh]**
	Lifetime of options can be different for both methods	

Note: "CU" is used for currency units instead of US$ or Euro

Let's point out again:

- Present values of investment options are comparable only if their lifetime is the same
- Annuities of options with different lifetime are comparable; annuity method does not require equal lifetime of options
- However, specific levelized electricity generation cost (LEC in CU/MWh), calculated with the present value method or the annuity method, can be compared also for projects with different lifetimes. This is because both, numerator and denominator of equation (2.13), include the same present value factor that is eliminated. This is also demonstrated in Example 2-3 and Example 2-4 below.

An alleged disadvantage of the annuity method is that the escalation rates for cost items are not considered in the calculation, while escalation rates can be considered in the present value method. The calculation must be conducted however, on year-by year basis.

This means, if we have to compare the present values of 5 options with lifetimes up to 35 years, we would need five spread sheets with at least 35 columns each. This is not a problem for Excel, but the calculation is not transparent and time consuming. With the annuity method the calculation can be conducted in one spreadsheet with five columns only.

In order to avoid drawbacks, the author of this book has modified the functions for annuity method and present value method so that the calculations can be conducted in transparent manner in just one column, including escalation of cost items (see section 2.3.1.). These modified functions are available as Add-Ins or Macros on the author's website; they have been used for the calculation in Example 2-3 and Example 2-4.

Example 2-3: Calculation of LEC by applying Annuity Method

Item		Unit	Steam PP	CCGT PP
Power and energy balance				
Power output, net		MW	600	400
Equivalent operating hours		h / a	5,500	5500
Power generation, net		MWh /a	3,300,000	2,200,000
Efficiency, net		-	42%	55%
Fuel consumption		MWh_t /a	7,857,143	4,000,000
CAPEX, incl. IDC+reinvest		**mln. US$**	**1,343**	**433**
Discount rate, in real terms		% /a	6.5%	6.5%
Lifetime		a	35	25
Fixed OPEX		% Capex	2.50%	1%
Fuel price LHV, w.o. escalation [1)]		US$ / MWh_t	12.29	27.5
Annual OPEX, at start of operation				
OPEX, fixed		mln. US$ /a	33.56	4.33
OPEX variable [2)]	10%	mln. US$ / a	106.18	121.00
Annual costs, incl. escalation for OPEX				
Annualized CAPEX		mln. US$ / a	97.76	35.37
Fixed OPEX [3)]	0.5%/a esc	mln. US$ / a	35.68	4.55
Variable OPEX [3)]	1.5%/a esc	mln. US$ / a	128.26	140.88
Total annual costs		**mln. US$ / a**	**261.71**	**180.80**
Levelized electricity cost		**US$ / MWh**	**79.30**	**82.18**

1) price coal 100 US$/tce natural gas LHV 25 US$/MWh
2) fuel +.10% non-fuel costs 3) escalation in real terms

Example 2-4: Calculation of LEC by applying Present Value Method

Item		Unit	Steam PP	CCGT PP
Power and energy balance				
Power output, net		MW	600	400
Equivalent operating hours		h / a	5,500	5,500
Power generation, net		MWh /a	3,300,000	2,200,000
Efficiency, net		-	42%	55%
Fuel consumption		MWh_t /a	7,857,143	4,000,000
CAPEX, PV total		**mln. US$**	**1,343**	**433**
Discount rate, in real terms		% /a	6.5%	6.5%
Lifetime		a	35	25
Fixed OPEX		% Capex	2.50%	1.00%
Fuel price LHV, w.o. escalation [1]		US$ / MWh_t	12.29	27.50
Annual OPEX, at start of operation				
OPEX, fixed		mln. US$ /a	33.56	4.33
OPEX variable	10%	mln. US$ / a	106.18	121.00
Costs present values, incl. escalation for OPEX				
CAPEX		mln. US$	1,343	433
Fixed OPEX [3]	0.5%/a esc	mln. US$	490	56
Variable OPEX [3]	1.5%/a esc	mln. US$	1,761	1,723
Total		**mln. US$**	**3,594**	**2,211**
PV of electricity production		**MWh**	**45,317,593**	**26,908,509**
Levelized electricity cost		**US$ / MWh**	**79.30**	**82.18**

1) price coal 100 US$/tce natural gas LHV 25 US$/MWh
2) fuel +.10% non-fuel costs 3) escalation in real terms

2.3.3 Capacity cost, energy cost and composite cost

In this book it is strictly distinguished between *capacity cost* (CU/kWa) and *energy cost* in CU/kWh (the latter are often called volume costs). The capacity costs reflect the fixed costs that are independent from the production while the energy costs reflect the variable costs that directly depend on the production. The composite levelized electricity costs (CU/MWh) are only valid for the full capacity hours (h/a) or respectively capacity factor (CF) for which they have been calculated. They are calculated with the following formula:

$$LEC = \frac{Capacity\,cost\left[\frac{\mathrm{CU}}{\mathrm{kW\,a}}\right]}{Full\ capacity\ hours\left[\frac{\mathrm{h}}{\mathrm{a}}\right]} + Energy\ cost\left[\frac{\mathrm{CU}}{\mathrm{kWh}}\right] \qquad \left[\frac{\mathrm{CU}}{\mathrm{kWh}}\right] \tag{2.17}$$

Following figure depicts the LEC versus full load operation hours of a base load and a peak load power plant.

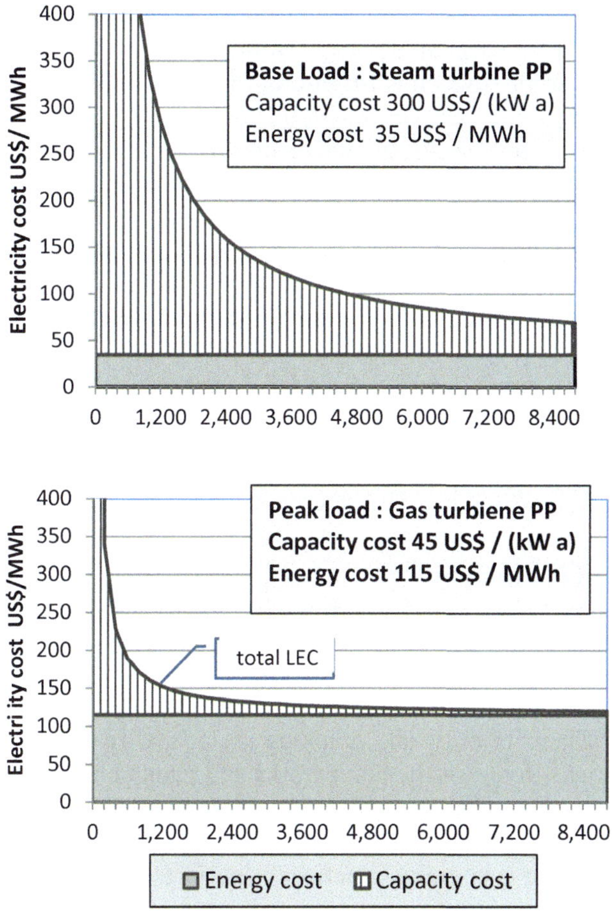

Figure 2-2: Electricity generation cost vs. full load operation hours

The base load power plant is characterized with high capacity cost and low energy cost, while the opposite is the case for the peak load power plant. Up to about 1200 h/a, the composite LEC of the peak load plant are lower and afterwards remain almost constant with high energy cost. The LEC of the base load plant are steadily decreasing with increasing full load hours.

2.3.4 Cash flow analysis and risk assessment

The cashflow analysis[9] is actually a projection of the profit and loss (P&L) statements over the lifetime of the project. The P&L statement gives all payments series, how the project incurs its revenues and expenses, through its operating activities; it shows the profit or loss and the cashflow over a specific accounting period, typically over a fiscal year. The *cashflow* is the actual outcome of the analysis. It is the cash ($/a) remaining after all costs, including repayment of the annual installment for loans, are deducted from the revenues. This is the available amount for payment of dividends for equity investors and for building reserves for future expenses.

Project developers use cashflow models in order to convince creditors and investors that the project will generate sufficient cashflows to repay loans and to obtain adequate returns on the investors' equity capital.

An integral and indispensable part of bankable feasibility studies for new projects is a risk assessment and mitigation process. This includes risk identification, analysis of the consequences in case of occurrence and risk management and mitigation.

An overview on project analysis under consideration of uncertainties and probable project risks is included in the mentioned book "Engineering Economics)[10], along with description of methods and tools such as sensitivity analysis, risk analysis and risk mitigation measures.

2.3.5 Forms of contracting

There are two main contracting forms[11]: *EPC-contracting* (Engineering Procurement Construction) and *Lot´s wise contracting.*

The EPC contractor has to deliver a turnkey plant and assumes all the construction risks. There is only one contract. A lot's wise contract comprises a number of contracts (lots) for the main works and plant components such as civil works, steam generator, flue gas cleaning system etc. The number of lots for power plant projects is between 10 and 40. The owner bears all the construction risks. EPC contracting is common practice for projects in the power sector.

---

[9]) Power & Energy Systems Engineering Economics, Chapter 5 – Financial and economic Analysis of Projects, same author

[10]) Same book, Chapter 7 –Project analysis under uncertainties

[11]) See chapter 10 "Development of Energy System Projects", this book

3 Thermal Power Plants Fired by Fossil Fuels

3.1 Steam Power Plants

3.1.1 General configuration and function

Steam power plants are currently the most common in the power industry. A simplified heat flow diagram, including the main plant components, is depicted in Figure 3-1 below.

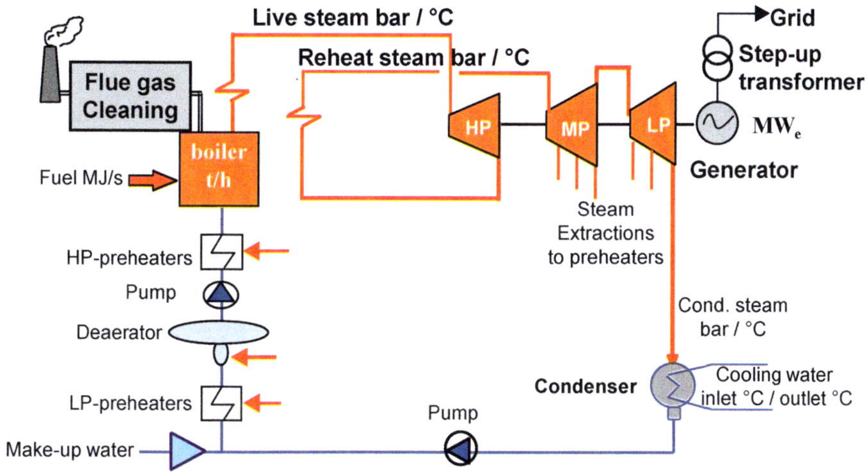

Source: Technologies & Economics, Author's own illustration

Figure 3-1: Simplified heat flow diagram of a steam power plant

The energy conversion from fuel to electricity is based on the Clausius Rankine thermodynamic cycle (Rankine cycle). The cycle includes the following thermodynamic processes: The chemical energy of the fuel is converted into thermal energy in the form of high pressure and temperature steam in the steam generator, known as the boiler. The steam is led to the steam turbine where its pressure and thermal energy is turned into mechanical energy by steam expansion. The generator converts the mechanical energy into electricity at a voltage between 6 kV to 30 kV. A step-up transformer increases the voltage level to that of the grid

P. Konstantin and M. Konstantin, *The Power Supply Industry*,
https://doi.org/10.1007/978-3-319-72305-1_3

(380 kV or higher). The exhaust steam leaves the turbine with a pressure considerably lower than the atmospheric pressure and is liquified in the condenser and pumped via the boiler feed water system to the boiler for re-evaporation. The combustion flue gases of the boiler pass the flue gas cleaning system, consisting of a denitrification plant (DeNOx), an electrostatic particulates' precipitator (ESP) and a desulfurization plant (FGD).

3.1.2 The Clausius Rankine steam cycle

The working principle of power generation in steam power plants is the *Clausius Rankine cycle*. The basic processes of the cycle are depicted in Figure 3-2 in an Enthalpy-entropy (h-s) diagram and in a Temperature-entropy (T-s) diagram. In h-s diagrams, energy amounts are depicted as lengths of line; this diagram is used for cycle calculations (Mollier h-s Diagram). In T-s diagrams, energy amounts are presented as areas; this diagram is suitable for visualization and helps to better understand the process flow.

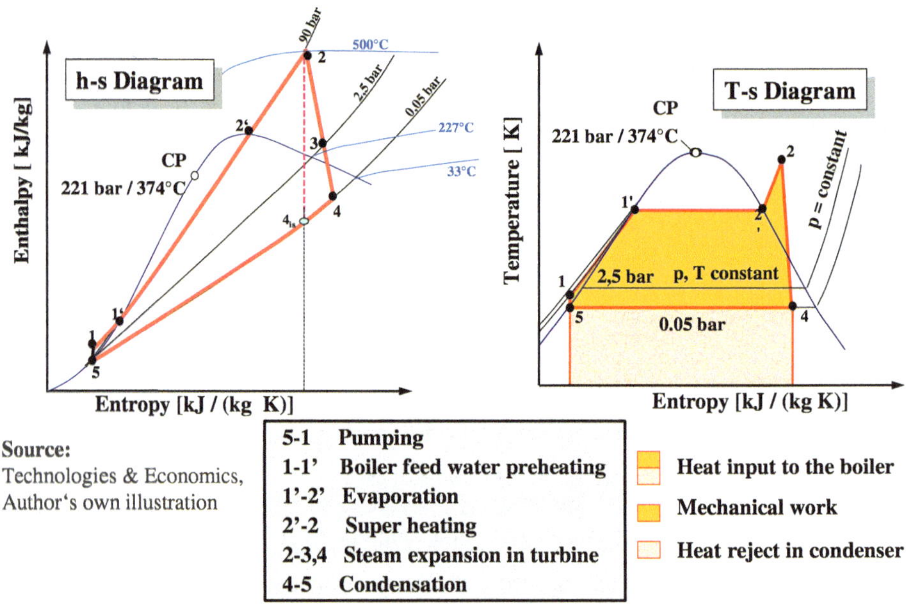

Figure 3-2: Simplified Rankine cycle illustration in h-s and T-s diagram

Typically, new steam power plants are operated in base load mode (full power output throughout the year). Their electrical output and efficiency depend on the live steam conditions (pressure and temperature) and the exhaust steam conditions (cooling medium temperature). Most of the existing power plants are designed for sub critical (SubC) steam parameters. The need of higher fuel efficiencies, in order to offset rising fuel prices, have forced manufacturers to increase steam parameters to super critical (SC) or even ultra-super critical (USC), as shown in Table 3-1 and Figure 3-3.

Table 3-1: Typical parameters for utility size steam power plants

Item	Unit	SubC	SC	USC
Typical rated power capacity	MW	600	800	800
Steam generation	t / h	1,850	2,455	2,040
Live steam parameters	bar / °C	160 / 535	240/ 540	285 / 600
Reheat steam parameters	bar / °C	42 / 535	48 / 540	59 / 620
Cooling water temperature *)	°C	31	31	31
Electrical efficiency, gross	-	42.0%	45.5%	48.5%
Heat rate	kJ / kWh	8,571	7,912	7,423

*) cooling tower, Central Europe

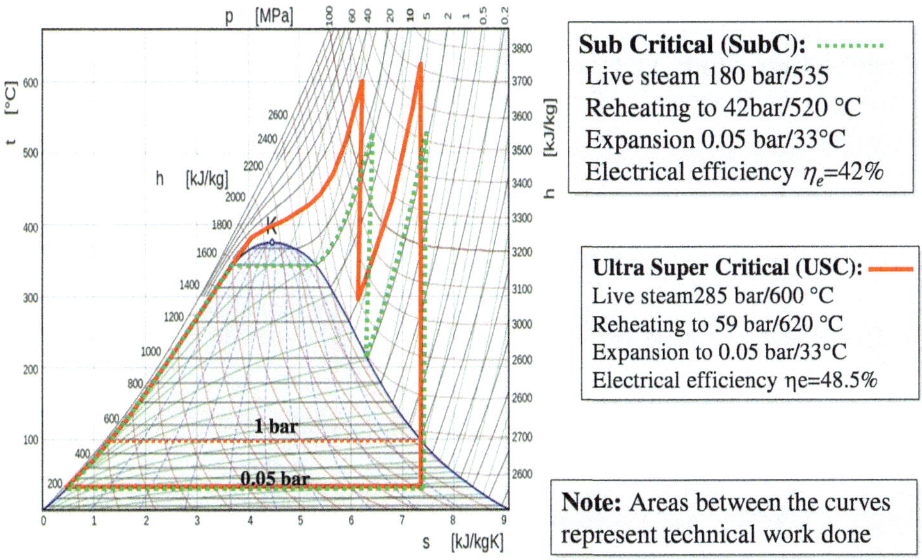

Source: Technologies & Economics, Author's own illustration

Figure 3-3: SubC vs. USC Rankine cycle in T-s diagram

Recommendation: The measures for a steady performance improvement of the Rankine cycle since its first industrial application till today are visualized in **Case Study 11.3.**

3.1.3 Modelling and simulation of thermodynamic cycles

Thermodynamic cycles are modelled and simulated with very sophisticated software tools available on the market. The author of this book is familiar with the software tool KPRO® [12]. The modelling of the power plant with KPRO® is achieved by means of "elements" (gas turbine, steam generator, steam turbine and heat exchangers) and connecting pipelines. The associated heat flow diagram is directly drawn on the screen with the aid of a graphic input system (Example see **Annex 12**). The existing elements and their connections with each other are derived automatically from the flow diagram.

It is pointed out however, that such kind of software tools requires a very strong background on thermodynamics and intensive training of the users. Experience has shown that users before starting using complex software tools should be able to calculate cycles with the traditional, conventional tools such as Mollier h-s diagram and steam tables or steam table software, otherwise there is a danger of erroneous application of the tools. This is especially recommended for students and young engineers.

For this purpose models and calculations of simple Rankine cycles are presented in **Case Study 11.1** and in **Case Study 11.2**. The latter is conducted for the simple cycle shown in Figure 3-4. Both are also available as softcopies on the author's website.

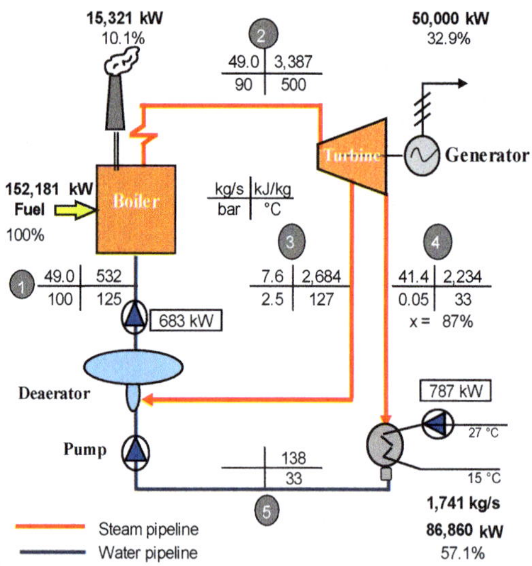

Figure 3-4: Heat flow diagram of a simplified Rankine cycle

A step-by step instruction for calculation of thermodynamic cycles of power plants is provided in **Tool Guide 3** in the Toolbox section of the book.

3.1.4 Main components of steam power plants

Annex 11 depicts an image of a modern steam power plant with all the main components.

3.1.4.1 *Steam generators*

There are two basic types of *steam generators (or boilers)* regarding the arrangement of their heat exchange surfaces, one-pass and two-pass boilers. The term *pass* refers to the flow direction of the flue gases. In one-pass boilers superheater reheater, evaporator and economizer are placed on top of the combustion chamber, while in two-pass boilers these are placed in the second pass. The former type is more common in Europe, the latter in Asia. Steam boilers are fired with a variety of fuels such as coal, heavy fuel oil, biomass and even municipal waste is used in waste-to-energy power plants.

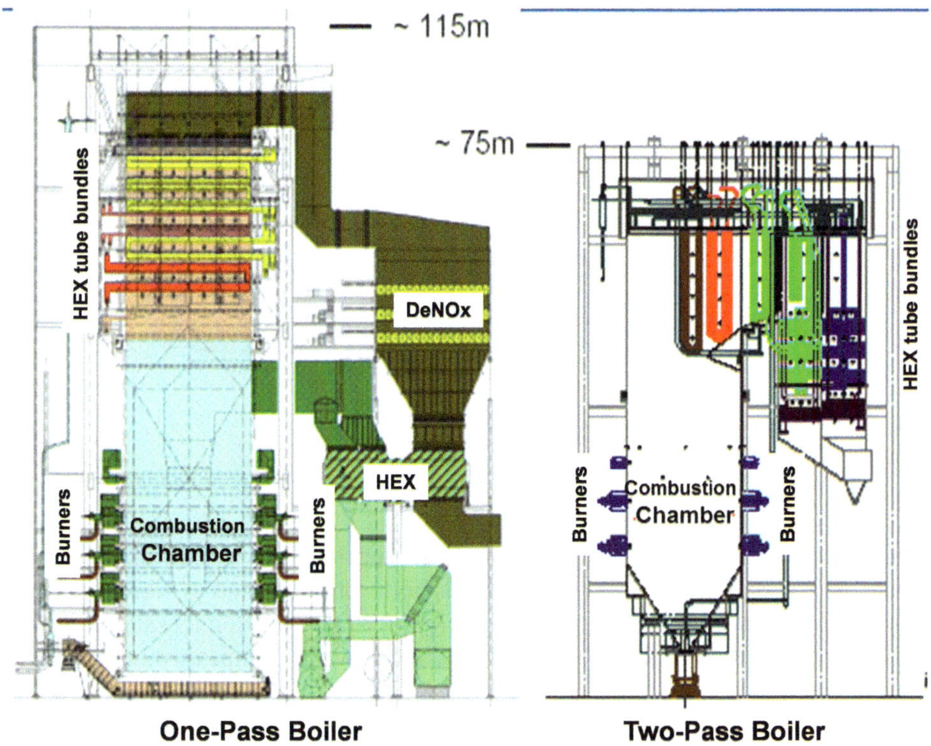

Figure 3-5: Boiler type configuration for a 600 MW_e power plant

3.1.4.2 *Steam turbines*

The steam turbine extracts thermal energy from pressurized steam coming from the boiler and expands it to generate mechanical work on a rotating shaft for driving the electrical generator. Steam turbines of utility size power plants usually are of condensing-reheat type design consisting of a high pressure (HP), intermediate pressure (IP) and low pressure (LP) part. The turbines are also equipped with steam extractions for the feed water preheaters. The steam volume passing the turbine depends on the pressure and determines the size and arrangement of the turbine. In Figure 3-6 the HP-part is of one-casing, one outflow, the IP-turbine of one casing two-outflow design. The volume of the steam in the LP part is very large, therefore the LP part consists of three double flow casings.

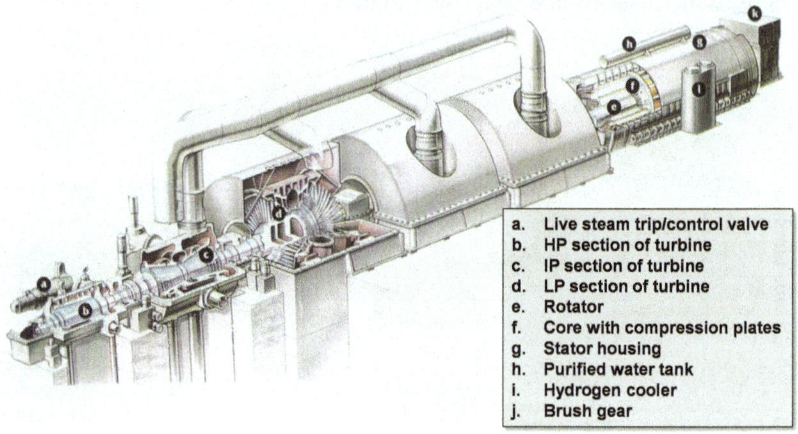

Figure 3-6: Steam turbine-generator set

3.1.4.3 *Steam–water cycle*

The HP steam entering the turbine leaves the turbine after expansion as exhaust steam at a pressure lower than the atmospheric pressure and is liquified in the condenser and pumped to the deaerator (Figure 3-1) by the condensate pumps. At the deaerator exit the water is pumped by the feedwater pumps through the boiler for re-evaporation.

The electrical efficiency depends on the live steam parameters (bar/°C) and on the cooling medium temperature for the condenser. The cooling medium can be sea or river water, where available, or circulating water from a cooling tower. In areas where no water source is available, ambient air is used as cooling medium in dry air cooling towers.

The temperature of the cooling medium has a significant influence on the efficiency of the power plant. In the case of water this can be estimated with the following approximation formula (t_{cw}: actual cooling water temperature):

$$\eta_{gross} = 0.442 \times (-\ 0.0028 \times t_{cw} + 1{,}0426)\ [\%] \qquad (3.1)$$

Example 3-1: Fuel costs vs. electrical efficiency

Calculate the net electrical efficiency and the specific fuel cost per MWh electricity for a steam power plant with SubC steam parameters located at the North Sea and at the Gulf.

Item		Unit	Location	
			North Sea	Gulf
Technical parameters				
Rated capacity, gross		MW	600	
Steam parameters *)		bar /bar /°C	160 / 42 / 535	
Cooling water temperature		°C	15	35
Electrical efficiency, gross **)		-	44.23%	41.75%
Heat rate		MJ / kWh	8.14	8.62
Energy balance				
Equivalent operating hours		h / a	7,500	
Annual electricity generation		GWh_e /a	4,500	
Annual fuel consumption		GWh_t /a	10,175	10,778
Fuel costs				
Fuel price		US$ / t_{ce}	120	
Fuel heat price	8.14 MWh/tce	US$ / MWh_t	14.74	
Annual fuel costs		**mln US$ / a**	**150**	**159**

*) Live steam pressure / reheat pressure / live steam temperature
**) Approximation formula η=0.442 x (-0.0028 x tcw + 1.0426)

3.1.4.4 *Flue gas cleaning technologies*

In order to meet emission standards, steam power plants, fueled with coal or heavy fuel oil, are equipped with flue gas cleaning systems including electrostatic precipitators (ESP) for particulates, flue gas desulfurization (FGD) plant for sulfur dioxide emissions and denitrification (DeNOx) plant for nitrogen oxide emissions. Following three main technologies are in application (Schematics are shown in **Annex 13, Annex 14** and **Annex 15**).

- Seawater FGD (with ESP and DeNOx if necessary)
- Limestone-Gypsum FGD (with ESP and DeNOx if necessary)
- Integrated WSA/SNOX System

In the *seawater FGD-Process* seawater is used to absorb and neutralize sulfur dioxide from the flue gas of fossil-fired power plants. Some 21,000 m³ of seawater are needed for desulphurization of 1 t SO_2. Seawater can be taken from the main condenser outlet. For fuels with a high sulfur content, additional water is required that must be taken directly from the seawater intake.
The *wet limestone/gypsum* process is currently the most common technology used in power plant flue gas scrubbing. The byproduct gypsum can be sold to the construction material industry or disposed of. The high amounts of limestone and gypsum require appropriate infra-structure and logistics for transport and disposal, especially for fuels with high sulfur content as shown in the table below.

Item	VR	HFO-380	AH	AL
Sulfur in fuel	**5.70%**	**3.70%**	**2.96%**	**1.42%**
Limestone consumption [t/h]	23.48	15.08	11.97	5.50
Gypsum production [t/h]	40.39	25.94	20.59	9.46

Note: Reference power plant: electrical output 600 MWe, fuel input 1,420 MWt
VR: vacuum residue; HFO: heavy fuel oil; AH: Arabian heavy; AL: Arabian light

The *integrated WSA/SNOX* process is a catalytic flue gas cleaning process, removing up to 98% of SO_2 and SO_3, and up to 96% of NOx. The sulfur is recovered as commercial grade concentrated sulfuric acid (H_2SO_4), while NOx is reduced to N_2. No chemicals or additives are required other than ammonia for NOx reduction. Due to the very efficient heat recovery concept the overall efficiency of the power plant can be increased up to 1.5 percentage points.

The specific cost for desulfurization vs. sulfur content and technology is shown in Figure 3-7 below.

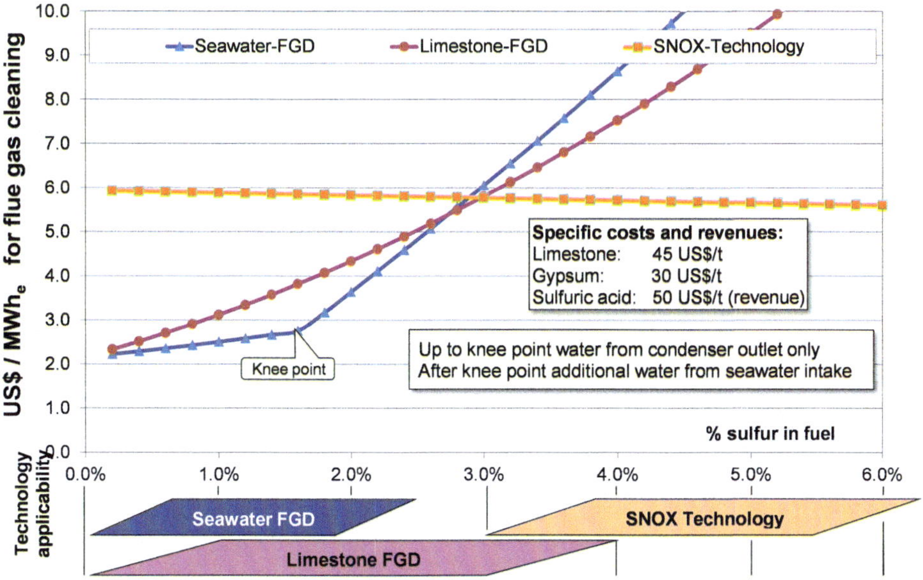

Figure 3-7: Specific cost for desulfurization vs. sulfur content in fuel

Table 3-2: SWOT analysis of flue gas cleaning technologies

Item	Seawater	Limestone/ Gipsum	SNOX
Technology	simple	complex	most advanced
Status of technology	proven and mature	proven and most mature	proven in chemistry industry
References	numerous	large number	few in power plants
SO_2 removal rate, maximum	98%	98%	98%
SO_3 removal rate, maximum	zero	zero	96%
NO_x removal rate	zero, DeNOx plant required	zero, DeNOx plant required	98%
Operation and maintenance	easy	complex	complex
Reagent	seawater *)	lime stone	only ammonia for NOx
Byproduct	no	gypsum	sulfuric acid
Market potential of byproduct	n.a.	limited	need of a long- term contract
Waste water and solid waste	no	yes	no
Electricity consumption	for pumping	relatively high	highly efficient heat recovery
Cycle efficiency **)	small reduction	reduction up to 1.5 p.p.	increased up to p.p.
Capital expenditures ***)	lower	Benchmark	about 40% higher, however DeNOx included

*) for fuels with 1.5% sulfur content seawater comes from the condenser, additional direct from seawater intake
**) p.p. percentage points referred to normal cycle efficiency w.o. FGD
***) referred to CAPEX of limestone gypsum FGD

3.2 Simple Cycle Gas Turbine Power Plants

3.2.1 General configuration and function

The basic arrangement of a gas turbine power plant is depicted in Figure 3-8 along with a section of a gas turbine. Typical for gas turbine power plants are the sizeable air inlet structures and the large chimney, compared to the gas turbine generator set. This is because gas turbines are operated with high excess air (AF~3, boilers ~1.3).

The most common fuel is natural gas; distillate oil is usually used as backup fuel. In the Arabian Peninsula heavy fuel oil and even crude oil has been used, with adverse effects on the output, efficiency, availability and lifetime of gas turbines. Unconventional fuels as e.g. blast furnace gas, converter gas, landfill gas, and different biogases are used.

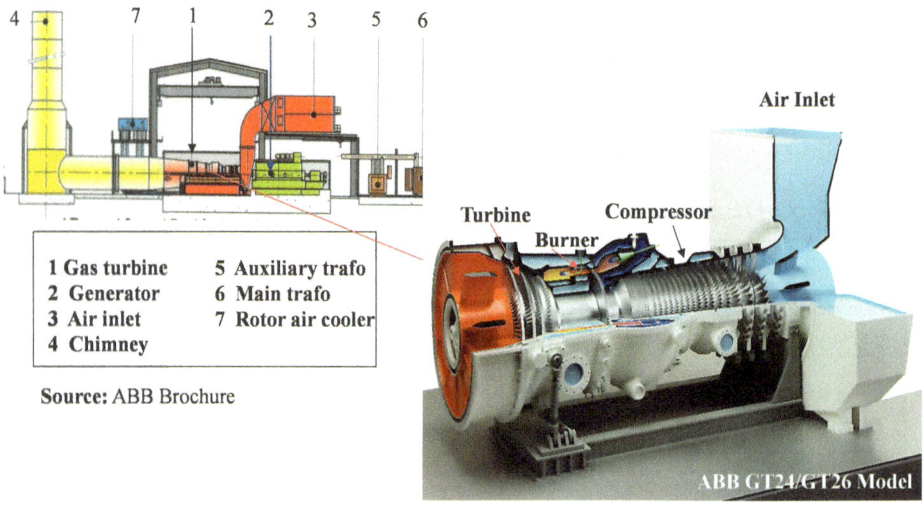

Figure 3-8: Schematic gas turbine power plant and gas turbine section

The main advantages of gas turbines include relatively quick start-up capabilities and high rates of load change. A hot startup takes about 17 to 20 minutes from zero to full output, depending on the type of the gas turbine. Therefore gas turbines are operated for peaking duty and as short time reserve. On the other hand, power output and electric efficiency drop steeply in part-load operation; so gas turbines are usually not operated under 60% load.

Note: The performance parameters of gas turbines are referred to as ISO standard conditions, these being:
→ Ambient temperature: 15°C, relative humidity: 60%
→ Pressure level: 1.013 bar (sea level)

Two types of gas turbines are used for power generation:

- Heavy duty gas turbines which are derived from steam turbines at a capacity range from 3 MW up to about 375 MW.
- Aeroderivative gas turbines which are derived from aircraft engines at a range up to about 60 MW.

Large, state-of-the-art gas turbines obtain remarkably high electric efficiencies up to 40%, based on lower heating value of the fuel. The most powerful gas turbine is currently the Siemens SGT5-800-H with a power output of 375 MW at 40% efficiency and the Mitsubishi M701J with 470 MW at 41% efficiency. Technical performance parameters are available in [13] and [14].

3.2.2 The gas turbine cycle

A gas turbine is a combustion engine consisting of a combustion air compressor, combustion chamber and a turbine –Figure 3-8.

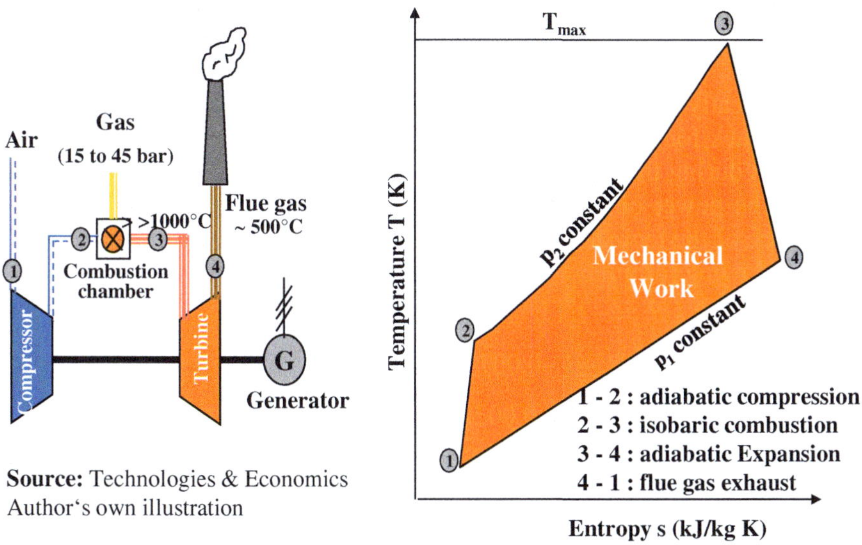

Source: Technologies & Economics
Author's own illustration

Figure 3-9: The basic gas turbine cycle

Compressor, gas turbine and generator are usually mounted on a common shaft. The gas turbine operation principle is called the *Joule cycle* in the European or *Brayton cycle* in the English literature; its working fluids are air – Figure 3-9.

The cycle, presented in Figure 3-9, is in *open cycle* configuration; the working fluid passes through the turbine only once and after expansion is released in the environment. *Closed cycle*, types where the working fluid is cooled down in a heat exchanger and recirculated are barely applied for commercial power generation.

Gas turbines are used in power plants in *simple cycle (SCGT)* configuration and *combined cycle (CCGT)* configuration. The latter is a combination of a gas turbine and a steam cycle.

The main parameters, influencing the performance of the cycle, are:

- the air inlet temperature to the compressor
- the pressure ratio ($\pi = p_2/p_1$)
- the inlet temperature to the gas turbine
- the load during operation

About 2/3 of the gas turbine's power output is used for driving the compressor and the remaining is the actual power output.

3.2.3 Performance vs. air inlet temperature

Gas turbines are volumetric devices. They draw a constant volume flow m³/s at a certain load while the power output is proportional to the mass flow rate kg/s. Hot air is less dense resulting to lower mass flow and power output. Therefore, site conditions, regarding ambient temperature and pressure (depending on site elevation), greatly influence the output and performance of gas turbines. This is a major drawback for regions with hot climate because the annual on-peak of grid loads occurs during the hot season, due to high power demand for air conditioning, while the power output of gas turbines drops. Rough estimations regarding the influence of the temperature and altitude on the performance, can be obtained with the following formulas:

Elevation *h*: $$P_h = -1,11 * 10^{-4} \times h + 1 \quad [\mathrm{kW}] \qquad (3.2)$$

Temperature *t*: $$P_t = P_{ISO} \cdot (-0.007 \cdot t + 1.105) \quad [\mathrm{kW}] \qquad (3.3)$$

$$\eta_t = \eta_{ISO} \cdot (-0.0022 \cdot t + 1.033) \quad [\%] \qquad (3.4)$$

Where "*h*" is the elevation in m and "*t*" is the ambient temperature in °C.

Example 3-2: Power output and efficiency vs. site conditions

Item	Unit	Values	
Technical parameters			
Gas turbine type		STG6-5000F	
Fuel	-	natural gas	
Rated power output, ISO	MW	208	
Electrical efficiency, ISO	-	38.10%	
Location (Saudi Arabia)		**Jazan Red Sea**	**Riyadh Central**
Elevation	m	7	608
Design temperature	°C	38	44
Correction factors:			
Elevation factor for power	-	0.999	0.933
Temperature factor for power	-	0.976	0.972
Temperature factor for efficiency	-	0.949	0.936
Actual power output	**MW**	**203**	**189**
Actual electrical efficiency	**-**	**36.2%**	**35.7%**

In order to prevent drop of output due to increased ambient temperatures in hot climates, the *inlet air cooling* is often applied either by injection of decarbonized water in the air inlet, that evaporates and cools down the combustion air, or by chilling the inlet air by compressor chillers. The most common is evaporative inlet cooling as shown in Figure 3-10 and Figure 3-11.
In cold climates, in contrary, icing of parts of the inlet path to compressor may occur, therefore the inlet air temperature is kept above 7°C by appropriate heating devices, see Figure 3-11.

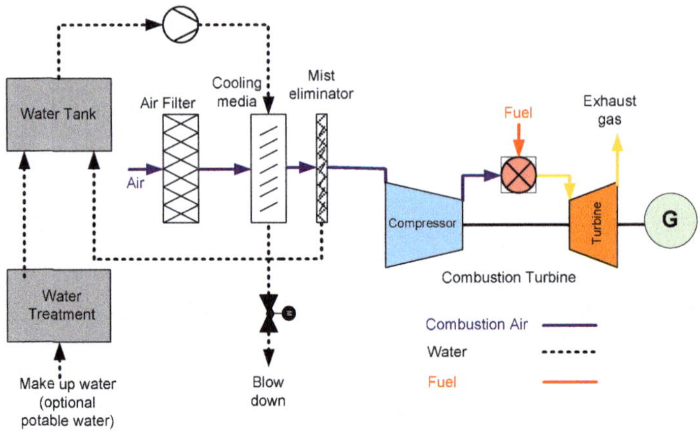

Figure 3-10: Schematic of evaporative inlet air cooling system

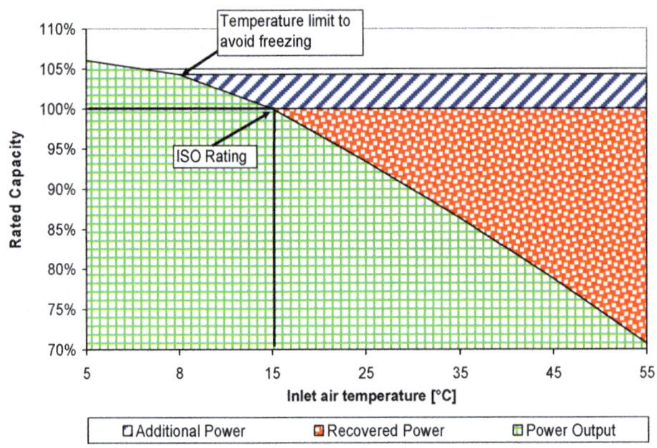

Figure 3-11: Power output vs. inlet temperature with evap. cooling

3.2.4 Pressure ratio and fluid inlet temperature into the turbine

A high pressure ratio $\pi = p_2/p_1$ (see Figure 3-9) increases the cycle efficiency and the specific power output referred to mass flow kW/(kg s). Aeroderivative gas turbines are designed for high pressure ratios in order to minimize weight and frontal area.

It is evident that inlet temperature to the turbine (Figure 3-9) has a major impact on increasing power output; there are, however, material limitations. As gas turbine material have developed, modern gas turbines obtain inlet temperature of about 1200°C and higher.

3.2.5 Reheating or sequential combustion

A possibility for increasing power output and efficiency is reheating also called sequential combustion. The hot gases leaving the HP part of the gas turbines are reheated and expanded in the LP part of the turbine – Figure 3-12.

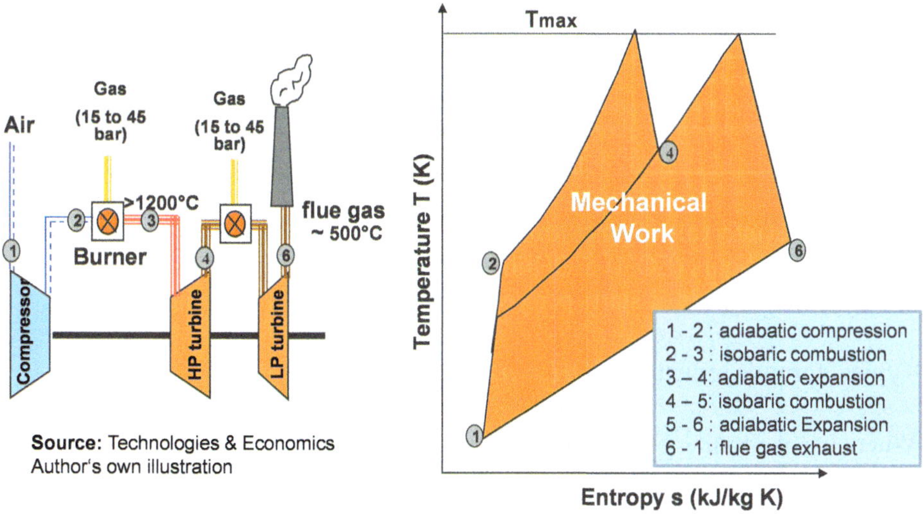

Figure 3-12: Gas turbine cycle with sequential combustion

3.2.6 Emission control

Sulfur dioxide and particulates' emissions are negligible for gas turbines fired by natural gas. Modern gas turbines are equipped with low-NOx burners and meet even strictest emission standards (NO_2<50 mg/m^3 for gas and 120 mg/m^3 for distillate oil, at 15% reference O_2). During operation with distillate as backup fuel, steam or water injection is applied to reduce NOx formation in the flue gases.

3.2.7 Maintenance of gas turbines

Performance and lifetime of gas turbines highly depend on a proper maintenance concept. The overall objective of maintenance is to ensure reliability of operation and optimization of performance at minimum lifetime cost. An optimal reliability centered maintenance (RCM) concept is a mix of three main components namely: corrective, preventive and predictive maintenance.

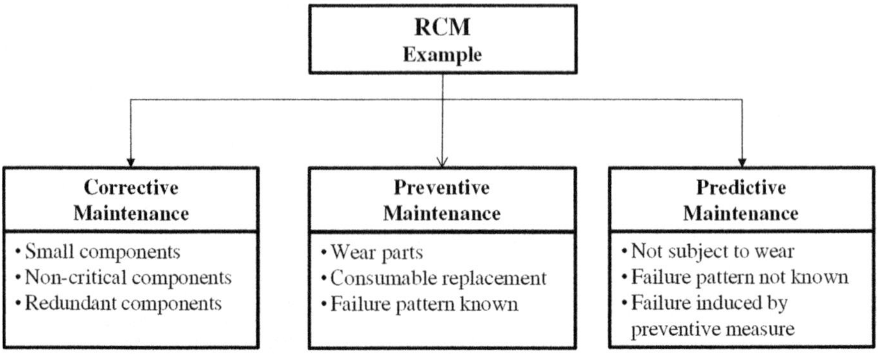

Maintenance intervals of gas turbines are commonly scheduled based on equivalent operating hours (EOH). EOH include operation time, load factor, fuel change and type and number of starts per year such as hot, warm and cold [15] [16]:

Typical formula structure: $EOH = a_1 t_1 + a_2 t_2 + a_3 n_1 + a_4 n_2$ (3.5)

Where (numbers differ based on experience and type of GT):
$a_1=1, t_1$ Full load operating hours; $a_2=4$, t_2 peak load operating hours
$a_3=10$, n_1: Number of startups/a; $a_4=10$, n_2: number of rapid load change
Note: Estimated startups: base load=10; intermediate load=50; peak load =300

Table 3-3: Guide values for maintenance of gas turbines

Type of Maintenance	Inspection	Interval EOH	Duration days
Short inspection	Boroscope inspection	12,000	5
Medium inspection	Hot gas path inspection	24,000	20
Major inspection	Inspection & overhaul	48,000	30

Maintenance is usually outsourced to external contractors based on longterm service agreements (LTSA), usually for 10 to 12 years duration. Maintenance contracts are a significant cost factor. The contracts commonly are composed of a fixed component (US$/a) and a variable component based on the EOH (US$/EOH). Especially for peak load gas turbines the EOH are becoming very high due to daily startups.

3.3 Combined Cycle Gas Turbine Power Plants

3.3.1 General configuration and function

The schematic in Figure 3-13 shows the configuration of a combined cycle gas turbine (CCGT) power plant, comprising gas turbine(s) with downstream heat recovery steam generator(s) (HRSG) and a common steam turbine generator set. HRSGs can be of single, double or triple pressure design, they may also be equipped with supplementary firing as shown in the schematic. The gas turbine and the steam turbine each have their own generator. There are also "single shaft arrangements" where the gas turbine, steam turbine and generator are arranged on one common shaft. Configurations with two gas turbines and one single steam turbine are most common in order to increase availability of the plant. An image of a CCGT power plant is shown in **Annex** 16.

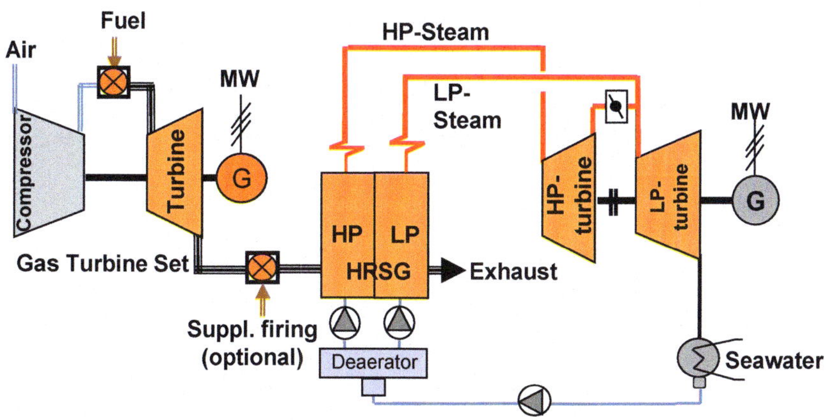

Source: Technologies & Economics, Author's own illustration

Figure 3-13: Simplified CCGT heat flow diagram, double pressure HRSG

3.3.2 Thermodynamic cycle

CCGT thermodynamic cycles consist of a combination of the Brayton (gas turbine) and the Rankine (steam) cycle. The heat from the exhaust gases of the gas turbine at about 500°C is utilized to generate high pressure (HP) and partly low pressure (LP) steam in the heat recovery steam generator which is then expanded in a steam turbine to generate additional electricity. The total electrical generation capacity is made up of the contributions from the gas turbine and the steam turbine. As rule of thumb about 2/3 of the power output is provided by the gas turbine(s) and 1/3 by the steam turbine. The electric cycle efficiency of CCGTs is considerably higher, compared to those of simple cycle gas turbines or steam power plants. Gross efficiencies of 60 percent at ISO conditions are common for new CCGT plants – see KPRO® model **Annex 17**.

Figure 3-14 and Figure 3-15 depict an advanced design of CCGT plants with dripple pressure of live steam. Typical designs have a power output of 400 MW ISO.

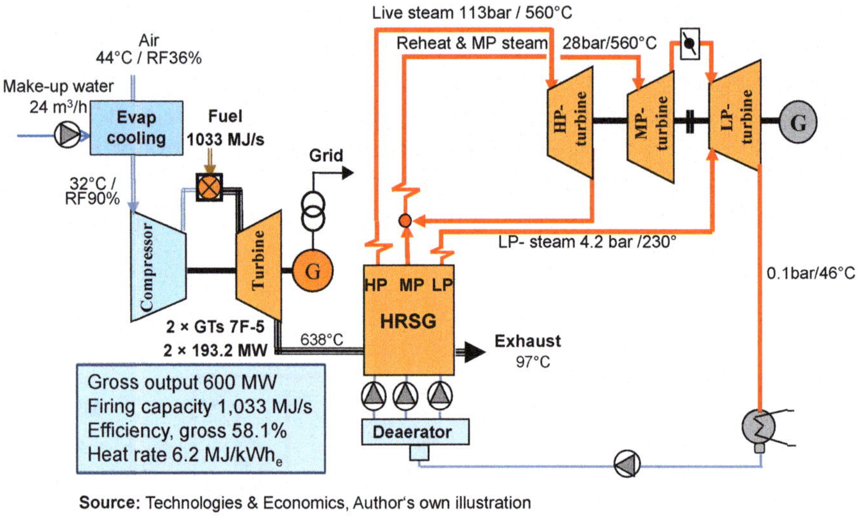

Figure 3-14: Simplified CCGT heat flow diagrams, triple pressure, reheat

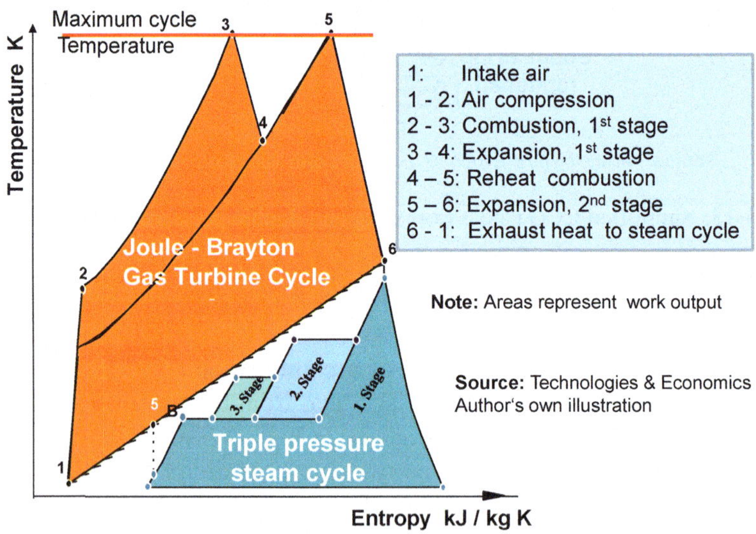

Figure 3-15: CCGT cycle T-s diagram, GT-reheat, triple pressure HRSG

CCGT power plants are operated in base and intermediate load mode, depending on the gas price. They are also operated to balance transient power supply from wind and solar power plants.
The most powerful plant, equipped with the Siemens SGT5-800H gas turbine, has a power output of 578 MW and gross efficiency of 60.75% at ISO conditions (first plan in operation in Irsching/Bayern/Germany since 2011, [17]).

The power output and electrical efficiency of the CCGT plants depend on the ambient temperature and to a lesser extent on the cooling water temperature. In order to offset the ambient temperature influence, gas turbines in hot climates are often equipped with inlet air cooling systems.

3.3.3 Selection of gas turbines

Gas turbines, suitable for CCGT applications, must fulfill the following main criteria:

- Advanced well proven design
- High electrical efficiency (low heat rate)
- High flue gas mass flow and temperature.

Heavy duty gas turbines are the predestinated technology for CCGT application. Heavy duty gas turbines with ISO output in the range of 250 MW_e obtain electrical efficiency up to 40% (heat rate 9.0 MJ/kWh) in H-class, and 38% (9.5 MJ/MWh) in F-class and flue gas temperature of 600 °C and higher.

Aeroderivative gas turbines offer the highest efficiency in simple cycle mode (up to 44.5% or 8.13 MJ/kWh heat rate) and can be operated more flexible. These GTs are, however, only available as rather small units in the range of up to about 100 MW. Furthermore their flue gas temperature is too low (~420 °C). This results to a significantly lower efficiency in a combined cycle mode compared to heavy duty industrial GTs.

3.3.4 Heat recovery steam generators

A major component of the CCGT power plants is the heat recovery steam generator (HRSG). HRSGs are actually heat exchangers which utilize the heat of the gas turbine exhaust gas to generate live steam of up to three different pressure stages, namely: high pressure (HP), medium pressure (MP) and low pressure (LP) steam.

In the CCGT cycle, depicted in Figure 3-14, the hot exhaust gases of the gas turbines enter the HRSG with a temperature of 638°C. The generated HP live steam is led to the HP steam turbine and after expansion is reheated in the HRSG and returns to the intermediate pressure steam turbine for further expansion. For better utilization of the exhaust heat, the HRSG generates also medium and low pressure steam. After evaporation the generated medium pressure steam is superheated in the reheat circuit of the HRSG. The low pressure steam is directly supplied to the low pressure steam turbine. After full utilization of the exhaust heat the flue gases leave the HRSG at a temperature of around 100°C.

3.4 Internal Combustion Engine Power Plants

3.4.1 Technologies and function

Power generating units with internal combustion engines (ICE) as the prime mover are a highly advanced and widespread technology. They are widely applied for captive power generation in industrial, commercial and institutional facilities as well as for isolated or small and medium sized power systems. In general, they are built in multi-unit configuration allowing high operational flexibility, reliability and availability. Their main advantages are quick start-up capability, high efficiency also in part load and in higher altitudes and ambient temperatures and fuel flexibility.

There are two basic thermodynamic cycles of internal combustion engines, the *Otto cycle* and the *Diesel cycle* [18]. The main mechanical components of the two cycles are the same. The expanding combustion gases move a piston inside of the cylinder. The piston is connected to a crankshaft that transforms the linear motion of the piston into the rotary motion of the crankshaft. Most engines have multiple cylinders that are connected to a single crankshaft. The main difference between Otto and Diesel cycles is the method of igniting the fuel. Otto-cycle engines use a spark plug to ignite a pre-mixed air fuel mixture introduced into the cylinder. Diesel-cycle engines compress the air introduced into the cylinder to a high pressure, raising its temperature to the auto-ignition temperature of the fuel. The fuel is injected at high pressure.

There are *2- or 4-stroke engines.* 2-stroke engines complete their operating cycle in two strokes of the piston during one revolution of the crankshaft; 4-stroke engines complete their operating cycle in four strokes of the piston during two revolutions of the crankshaft.

Engines are further categorized by *crankshaft speed (rpm* – routes per minute). The speed of the engine basically determines its weight and size and also the capital costs. Power gen sets are medium and low speed engines. The speed ranges for power generator sets (gen sets) typically are:

- High speed > 1000 rpm; < 2 MW
- Medium speed 400 to 1000 rpm; 1 to 10 MW
- Low speed <400 rpm; 3 to 80 MW

Natural gas spark ignition engines are offered up to 20 MW. Their efficiencies are typically lower than those of diesel engines because of their lower compression ratios. Modern spark ignition ranges fired by natural gas obtain efficiencies 42 % to 47%.

For power generation diesel engines are commonly used. Large four-stroke diesel engines, fired with liquid fuels, are offered in sizes up to 20 MW and obtain efficiencies between 45 and 50%. Two-stroke engines are available up to 80 MW.

Single and *dual fuel* gas-diesel engines cover a power range up to about 20 MW. They utilize gas as primary fuel with a small portion of diesel as a pilot fuel for ignition. They can switch from gas to 100% diesel during operation. Multi fuel designs utilize heavy and distillate fossil fuels and a range of bio fuels.

3.4.2 Operational characteristics of engine power generation

Internal combustion engine (ICE) based power plants have not been the primary choice for utilities of large power systems, The most preferred technologies have been large steam power plants, usually fired by coal for base load, combined cycle gas turbine power plants for intermediate load and single cycle gas turbines for peak load. These technologies are at an advanced state-of-the-art and ensure a safe and cost efficient electricity supply.

Fluctuations of demand have been predominately served by gas-fired simple or combined cycle gas turbine units that are synchronized to the grid, but are operated at part load. The situation is changing with the rapidly increasing penetration of power generation from renewable energy in the systems. The intermittency of solar and wind power requires the availability of appropriate backup conventional power capacities with quick response to fluctuation of demand and fast startup ability from zero to full output. Growing shares of renewables require also appropriate conventional backup capacities. Power generation, relying on part load operation e.g. of gas turbines, will reduce considerably the efficiency of the whole system. In this respect, the existing power generation systems need to be complemented by dispatchable, dynamic capacity with the capability of handling frequent fast starts, stops and load ramps.

Transmission system operators have to maintain the stability of the systems despite the intermittency of wind and solar power supply. This variability is managed with redundant generating capacity that can quickly respond to fluctuations in demand. ICE power plants provide flexible power generation that can be rapidly brought online avoiding inefficiency due to part load operation.

Table 3-4: Start-up time of different power generating technologies **[19]**

Type mover	Starting Conditions	Full load in min
Gas engine, Wärtsilä,34SG , 9.3 MW *)	hot start, 70°C cooling water, prelubrication of engine and generator bearings	5
Gas engine, Wärtsilä, 50SG, 18.8 MW *)		7
Gas turbine aeroderivative **)	hot start conditions	10
Gas turbine heavy duty **)	hot start conditions	15

*) **Source**: Wärtsilä brochure power plant solutions 2013

**) Authors own researches

The *modular multi-unit* configuration of ICE power plants provides (Figure 3-16), among others, the following advantages:

- A short construction time with prefabricated components is a crucial advantage, especially for power systems with high growth rates of demand, so that installation of new power generation capacities can better be adapted to the growth of demand
- Superior operational flexibility, due to modular multi-unit configuration, as power output can follow the actual demand by switching units off and on avoiding lower efficiency due to part load operation. Practically, there is no part load operation
- High plant availability and reliability close to 100% in n+1 or n+2 configuration of multi-unit plants
- Operational flexibility due to modular multi-unit design, very fast start-up capability from zero to 100% load within few minutes
- Fuel flexibility, multi-fuel units enable the choice of the most feasible gaseous and liquid fuels
- High energy efficiency up to 50 percent also in part low of units
- Full plant output also at high altitudes and hot and dry ambient conditions
- Minimal water consumption due to close-circuit dry air cooling
- Electrical efficiency remains high also in hot summer; the efficiency drop at 50 °C is only about one to two percentage points.

Figure 3-16: Engine power plant in multi-unit arrangement, Wärtsilä **[20]**

3.4.3 **Maintenance**

Maintenance is typically defined as additional expenses caused by "wear and tear" during startups and normal operation of the plants. As start-ups with engine technology are considerably shorter than with any other thermal technology, fuel and electricity consumption is negligible and there is no "wear and tear" effect in engine technology. It can be concluded that there are no additional start-up expenses with engine power plants. Engines can be started and stopped unlimited times per year.

Fees in *maintenance contracts* are typically calculated based on equivalent operating hours (EOH). These include the actual operating hours plus additional equivalent hours for each startup. This results to a significant EOH amount, especially for peaking plants with frequent startups.

Important note: For ICE power plants with multi-unit modular configuration equivalent operating hours (EOH) are equal with running hours. There are no additional EOH for startups.

All *maintenance* can be effectively performed on site. In a multi-unit plant, only one of the engines is maintained at a time, without affecting the operation of the other units of the plant. This allows scheduling the maintenance unit-by-unit, thereby maximizing the available power generation capacity. Ideally, the maintenance is scheduled at periods of lower power demand.

3.4.4 Emission control

Otto gas engines use two different primary emissions reduction methods for nitrogen oxides: The *Non-Selective-Catalytic-Reduction* (NSCR), called also three-way-catalyst, reduces simultaneously nitrogen oxides (NO_x), carbon monoxide (CO) and hydrocarbons (C_xH_y) to very low concentrations. But it is applied for engines up to about 1 MW only, and the engines must be operated at stoichiometric combustion process with zero excess air.

Larger Otto gas engines use the *lean-burn combustion* process. In this process, natural gas and air are premixed in an air/fuel ratio of 1.15, respectively with about 15% excess air, before being fed into the cylinders. Sulfur dioxide and particulates emissions are negligible for engines fired by natural gas.

Depending on the type of fuel (e.g. HFO), diesel engines may be equipped with flue gas cleaning systems as required by the permitting legislation, including electrostatic precipitators (ESP) for particulates, flue gas desulfurization (FGD) and denitrification (D_eNO_x). These systems are capital intensive for installation and also their operating costs are relatively high.

3.5 Economics of Fossil Fuel Fired Power Plants

Case Study 11.4, presents an integrated techno-economic model for calculation of electricity generation costs of fossil fuel fired power plants. Its general structure is based on the model already described in section 2.3.2 also shown in Figure 2-1. The model includes six selected types of power plants fired with fossil fuels. It has been developed in MS-Excel and consists of eight spreadsheets, the contents of which are described below.

The Model is depicted in Chapter 11, Case Studies as hardcopy and is available as softcopy on the author's Website.

The first spreadsheet of all case studies in this book presents the *summary of results* for a quick overview.

The spreadsheet *Input Technical Parameters* includes the key technical parameters of the selected plants, as far as they are required for cost calculations. The selected plants include: For base load, two Rankine cycle steam power plants, fired with hard coal, are selected. They consist of one unit each with typical power output for this type of plants. For intermediate load, a gas fired combined cycle gas turbine (CCGT) power plant and an internal combustion engines (ICE) multi-unit power plant, fired with heavy fuel oil. For peaking mode, a combustion engine and a gas turbine plant. The selection of the plants is of explanatory nature and does not represent any preference.

The performance of the plants at rated conditions is defined by gross power output and efficiency (or heat rate). Note, however, that output and efficiencies depend on the ambient temperature and the temperature of the cooling medium for the condenser at the site. Therefore, the technical parameters shall reflect the reference site conditions (RSC). In our case the calculation is based on ISO conditions because there is no specific site. The annual average efficiency is lower than the rated efficiency due to wear, degradation and also part load operation. This is considered with a deterioration allowance.

In the spreadsheet *Input Financial*, the operational and financial parameters for the calculation of the generation cost are defined, for most of the parameters typical values based on experience from real projects. For maintenance of gas turbines and engines usually O&M contracts with manufacturers or other bidders of such services are closed. These usually consist of a fixed and a variable tariff component. Fees for O&M contracts are significant cost items. The variable O&M cost are referred to the equivalent operating hours (EOH) that include the running hours and equivalent hours for startups.

An essential parameter for the calculation of electricity generation costs are the *fuel prices* of the different fuels and their relation to each other. In the author's book Engineering Economics [21] it is shown, that the ratio of the prices of the different fuels to the crude oil price can be assumed to remain relatively constant for the medium and longterm evaluation of investments. This seems to be a practical approach considering the lifetime of power generation plants. Hence, the *prices of the different fuels* are defined as a fixed ratio based on the crude oil price.
The discount rate is based on the weighted average cost of capital which is depicted in *spreadsheet WACC*.

Spreadsheet *EOH-O&M* includes inputs and some intermediate calculations required for the actual generation cost calculation. The annual electricity generation is calculated from the net power output multiplied with full load hours. Typical values for full load hours are assumed for the load segment for which the plants are designated. However, due to forced outages during operation, the actual full load hours deviate from the typical values. This is considered with typical availability factors for the different technologies.

Spreadsheet CAPEX: Consulting companies usually estimate capital expenditures (CAPEX) based on bids for recent projects. Annually updated cost estimates[12] for several types of power plants are available [22]. Guide values of CAPEX for different technologies are also shown in **Annex 18**. Updated prices for gas turbines and CCGT are annually published in gas turbine world hand book [13], see

[12] eia - Updated Cost Estimates for Utility Scale Electricity Generating Plants and Gas Turbine World Handbook

Annex 19, **Annex 20** and **Annex 21**. If projects are in an advanced stage, *budget offers* from manufacturers for the main components should be acquired. In feasibility study phase a margin of ±25% is expected in most cases.

The cost estimates for CAPEX in this case study are based on specific cost per kW known from different EPC offers (Engineering Procurement Construction) for similar projects. They consist of three main items namely: Overnight costs, owner's costs and interest during construction. *Overnight EPC costs* include civil works, delivery and installation costs for mechanical equipment, electrical instrumentation and control equipment and project indirect costs. EPC prices usually do not show a detailed cost breakdown. *Owner's costs* mainly include costs for studies during project development, legal fees for construction permits, infrastructure and provision of electricity and other utilities during construction. *Interest during construction IDC* is due for dispersed installments for bank loans during construction.

The complete calculation of the electricity generation costs, including also the main techno-economic parameters, is depicted in the spreadsheet *GenCostCalculation*. The annual costs are calculated on real terms, excluding inflation, using the annuity method (annual equivalent amounts). The calculation of the O&M costs can be conducted with or without escalation rates (*yes* or *no* input in the spreadsheet "Input financial"). For calculation with escalation the Add-In "ANesc" is needed. The final output of generation cost calculations are levelized electricity generation costs (LEC) broken down in capacity cost (US/kWa), energy cost (US$/MWh) and composite cost (US$/MWh). A significant cost component for fossil fueled power plants is the costs for carbon emission (CO_2) allowances. They are calculated separately as prices for CO_2-certificates and are very volatile on the market.

Table 3-5 below depicts the results of the calculation of the case study in a compressed manner.

The case study includes also two graphs shown in Figure 3-17 and Figure 3-18. The graphs depict the *merit order* of power plants. Merit order refers to the operational ranking of power plants for dispatching to deliver power into the grid. Power plants with lower marginal specific cost are dispatched first, until the demand in the related load segment is covered, followed by the plants ranked next.

Figure 3-17 depicts the total (fixed plus variable) power generation cost vs. full load hours. They are also known as the long run marginal cost – LRMC. This information is relevant for defining the type of required new power plants in a power generation expansion plan. If, e.g., there is a shortfall in base load segment (full load hours 7,000 h/a) steam PPs or CCGT PPs are the most cost effective. For peak load with less than 1000 h/a, gas turbines or ICE PPs have the lowest cost.

Table 3-5: Techno-economics of fossil fired PPs, Summary of Results

Item	Unit	Steam USC coal	Steam SubC coal	CCGT nat. gas	IC Engine HFO	IC Engine LFO	GT LFO
Energy balance							
Number of units	-	1	1	1	24	20	2
Power output net	MW	744	555	404	402	335	329
Net electricity production	GWh_e /a	5,566	4,152	1,959	2,009	502	478
Fuel consumption	GWh_t / a	11,843	10,127	3,320	4,481	1,120	1,462
Financial constraints							
Life time	a	35	35	25	25	25	20
Construction time	a	5.0	5.0	2.5	2.0	2.0	1.5
Discount rate (WACC), real terms	% / a	6.5%	6.5%	6.5%	6.5%	6.5%	6.5%
Fuel price	US$ / MWh_{LHV}	19.3	19.3	39.5	38.0	61.3	61.3
CAPEX, 2014 US$, ±25%	**mln US$**	**2,440.5**	**1,564.2**	**404.3**	**415.0**	**282.3**	**120.6**
Annual power generation costs	**mln US$ / a**	**462.6**	**348.5**	**179.7**	**227.5**	**101.7**	**109.9**
Annualized CAPEX	mln US$ / a	177.7	113.9	33.1	33.9	23.1	10.9
OPEX fixed	mln US$ / a	51.2	35.4	8.6	8.0	6.1	3.2
OPEX variable, incl. fuel costs	mln US$ / a	233.6	199.2	138.0	185.6	72.5	95.8
Power generation cost, levelized							
Capacity (fixed OPEX + Annualized capex)	US$ / (kWa)	307.7	269.0	103.2	104.3	87.1	42.8
Energy (variable cost)	US$ / MWh	42.0	48.0	70.5	92.3	144.3	200.4
Composite cost, excl. CO_2-cost *)	**US$ / MWh_e**	**83.10**	**83.93**	**91.75**	**113.21**	**202.35**	**229.85**
Composite cost, incl. CO_2-cost *))**	**US$ / MWh_e**	**86.74**	**88.10**	**93.46**	**116.34**	**205.32**	**233.92**
*) referred to full load hours	h /a	7,481	7,481	4,850	5,000	1,500	1,455
**) Spec. Emission cost	US$ / t_{CO2}	5.0	(for example, spreadsheet Input financial)				

Existing fossil fueled power plants are scheduled for deployment based on their marginal costs which actually are the variable costs. The marginal costs of fossil fueled PPs are shown in Figure 3-18. They are also known as the short run marginal cost- SRMC.

Hydro, wind or solar PPs are, so-called, must-run plants; they have priority for dispatching because their marginal costs are almost zero although their total costs (including capital and fixed costs) are high.

It is pointed out, however, that the electricity generation cost is not the only criterion for deployment of power plants. A short start-up time is, e.g., also an important criterion, especially for peaking and backup power plants.

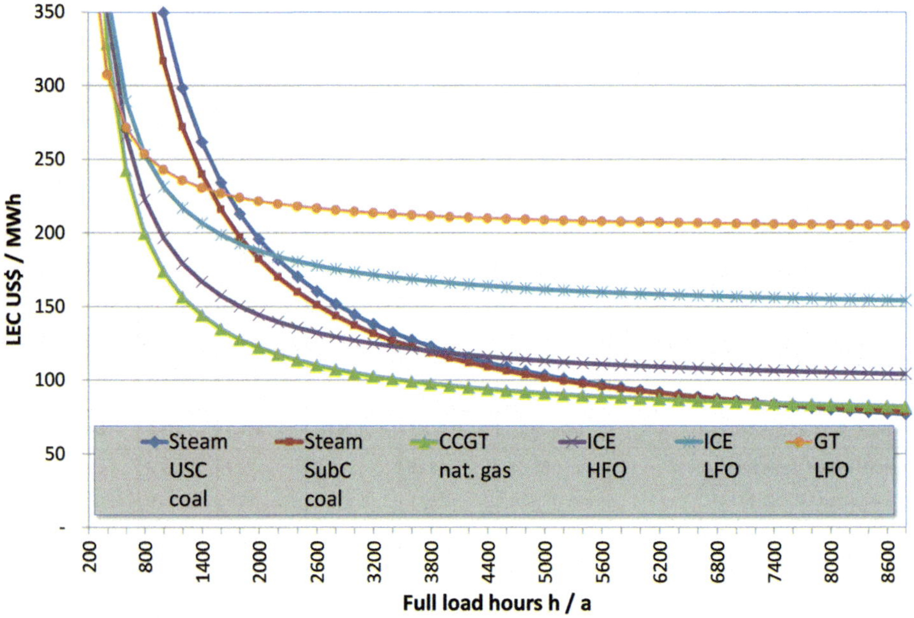

Figure 3-17: Electricity generation cost vs. full load hours

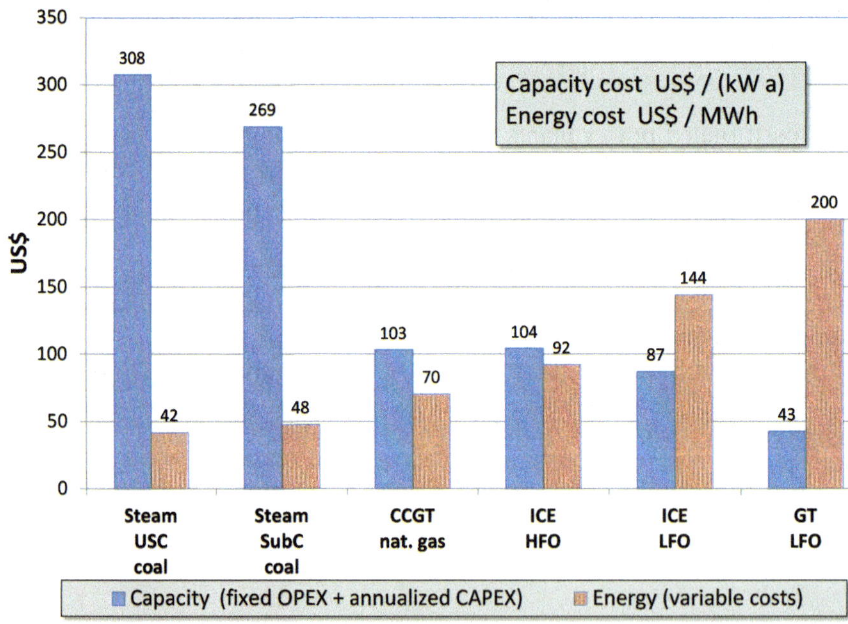

Figure 3-18: Structure of the electricity generation cost

3.6 Fossil Fuels and Climate

Combustion of fossil fuels releases carbon dioxide and other greenhouse gases into the atmosphere, which contribute to the global warming and climate change. Climate change is of vital concern worldwide and a serious problem, requiring the community of states to act and set up strategies to mitigate consequences. Serious efforts in this respect have been started at the Conference on Environment and Development of Rio in 1992, presented in the United Nations Framework Convention on Climate Change (UNFCCC). Based on the convention, the Kyoto Protocol from 1997 commits its signatory Parties by setting internationally binding emission reduction targets. A number of conferences followed. The latest took place in Paris in December 2015 and concluded with an agreement of the Parties to reduce their CO_2 emissions and to do their best to keep global warming well below 2°C.

As shown in Figure 3-19, the carbon footprint depends on the fuel burned and the type and electrical efficiency of the power generation plant. The technical means to obtain the above goal are: increasing electrical efficiency, shifting to fuels with lower emissions and accelerated development of renewable energy.

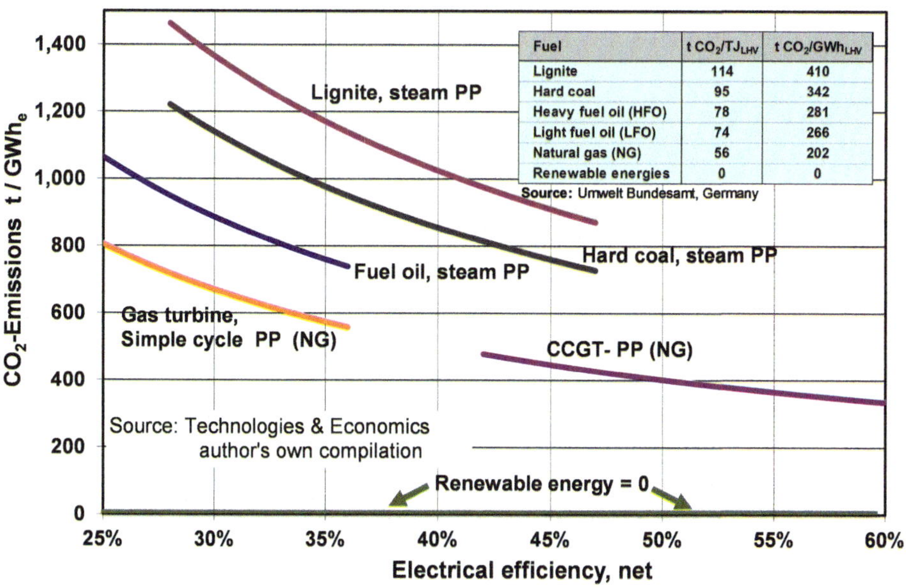

Figure 3-19: CO_2 Emissions of electricity vs. PP electrical efficiency

It becomes evident from the figure that lignite and hard coal have the most serious problem while natural gas in combination with CCGT is less problematic.

Biomass is considered to generate zero CO_2 emissions. This is due to the fact that plants absorb as much carbon dioxide during their growth as emitted into the atmosphere during combustion. The carbon footprint of solar and wind power is zero.

4 Nuclear Power Plants

4.1 Technology Description

Nuclear power plants are thermal power plants similar, to those burning fossil fuels. The main difference lies within the source of heat used to generate steam. In conventional power plants fossil fuels are burned in the boiler to generate high pressure steam that is used to drive the turbines which produce electricity. Nuclear power plants use the heat released by the continuous nuclear fission of atoms in the nuclear reactor. Fission is the process of splitting the nucleus of uranium atoms in certain elements.

There are several types of nuclear reactors. The most common is the pressurized water reactor (PWR) and the less numerous boiling water reactor (BWR). Both types use water as coolant and moderator. Since water normally boils at 100°C, they have robust steel pressure vessels or tubes to enable the higher operating temperature.

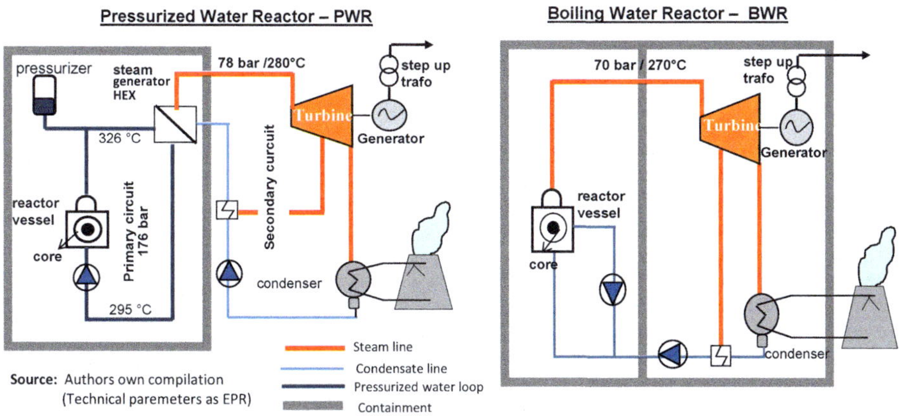

Figure 4-1: Simplified schematics of nuclear power plants

Basic fuel is usually uranium dioxide (UO_2), low enriched (3% to 4%) with the uranium isotope-235, arranged in tubes to form fuel rods. The rods are arranged into fuel assemblies which are submerged in the reactor core. The heat value of the nuclear fuel is very high compared to that of fossil fuels. For comparison, the

P. Konstantin and M. Konstantin, *The Power Supply Industry*,
https://doi.org/10.1007/978-3-319-72305-1_4

heat value of hard coal is 29.3 MJ/kg ce (coal equivalent), while that of uranium dioxide is 3,888,000 MJ/kg (based on UO_2 enriched to 3.5% and 45,000 MWd/t burnup in light water reactors, see also 4.1.4).

4.1.1 Main components

The main components of a nuclear reactor are [23][13] the pressure vessel, the moderator, control rods, the coolant, the containment and the steam turbine generator.

The *Pressure vessel* is usually a robust steel vessel containing the reactor core and moderator/coolant. The *moderator* is a material in the core which slows down the neutrons released from fission so that the chain-fission can take place. It is usually water, in certain types of reactors it may be heavy water or graphite.

The *Coolant* is a fluid circulating through the core to transfer the heat from it to the steam generator. In light water reactors, the water moderator functions also as primary coolant. Power reactors moderated and cooled by water are called light water reactors – LWR. In BWRs, there is a secondary coolant circuit where the water becomes steam.

The *Control rods* are made with neutron-absorbing material and are inserted or withdrawn from the core to control the rate of reaction, or to halt it.

The *Containment* is the structure around the reactor and associated steam generators which is designed to protect it from outside intrusion and to protect those outside from the effects of radiation in case of any serious malfunction inside. It is typically a meter-thick concrete and steel structure.

The *Steam generator*, (essentially a heat exchanger – HEX) is part of the cooling system of PWR (not in BWR) where the high-pressure primary coolant, bringing heat from the reactor, is used to make steam for the turbine, in a secondary circuit. Reactors may have up to four 'loops', each with a steam generator.

4.1.2 Nuclear reactors on the market

Several generations of reactors are commonly distinguished [24][14]. Generation I reactors were developed in the 1950-60s; they are not in operation any more. They mostly used natural uranium fuel and graphite as moderator. Most reactors still in operation are of Generation II. They typically use enriched uranium fuel and are mostly cooled and moderated by water. Generation III are the advanced reactor developments of the second generation with enhanced safety. However,

[13] WNA – World Nuclear Association, Nuclear power reactors
[14] WNA – World Nuclear Association, Advanced nuclear power reactors

there is no clear distinction between Gen II to Gen III because there was a steady development. Generation IV designs are still on the drawing board and will not be operational before 2020 at the earliest, probably later.

Currently there are several types of nuclear power plants offered on the market. The following are known to be under construction:

The European Pressurized Reactor – EPR: The French manufacturer Areva NP has developed a large pressurized water reactor (typically 4590 MW_t, 1750 MW_e gross and 1630 MWe net), called European Pressurized Reactor (EPR). This is a 4-loop design derived from the German Convoi types. It will operate flexibly to follow loads, have fuel burnup of about 65 GWd/t and an electrical efficiency of 37% gross and of 36% net. Availability is expected to be 92% over a 60-year service life.

It has double containment with four separate, redundant *active safety systems*, and boasts a core catcher under the pressure vessel. Active means, urgency cooling is provided by pumping water. The safety systems are physically separated through four ancillary buildings on the same concrete raft, two of them are aircraft crash protected. The primary diesel generators have fuel for 72 hours, the secondary backup ones for 24 hours, and tertiary battery back-up lasts 12 hours. It is designed to withstand seismic ground acceleration of 600 Gal without safety impairment.

The first EPR unit is being built at Olkiluoto in Finland, the second at Flamanville in France, the third will be at Penly in France, and two further units are under construction in China.

The Toshiba/Westinghouse AP1000: AP1000 is a 2-loop PWR which has evolved from the smaller AP600, one of the first new reactor designs certified by the United States Nuclear Regulatory Commission (US NRC) in 2005. A major design objective of the AP1000 was simplification of overall safety systems, normal operating systems, the control room. Construction techniques, instrumentation and control systems provide cost savings with improved safety margins.

It has been designed with a *passive safety concept*. It has a core cooling system including passive residual heat removal by *convection* instead of pumping, improved containment isolation, passive containment cooling system to the atmosphere and in-vessel retention of core damage (corium) with water cooling around it. No safety-related pumps or ventilation systems are needed. It is being built in China, quoted as 1200 MW_e gross and 1117 MW_e net and thermal capacity 3400 MW_t. The first ones being built in China are on a 51-month timeline to fuel loading, or 57-month schedule to grid connection.

South Korea's APR1400: The APR1400 advanced PWR has a capacity of 1,455 MW_e gross and 1,355 MW_e net, heat generation 3,983 MW_t with 2-loop primary circuit. The first of these is under construction - Shin-Kori-3 & 4 in South Korea. Fuel will have up to 55 GWd/t burn-up, refueling cycle about 18 months, outlet temperature 324°C. Construction time is expected to be 48-months which is quite ambitious. It is designed for a plant life of 60 years. It has been chosen as the basis of the United Arab Emirates nuclear program for 4 nuclear power plants, and is also being offered in Finland.

The Russian VVER-TOI developed by Atomenergoproject [25] has a rated capacity of 1,255 MWe gross, heat generation of 3,312 MWt and gross efficiency 37.9 %. It is designed with a passive safety system with 72 hours grace period requiring no operator intervention after shutdown. The availability factor is 93% based on a fuel cycle of 18 months. The construction period is expected to be 48 months and planned service life 60 years. The first units are planned in Nizhny Novorod in Russia and in Akkuyu in Turkey.

4.1.3 Nuclear fuel

Uranium deposits consist of the two uranium isotopes namely U-238 of about 99.3% and U-235 of about 0.7%. Only the isotope U-235 is fissile and can generate huge amounts of thermal energy in a fission chain. At the mines the natural uranium is separated from waste residues and comes to the market place in form of *Uranium Oxide* U_3O_8 also called yellow cake. This undergoes several process steps to become *Uranium dioxide* UO_2 that is the actual nuclear fuel.

The nuclear fuel used for power generation is enriched Uranium; its content of the isotope U-235 is increased from 0.7% to about 3% to 4%. This is compiled in nuclear fuel assemblies that come into the reactor core of the power plants.

4.1.4 Performance parameters of nuclear power plants

The performance parameters introduced for fossil power plants in 2.2 are also applicable for nuclear power plants. A new parameter is the so-called fuel *burnup* This is equivalent to the heating value of fossil fuels; however, the unit is different. Instead of kJ/kg or kWh/kg it is given as MWd/kg. It is defined with the following formula:

$$\text{burnup:} \quad q_{nf} = \frac{P_t\left[\mathrm{MW_t}\right] \times t\left[\frac{\mathrm{d}}{\mathrm{a}}\right]}{m_{nf}\left[\frac{\mathrm{kg}}{\mathrm{a}}\right]} \quad \left[\frac{\mathrm{MW_t\ d}}{\mathrm{kg}}\right] \tag{4.1}$$

Where:

P_t : Thermal power of the reactor MW_t

t : Full load operating time d/a (not h/a)

m_{nf}: Nuclear fuel consumption during operating time

The burnup rate is given by the manufacturers; for cost calculations the mass flow of fuel during the operating time is needed.

$$\text{Fuel mass flow:} \quad m_{nf} = \frac{P_t\left[\mathrm{MW_t}\right] \times t\left[\frac{\mathrm{d}}{\mathrm{a}}\right]}{q_{nf}\left[\frac{\mathrm{MW_t\ d}}{\mathrm{kg}}\right]} \quad [\mathrm{kg/a}] \qquad (4.2)$$

Example 4-1 : Calculating equivalent heating value of nuclear fuel

The equivalent heating value (LHV) of nuclear fuel can be determined by multiplying the fuel burnup with the hours of the day (24 h). In the following example it has been calculated for standard conditions (burnup 45 MWh/kg and 3.5 % enriched UO_2) as well as based on the burnup of the Areva's European pressurized reactor (EPR).

Item		Unit	Standard	EPR
Fuel burnup		MW_t d / kg	45	65
Time		h /d	24	24
Equivalent heating value		MWh /kg	1,080	1,560
	3600	MJ /kg	3,888,000	5,616,000

Example 4-2: Fuel consumption in comparison

In the example the annual fuel consumption (GWh_t and t/a) of a conventional steam power plant and that of advanced reactors is calculated. The latter is calculated with two methods using the equivalent heating value and based on the burnup rate. The conventional steam PP's consumption mass flow is about 150 times, that of the nuclear PP.

Item	Symbols	Unit	Conventional Steam PP	Advanced Nuclear PP	
Type	-	-	USC	EPR -Areva	
Fuel	-	-	hard coal	UO_2	
Number of units	-	-	2	1	1
Rated power output, per unit, gross	P_e	MW_e	875	1,750	1,750
Rated thermal power	P_t	MW_t	1,804	4,590	4,590
Electrical efficiency, gross		-	48.5%	38.1%	38.1%
Heat value LHV/ Fuel burnup	-	-	8.14 GWh/t	1,560 GWh/t	65 GWd/t
Equivalent full load hours	t	h /a	7,500	7,500	7,500
		d / a	n.a.	n.a.	312.5
Electricity generation	$W_e = P_e \times t$	GWh_e /a	13,125	13,125	13,125
Fuel consumption	$Q = W_e / \eta$	**GWh_t /a**	**27,062**	**34,425**	**34,425**
	m	**t /a**	**3,325**	**22.1**	**22.1**
Formula	$m =$	t /a	Q / LHV	Q / LHV	$P_t \cdot t / burnup$

4.1.5 Decommissioning and waste disposal management

Decommissioning [26] of nuclear power plants after the end of their operational lifetime includes permanent shutdown, progressive dismantling of the plant, all cleanup of radioactivity and finally unrestricted site release for reuse.

About 95% of the high level radioactivity in nuclear reactors is associated with the fuel which is removed following permanent shutdown. Apart from some surface contamination of plant, the remaining radioactivity comes from activation products in steel which has long been exposed to neutron irradiation, notably the reactor pressure vessel. The depleted fuel is first stored under water in cooling ponds at the reactor site for several years. The concrete ponds and the water, covering the fuel assemblies, provide radiation protection, while removing heat generated during radioactive decay.

Decommissioning is a long-lasting and costly procedure conducted under supervision of the national regulatory authorities in strict compliance with the laws, standards and regulations. The International Atomic Agency (IAEA) defines three options [26] [27] for decommissioning:

- Intermediate dismantling (Decon)
- Safe enclosure (SafeStore)
- Entombment (Entomb)

Immediate Dismantling allows for the facility to be removed from regulatory control relatively soon after shutdown or termination of regular activities. The term immediate is however understatement; the whole process takes 10 to 20 years, depending on the complexity of the plant. An advantage of the option is that most of the operational staff can be retained, contributing with their experience and skills also in the decommissioning process.

Safe Enclosure means a deferred dismantling. The plant is placed into a safe storage configuration until the eventual dismantling and decontamination activities occur after residual radioactivity has decayed. This option postpones the final removal for a longer period, usually in the order of 40 to 60 years. The plant must be kept under surveillance during this period. This option is preferred if another nuclear power plant is in operation at the same site and the same staff can be employed for surveillance.

Entombment entails placing the facility into a condition that will allow the remaining on-site radioactive material to remain on-site without ever removing it totally. This option usually involves reducing the size of the area where the radioactive material is located and then encasing the facility in a long-living structure such as concrete shell, that will last for a period of time to ensure the remaining radioactivity to be no longer of concern. This option is applied after grave nuclear accidents e.g. in the case of Chernobyl.

4.2 Economics of Nuclear Power

The costs of nuclear power are difficult to quantify, especially the capital expenditures for building the plant that is the overwhelming cost component, the cost for nuclear waste management and the longterm cost for disposal high level radioactive waste. Nevertheless, we try to calculate the costs of nuclear power and analyze their economics and notably to provide more transparency in their costs structures, according to best of our knowledge and belief.

The **Model** for the calculation of the electricity generation cost is presented in **Case Study 11.5** as a hardcopy and is available as softcopy on the author's website. It includes the following spreadsheets:

- Summary of results
- Weighted Average Cost of Capital (WACC) used as discount rate
- Nuclear fuel cost
- CAPEX breakdown
- Costs for waste disposal and reserves for decommissioning
- Electricity generation cost calculation
- Electricity cost breakdown

In the following some explanations of the main technical-economic parameters for the cost calculations are summarized:

CAPEX: Manufacturers of advanced, third generation reactors were very optimistic at the beginning with regard to the construction costs and construction time.

Initially, cost estimates were in the range of about 1500 US$/kW and construction time of about 3 years. New contracts give cost of considerably higher than 5,500 US$/kW and construction time of plants under construction has been longer than five years in most cases. Significant costs overruns are also always very common. Moreover, there are few sources available and they do not specify exactly what kind of costs is included.

The U.S. Energy Information Administration [22][15] gives a capital cost estimate for two units Westinghouse AP 1000 nuclear power plant. This is in our opinion the most transparent construction costs estimate available we could find in the whole literature and therefore it is taken as basis for the cost calculations. The *discount rate* is based on the WACC. Taking into account the longevity rise of the investment, a higher rate of return on equity of 1.5 percentage points, compared to conventional power plants, is considered.

Costs of nuclear fuel: The model for the calculation of the cost of nuclear fuel (UO_2) is described in detail in the author's book "Engineering Economics [28]" and on the website of the World Nuclear Association. [29][16].

The cost of nuclear fuel is expressed in US Dollar per kg uranium dioxide (US$/$kg_{UO2}$). The prices of nuclear fuel are composed of the following costs components:

- Cost for Uranium oxide U_3O_8
- Conversion cost of Uranium oxide to Uranium hexafluoride UF_6
- Cost for enrichment of Uranium hexafluoride UF_6
- Cost of nuclear fuel fabrication to UO_2 assemblies

The nuclear fuel costs calculation is also depicted in brief in Table 4-1 below. The *service life* for the calculation is assumed to be 50 years and *construction time* 6 years. The calculation is conducted inflation adjusted, in real terms in US$ 2013.

The *cost for nuclear waste management* and disposal of depleted fuel during operation is set to 1.5 US$/$MWh_e$ (0.15 cent/kWh_e). For comparison, this charge is in the USA 0.1 cent /KWh, in Sweden 0.13 cent/kWh_e.

Cost and finance for decommissioning. A survey of OECD published in 2003 [30] reports costs for decommission in 2001 US$ of 200 to 500 per kWe for PWRs and 300 to 550 per kWe for BWRs. Projected in US$ 2013 [17)] this will be 259 to 649 for PWRs and 389 to 714 per kWe for BWRs.

Due to the longevity of the investment it is assumed in the case study that an amount totaling about 15% of the CAPEX in US$ 2013 must be accumulated in a

[15] eia – Updated capital cost estimates for utility scale power plants, April 2013
[16] WNA Website – Information library, The economics of nuclear power
[17]) USA CPI 2011= 81,2, CPI 2013=106.8

period of 30 years corresponding to 832 US$/kWe or a fee of about 2.15 US$/MWhe produced electricity. After 50 years of operation this amount will be about 38% or 2,100 US$/kW. Note, however, that for decommissioning of the 17 German nuclear reactors after phase out in 2022, about 38 bn€ are available corresponding to about 1,850 €/kW.

Table 4-1: Composition of the nuclear fuel cost components

Component	Explanation	Units needed for 1 kg UO_2*)	US$/Unit **)	Total
Uranium oxide U_3O_8	This is the form Uranium is offered in the market place. It includes 0.7% of the fissile isotope U-235.	8.90 kg U3O8	72.64	646
Conversion in UF_6	The Uranium oxide is converted in gaseous form in Uranium hexafluoride (UF_6).	7.50 kg U	7.92	59
Enrichment	Uranium hexafluoride (UF_6) is in enriched in centrifuges to enriched UF_6 with a concertation of 3 to 4 percent U-235.	7.30 kg SWU	91.83	670
Fuel fabrication	The enriched UF_6 is converted in Uranium dioxide (UO_2), the actual nuclear fuel, in form of powder. It is compressed in pellets and filled in thin pipes bundle up in fuel assemblies.	-	-	275
Nuclear fuel	Assemblies of nuclear fuel	**1 kg UO_2**	-	**1,651**

*) **Source:** World Nuclear Association, information library, July 2015
http://www.world-nuclear.org/info/Economic-Aspects/Economics-of-Nuclear-Power/
**) Power & Energy Systems Engineering Economics, average 2013

In Table 4-2 and below the spreadsheet "Summary of results" of the Model, including the key techno-economic parameters of the calculation and the main results, are depicted.

Table 4-2: Techno-economics of nuclear PPs, Summary for results

Item		Unit	Value
Power and Energy balance		-	
Rated power each, total, gross		MW_e	2,400
Thermal reactor power, total		MW_t	6,800
Electricity generation, net	7,500 h/a	GWh_e / a	16,740
Fuel consumption, in thermal units		GWh_t / a	47,430
metric tons nuclear fuel		t / a	35.4
Technical-financial constraints			
Service life for calculation		a	50
Discount rate, on real terms (WACC)		% / a	7.1%
Cost of nuclear fuel *)		US$ / kg UO_2	1,651
Reserve funds for decommissioning, waste disposal		US$ / MWh_e	3.65
Capital expenditures (CAPEX), US$ 2013 **)		**Mln US$**	**13,720**
Annual costs, US$ 2013		**Mln US$ /a**	**1,483**
Annualized CAPEX		Mln US$ /a	1,002
Fixed Operating expenses (fixed OPEX)		Mln US$ /a	362
Variable operating expenses (variable OPEX)		Mln US$ /a	120
Capacity cost ref. to net power		**US$ / (kW a)**	**611**
Energy cost, ref. to net electricity production		**US$ / MWh_e**	**7.14**
Composite cost		**US$/ MWh_e**	**88.60**

*) Average 2013, book, Engineering Economics

) **Source: eia - U.S. Energy Information Administration, updated capital cost estimates 2013,

**) including EPC price, owner's expenses and interest during construction

From the table above it becomes evident that the main cost item is by far the capital cost, while the share of the fuel cost is almost marginal.

4.3 Balancing Benefits and Risks of Nuclear Power

At the beginning of its introduction in the early 1960s, nuclear power was considered as a viable solution for the energy supply of the world and was met with a high level of approval. In the course of time, some nuclear accidents, especially those of Chernobyl and Fukushima, made apparent the high risks of this technology. Another concern remaining is the still unsolved problem of final disposal for highly radioactive waste.

Against these backdrops, views on nuclear power are meanwhile much divided and controversial.

Besides the above worrying aspects, governments have to take care of other crucial issues such as reliability and security of energy supply for their population and national economy. This is a particular concern and challenge for large countries with a high growth of power demand. Furthermore, importing countries endeavor to reduce dependency from imports of fossil fuels and deem nuclear power as an alternative. Besides costs considerations, it is obvious that renewable energy, except hydro, where it is available, cannot provide required base load power. In this respect, fossil fueled or nuclear power plants remain indispensable.

Nuclear power is also regarded as low-carbon energy and a viable option against climate change, alongside renewable energy.

Against this background of obvious risks, security of supply and imminent climate change, governments have adopted completely different approaches in their energy policy.

Just to mention a few examples: Italy. Austria and even Australia, being the major exporter of Uranium, have abstained from nuclear power. Germany is phasing-out nuclear power and will shut-down all existing nuclear power plants by 2022. In contrary, France, Finland, U.K, Russia, China, the USA and other countries rely on and plan to continue using nuclear power. There are also some newcomers, e.g., Abu Dhabi while Turkey and Saudi Arabia are candidates.

Another big challenge for the coming decades is the decommissioning and the likely replacement of a large number of nuclear power plants, which approach the end of their technical lifetime.

Against this background and without any prejudice, we can conclude that nuclear power will remain in use for the foreseeable future, until alternative technologies will be developed.

On the other hand, in some industrialized countries, especially in Central Europe, the growth of electricity demand is zero or even negative. They also have the technical and economic means to invest and expand renewable energy, especially offshore wind power (Denmark, Germany).

5 Power Generation from Renewable Energies

5.1 Hydroelectric Power Plants

5.1.1 Technology description

Hydro power is an age-old technology; the energy potential of water in motion has been used for centuries in flour mills or for pumping irrigation water. Hydroelectric power plants convert the kinetic and potential energy of water flows in mechanical and subsequently in electrical energy. The power output of hydroelectric generators is calculated by the formula:

$$P = \eta \times \rho \times g \times \dot{Q} \times h \quad [\mathrm{W}] \tag{5.1}$$

Where:

η: System efficiency (0,80 – 0,90)

ρ: Density of water (1000 kg/m^3)

g: Gravity acceleration (9,81 m/s)

$\dot{Q}$: Water flow rate (m^3/s).

h: Head of water (m)

By substitution of the three constants with their values (0,85x1000x 9,81/1000 = 8.34), the formula can be converted into a simplified numerical form:

$$P = 8.34 \times \dot{Q} \times h \quad [\mathrm{kW}] \tag{5.2}$$

The power output is proportional to the volume of water flow and the water head. Water head is the pressure of falling water due to gravity.

Hydroelectric power plants are categorized in several types, based on the water head and site conditions. The most important types of hydro power plants are: run-of-river plants, dam plants and pump storage plants.

Run-of-river power plants – Figure 5-1 – use the energy of flowing river water to drive hydro turbines to generate electricity. The water volumes of utility scale hydro power plants are usually large, while the elevation difference (water head) between upstream and downstream water levels is relatively small. Run-of-river power plants operate continually as base load plants. Their electricity production

P. Konstantin and M. Konstantin, *The Power Supply Industry*,
https://doi.org/10.1007/978-3-319-72305-1_5

depends, however, on the water flow of the river that is subject to seasonal variations. They normally produce more electricity in summer than in winter season.

Run-of-river *pondage hydropower* plants are capable to control the river water flow. They store river water upstream of the plant during off-peak times and release it during on-peak times:

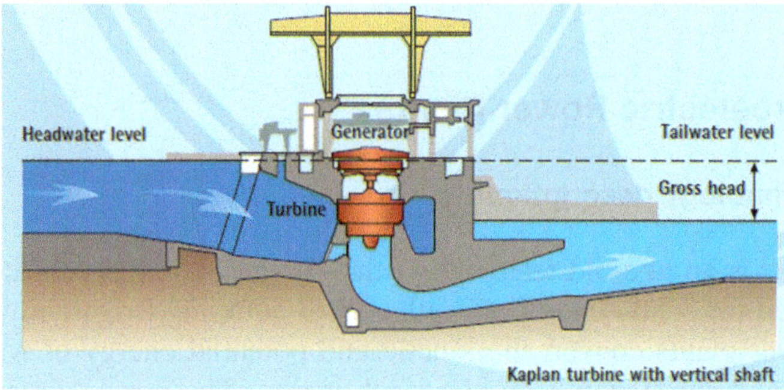

Figure 5-1: Run-of-river hydro power plant (Blue Danube Austria)[18]

Dam hydro power plants – Figure 5-2 –are typically large hydropower systems using a dam to store river water in a large upper reservoir. Water, released from the upper reservoir, flows through the turbine generators to produce electricity and is released into the lower reservoir. The water may be released either to meet changing electricity needs or to maintain a constant reservoir level.

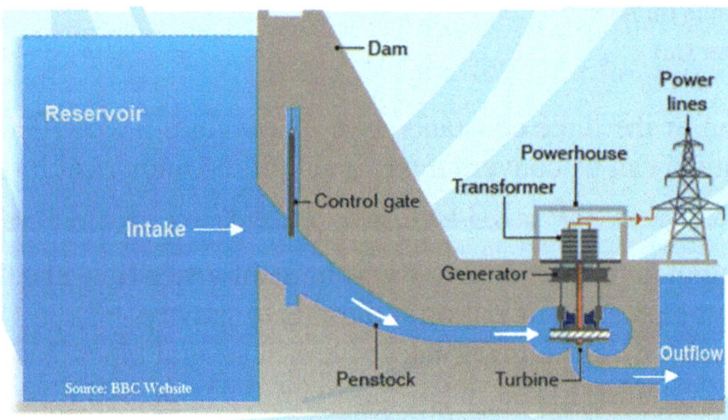

Figure 5-2: Cross-section of typical hydroelectric dam power plant

Pump storage hydro power plants – Figure 5-3 –are composed of an upper basin, the power house and a lower basin. The turbine in the power house can be operated either in power generating or in pumping mode. During surplus electric-

[18] Taken from http://www.qrz.com/db/OE7XWI

ity production, either of conventional or renewable power generators, water is pumped from the lower to the upper basin. During on-peak times, water is released from the upper basin via a pressure pipeline (penstock) to the power house to drive the turbine generators and is discharged afterwards to lower basin. The turbine can reach 100 percent output within seconds thus pump storage power plants ensure the fastest peak load capability of all other systems.

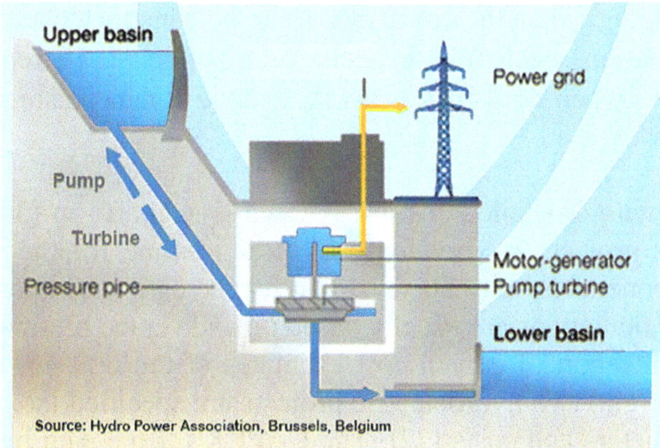

Figure 5-3: Pump storage hydro power plant **[31]**

The importance of pump storage hydropower is growing with the increasing penetration a power generation from renewable energy. Excess power production of wind farms or solar power plants can be used for pumping and accumulating water into large reservoirs that can later be used for power production in times of low power supply from renewables. An additional advantage is the extremely fast startup capability of turbines for balancing fluctuations of supply from renewable sources.

The three main types of hydro turbines [32] [33] used in hydro power stations and their typical applications are depicted in Figure 5-4.

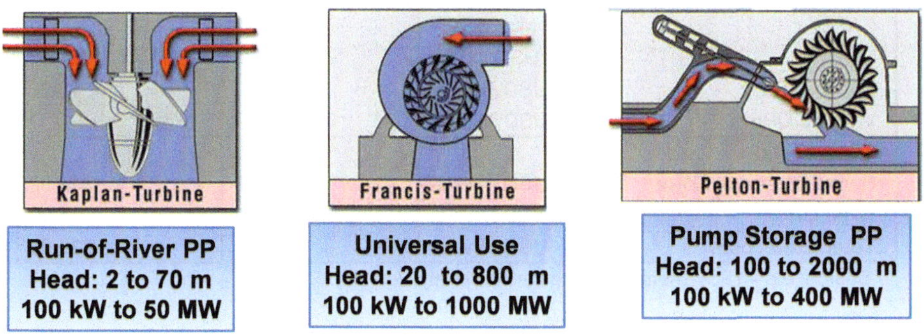

Figure 5-4: Hydro turbines and their typical applications

5.1.2 Economics of hydropower

Capital costs (CAPEX) of hydro power strongly depend on site conditions as water flow rate, head (m) between intake and discharge of water levels. Especially the costs, related to the extent of civil works, are very site specific. There is also a big costs difference depending on the type of (re)construction. A distinction is made between:

- New building, when the entire plant is to be constructed anew
- Modernization, when only the technical components are replaced and
- Extension, when construction work is done for increasing the water flow rate.

A good base[19] with a detailed cost breakdown is available in [34]. The life time of hydropower projects is very long. Typical lifetime for the mechanical and electrical components is 40 to 50 years and for the civil structures 80 years. Generation costs' calculations are usually based on an average lifetime of 50 years.

In the annual report of capital cost estimates of the U.S. Energy Information Administration (eia) two hydroelectric projects are specified including their main technical parameters and costs [22]. The CAPEX for the dam hydroelectric project is low compared to figures known in Europe. It is probably an extension project at an existing site – Table 5-1.

Table 5-1: Costs of hydroelectric power plants

Item			Unit	Hydroelectric Dam Power Plant	Hydroelectric Pump Storage Power Plant
Technical parameters					
Nominal capacity			MW	500	250
Type of turbine			-	2 xFrancis	2 xFrancis
Minimum head			m	200	not specified
Financial parameters					
CAPEX			Mio. USD	1,468	1,322
Specific			USD / kW	2,936	5,288
Fixed OPEX			USD / kW a	14.13	18.00
Variable OPEX			USD / MWh	0	0
Annual costs					
Capital costs	5.50%/a	50 a	USD / kW a	173.40	312.32
Fixed OPEX			USD / kW a	14.13	18.00
Total fixed costs			**USD / kW a**	**187.53**	**330.32**

*) Source: eia – U.S. Energy Information Administration 2013

[19] Cost base for hydropower plants, Norwegian Water Resources and Energy Directorate 2012

Due to mentioned uncertainties with regard to capital cost estimates a calculation of electricity generation cost for hydroelectric power plants is only feasible for concrete projects.

Nevertheless, a generation cost calculation for three types of power plants is presented in the following Example 5-1, based on technical-economic parameters that can be regarded as reasonable for project sites in Europe. As almost all the costs are capital cost (±25%), the insecurity margin is high. Due to the long life-time of the projects the electricity generation costs are relatively low compared to those of fossil power plants.

Example 5-1: Electricity generation cost of typical hydro power plants

Item			Unit	Run of River	Dam Hydro	Pump Storage
Technical Parameters						
Nominal power			MW	150	150	150
Load segment			-	base	Intermediate	peak
Typical full load utilization time			h / a	4,500	3,500	1,000
Annual electricity generation			GWh / a	675	525	150
Financial constraints						
Water head			m	20	200	250
Pumping electricity		eta=85.0%	GWh /a	-	-	176
Pump utilization time			h / a	-	-	1,000
Cost of electricity for pumping			€ / MWh	-	-	30
Life time			a	50	50	50
Discount rate, in real terms			% / a	4.58%	4.58%	4.58%
Fixed OPEX			% Inv./ a	1.0	1.1	1.2
CAPEX, estimate, US$ 2014, ±25%			**Mio. €**	**525**	**600**	**675**
specific			€ / kW	3,500	4,000	4,500
Annual electricity gen. Costs, in real terms						
Capital costs	4.58%/a	50 a	Mio. € / a	26.9	30.7	34.6
Fixed OPEX			Mio. € / a	0.5	0.7	0.8
Variable Cost			Mio. € / a	-	-	5.3
Total annual costs			Mio. € / a	**27.4**	**31.4**	**40.7**
Levelized Electricity cost, real terms			**€ / MWh**	**40.62**	**59.80**	**271.19**
Capacity cost			€ / (KW a)	179.28	204.89	230.50
Variable cost			€ / MWh	-	-	35.29

Note, however, that the electricity generation costs of renewable energy projects are calculated with a lower discount rate, 4.58%/a, instead of 6.5%/a, in real terms for conventional plants. See explanatory note in section 2.3.

It is also worth mentioning that the prices for on-peak electricity are sometimes very high, up to 1000 €/MWh. Thus pond and pump storage power plants can obtain very high revenues in a short time.

In Example 5-2 below the electricity generation costs are calculated according to eia technical parameters and cost estimates as shown in Table 5-1.

Example 5-2: Electricity generation cost of hydro power plants, eia estimates

Item	Unit	Run of River	Dam Hydro	Pump Storage
Technical Parameters				
Nominal power	MW	500	500	250
Load segment	-	base	Intermediate	peak
Typical full load utilization time	h / a	4,500	3,500	1,000
Annual electricity generation	GWh / a	2,250	1,750	250
Financial constraints				
Water head	m	20	200	250
Pumping electricity eta=85.0%	GWh /a	-	-	294
Pump utilization time	h / a	-	-	1,000
Cost of electricity for pumping	€ / MWh	-	-	30
Life time	a	50	50	50
Discount rate, in real terms	% / a	4.58%	4.58%	4.58%
Fixed OPEX	% Inv./ a	1.0	1.1	1.2
CAPEX, estimate, € 2014, ±25%	**Mio. €**	**1,950**	**2,600**	**1,723**
specific (acc. to eia 2013) US$/€ 1.30	US$ / kW	3,000	4,000	5,300
Annual electricity gen. Costs, in real terms				
Capital costs 4.58%/a 50 a	Mio. € / a	99.9	133.2	88.2
Fixed OPEX	Mio. € / a	2.0	2.9	2.1
Variable Cost	Mio. € / a	-	-	8.8
Total annual costs	Mio. € / a	**101.8**	**136.0**	**99.1**
Levelized Electricity cost, real terms	**€ / MWh**	**45.26**	**77.74**	**396.48**
Capacity cost	€ / (KW a)	199.77	266.36	352.92
Variable cost	€ / MWh	-	-	35.29

5.2 Wind Power Plants

5.2.1 Technology description

Wind turbines convert the kinetic energy of wind into mechanical energy. The theoretical wind power is proportional to the area that is subject to the wind flow and to the third power of the wind speed:

$$P = \frac{1}{2} \cdot \rho \cdot A \cdot w^3 \quad [\mathrm{W}] \tag{5.3}$$

However, the kinetic energy of upstream wind cannot be completely converted into mechanical energy. This is because the same air mass flow continues to flow downstream of the rotor with a lower speed. So the converter can only capture the energy difference between the upstream and downstream air flow. According to the Betz equation [35] there must be a wind speed change from the upstream to the downstream side of the converter, in order to extract energy from the wind. Betz proved with his equation that only a fraction up to a maximum of 59.3% of the kinetic wind energy can be converted into mechanical energy. This is called the *Betz performance coefficient* "c_{p_Betz}=0,593". The maximum energy yield is obtained when the wind speed downstream of the converter decreases to 1/3.

Modern wind turbines obtain performance coefficients between "c_p =0.45 to 0.55". After replacing the Area with $A=\frac{1}{4}\cdot\pi\cdot D^2$ and 1 W =1/1000 kW in (5.3) the equation becomes:

$$P = \frac{c_p}{8000} \cdot \pi \cdot \rho \cdot D^2 \cdot w^3 \quad [\mathrm{kW}] \tag{5.4}$$

Where:

c_p: Actual performance coefficient of the wind turbine
ρ: Air density [kg/m^3]
D: Rotor diameter [m]
w: Wind speed [m/s]

From the equation (5.4) it becomes evident, that the most important criterion for the selection of an appropriate location for wind power is the wind speed at the site. Wind atlases are maps containing data of wind speed and wind direction in a region. A climatological wind atlas indicates hourly averages of wind speed at a standard height of 10 meters over longer periods, usually 30 years. The wind potential is usually displayed with lines of constant wind speed.

Figure 5-5 depicts the cross section of a wind turbine including brief description of its main components and their function.

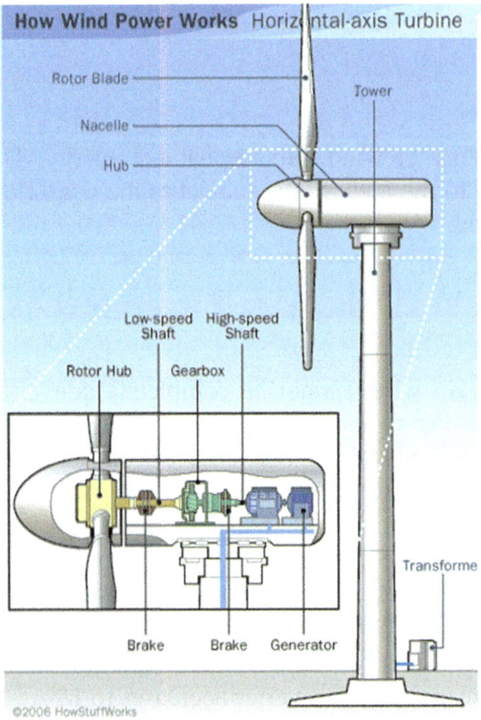

Source: http://www.howstuffworks.com/

The *rotor blades* capture wind's energy and convert it to rotational energy.
The *shaft* transfers the rotational energy to the generator.
The *Nacelle*-casing holds the gearbox, the generator, the electronic control unit, the yaw and the brakes mechanisms.
The control unit monitors the system and shuts down turbine in case of malfunction. The *yaw mechanism* moves the rotor to align with direction of wind.
T*he brakes* stop rotation of shaft in case of power overload or system failure.
The *tower* supports rotor and nacelle and lifts the entire setup to a higher elevation where the wind speed is higher and the blades can safely clear the ground.
The e*lectrical equipment* carries electricity from generator down through tower to the transformer.

Figure 5-5: Cross section of wind turbine with main components **[36]**

For the calculation of the annual energy yield of a wind turbine the following data are required as a minimum:

- The annual average wind speed
- The frequency distribution of the wind speed
- The performance curve of the wind turbine

The *wind speed* increases according to a logarithmic law with the elevation. For the calculation of the energy yield the annual average wind speed at the hub of the wind turbine is needed. It can be calculated with the formula depicted in Figure 5-6. The roughness length z_o describes the ground roughness of the terrain. Roughness classes (0 to 4) along with the related roughness lengths "z_o" in meters for different landscape forms are defined in the *European Wind Atlas WAsP*[20] [37] - see table **Annex 22**.

[20] See Website of Danish Wind Energy Association, Wind Energy Concepts

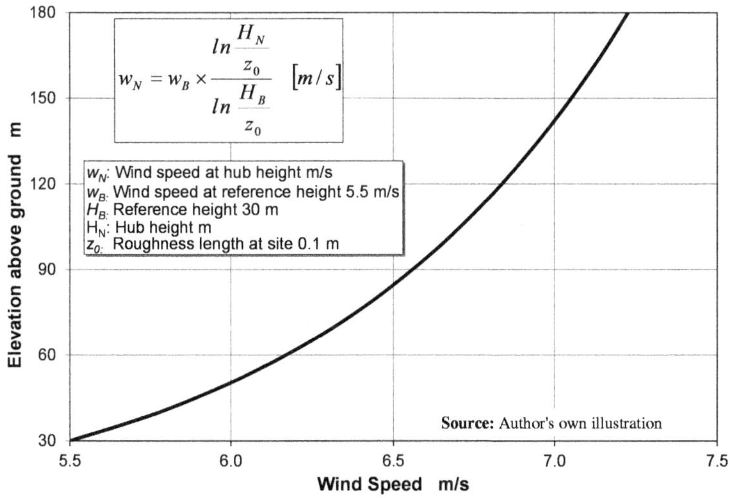

Figure 5-6: Wind speed vs. elevation

Besides the annual average wind speed also its *frequency of distribution* is required. This is the statistical duration in hours of each wind speed interval over the year. Usually, the measured wind speeds at a typical site show a Weibull distribution. For the purpose of calculation, a Weibull distribution curve is fitted to the distribution of the measured data by means of the scale and shape parameters. For onshore sites in Europe the Rayleigh- Model is used Figure 5-7, a subvariant of the Weibull-Model.

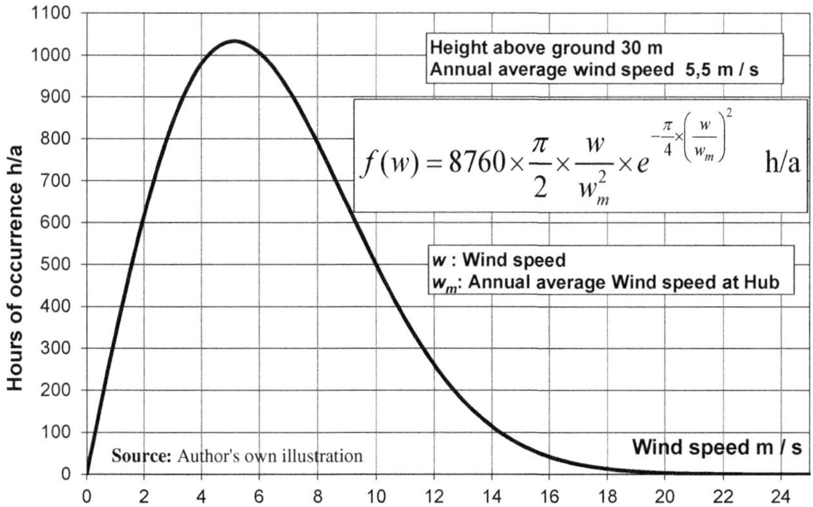

Figure 5-7: Rayleigh frequency distribution of the wind speed

Note: the original formula gives the h/a in percent. The author has included in the formula the 8760 h/a in order to get the distribution in h/a vs. wind speed interval, as it is helpful for practical application.

The *performance curve* of wind turbines is included in the technical data of manufactures or can be delivered upon request either as table or in digital form. The performance curve depicts the power output in kW and the performance coefficient c_p vs. wind speed. This is characteristic for each wind turbine type – Figure 5-8.

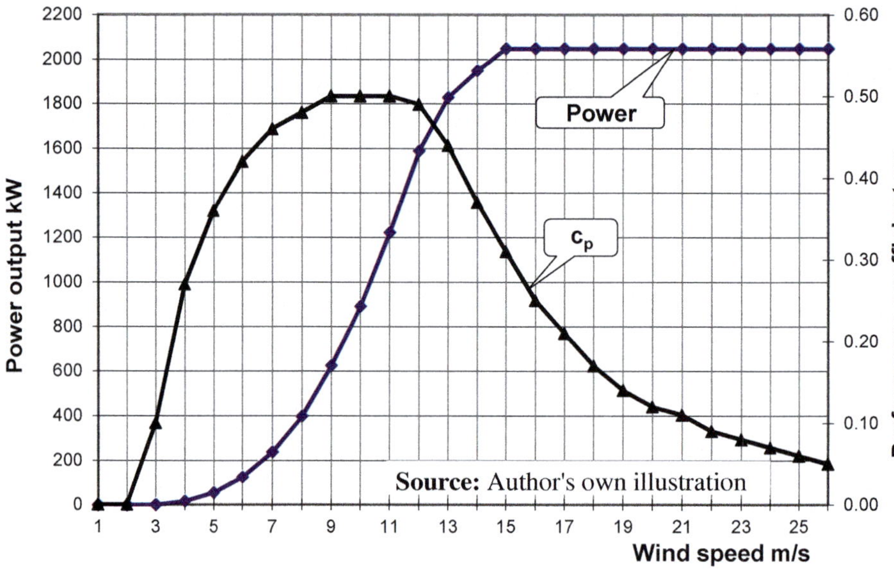

Figure 5-8: Performance curve of a wind turbine

There are following distinct areas of the performance curve. The turbine starts to generate power after reaching the *cut-in wind speed,* in this case about 4 m/s. Up to this speed the captured wind energy is not sufficient to overcome inertia and friction losses. Afterwards the power output increases with the third power of the wind speed up to the *rated wind speed,* in this case 16 m/s, the nominal power output of the generator is obtained. As soon as the *cut-out wind speed* is obtained, about 26 m/s, the wind turbine is shut-off in order to prevent damage of the plant due to exciding of the mechanical strength limits.

After the nominal power is obtained, the *power control* limits the wind energy input to the rotor at high wind speeds in order to prevent exceedance of the material's strength and damage of the rotor. This is in most wind turbines done by altering the angle of the rotor blades against the wind direction, known as *pitch control.* As soon as the cut-out wind speed is reached, the pitch control system brakes the rotor blades aerodynamically by moving the blades in feather position.

5.2.2 Classification of wind turbines

Depending on the wind potential of the site, different types of wind turbines are available on the market. They are classified according to the IEC 61400 (International Electro technical Commission) based on three parameters: the average wind speed, an extremely high wind speed that statistically can occur only once every 50 years and the turbulence intensity class – Table 5-2.

Table 5-2: Wind turbine classification acc. to IEC 61400

Parameter	Symbol	IEC-Wind Class			
		I	II	III	IV
Average wind speed	v_{ave}	10.0 m/s	8.5 m/s	7.5 m/s	6.0 m/s
Maximum, 10 minute average wind speed in a 50 year period	v_{50}	70.0 m/s	59.5 m/s	52.5 m/s	42.0 m/s
Air turbulence intensity at a wind speed of 15 m/s TI_{15}	A	18.0%			
	B	16%			
	C	14%			

Note: Wind speed at Hub height

Air turbulence is defined as air speed fluctuation during a certain period of time. It increases in locations with uneven terrain and a high ground roughness caused by obstacles such as vegetation, buildings etc. It is measured at a wind speed of 15 m/s.

A wind turbine of IA class, e.g., is suitable for sites with strong wind and high turbulence intensity, and it is particularly robust in design. A turbine of class III is suitable for low wind, low turbulence locations; its rotor blades are long and its hub altitude high to capture more wind energy.

There is also a distinction of wind turbines for onshore or offshore application. The rated power of the former ranges between 1.5 MW to 7.5 MW while that of the latter ranges between 3.6 and 8.0 MW.

Technical data of selected types of wind turbines are shown in **Annex 23** and **Annex 24.**

5.2.3 Backup capacities to balance fluctuating power supply

The intermittent power supply of wind turbines is a real challenge for power system operators. As the power output of wind turbines is proportional to the third power of the wind speed, even small changes of the wind intensity have significant impact on the power output. The capacity factor increases with larger installed capacity and share of offshore wind power. The average capacity factor, e.g., of the entire installed wind power capacity in Germany obtained about 21 percent in 2015 (41.6 GW onshore, 2.3 GW offshore) [38]. In general however, reserve capacity requirements with quick start capability are high. There is a distinction between *positive and negative reserves*. Positive reserves are required for times of weak wind. Negative reserves are required for times with strong wind and excess production. In the former case conventional power plants must be switched on to balance the deficit in power supply. In the latter case conventional power plants must be switched-off or be operated at reduced capacity to allow the wind power into the system.

In liberalized energy markets, system operators must buy both positive and negative reserve capacities from energy generators usually in auctions for the day ahead. System operators have even to pay consumers who have the potential to take-off excess production from wind power.

5.2.4 **Yield calculation of wind turbines**

For the preliminary calculation of the energy yield of wind turbines, the following parameters are required:

- The performance characteristic of the wind turbine in digital form
- The annual average wind speed at hub height at the site
- The wind distribution

The performance characteristic is provided by the manufacturers – Figure 5-8. The annual average wind speed, usually at 10 or 30 meters elevation, is taken from wind atlases offered by several suppliers on the market. The wind speed at hub height is calculated with the formula stated in Figure 5-6 and the speed distribution over the year is assessed based on the Weibull or Rayleigh model – Figure 5-7. Based on these, the calculation of the energy yield of a *single* wind turbine is demonstrated in Example 5-3 below.

Wind power is commonly produced in utility size onshore – **Annex 26** –and offshore – **Annex 27** –windfarms consisting of numerous wind turbines and installed capacities of many MW. The calculation of the energy yield of a *wind farm*, consisting of many wind turbines, in most cases of different types, is much more complex. There are additional parameters to be considered as type, elevation, hub height and efficiency of each turbine. Furthermore, the mutual interference of the wind conditions between the different wind turbines and many other factors must be considered. Offshore windfarms can obtain almost double as high annual yield compared to onshore windfarms.

For this purpose there are several software programs available as, e.g., WindPro and WindFarmer. Both programs are based on the Wind Atlas and Analysis Program (WAsP) of the RISØ National Laboratory, Roskilde, Denmark that is established as the standard software.

Example 5-3: Reference annual electricity yield of a single Wind turbine

The computation of the annual electricity yield of a single wind turbine is conducted by multiplication of the power output values from the performance curve with the corresponding values of duration of each wind speed from the Weibull or Rayleigh distribution over all wind speed intervals.

Annual average wind speed *w*:								5.5 m/s			30 m above ground				
Rated power output								3,000							
Availability								0.97							
Hub height H_N:								149 m							
Average wind speed at hub height w_N:						Z0=0.10		7.05 m/s			Roughness class 2				
w_N	m / s	0.0	1.0	2.0	3.5	4.0	5.0	6.0	7.0	8.0	9.0	10.0	11.0	12.0	**Sub - total**
P_e	kW	0	0	0	49	155	339	628	1,036	1,549	2,090	2,580	2,900	3,000	
t	h / a	0	273	520	799	861	933	941	894	806	693	570	450	341	
W_e	MWh	0	0	0	38	129	307	573	898	1,210	1,404	1,426	1,265	992	**8,243**
Continuation															
w_N	m / s	13.0	14.0	15.0	16.0	17.0	18.0	19.0	20.0	21.0	22.0	23.0	24.0	25.0	**Sub - total**
P_e	kW	3,000	3,000	3,000	3,000	3,000	3,000	3,000	3,000	3,000	3,000	3,000	3,000	3,000	
t	h / a	249	175	118	77	49	30	17	10	5	3	1	1	0	
W_e	MWh	724	509	345	225	142	86	51	29	16	8	4	2	1	**2,142**
P_e kW: Performance characteristic of the WT *)								**Grand total, annual yield MWh**							**10,385**
t h/a: Frequency distribution acc. Rayleigh Model								**Capacity factor**				**39.5%**			**3,462 h/a**

5.2.5 Exceedance probability

The energy production from renewable sources fluctuates from year to year, depending on the prevailing weather conditions. The base case energy production (yield) of renewable energy projects is calculated for a *reference* year, reflecting the statistical average climatic conditions of the recent 10 to 20 years.

According to the Gauß normal distribution function, there is a probability of 50% that the base case yield may be exceeded and 50% that it may fall below. In the normal distribution this is the mean value μ, and the exceedance probability is denoted with P50 (*P* stands for probability).

Investors usually base their decision for investment on *P50* exceedance probability while creditors usually require *P90*. This means, they base their decision for approving a loan on a yield which statistically will be exceeded in 90% of the years during the lifetime of the loan, and this is lower than the base case yield.

The theoretical background is explained in the author's book "Engineering economics" [39]. The exceedance probability can be calculated with a slightly modified MS-Excel function "NORM.INV" as shown below:

$$X_{PXX} = 2 \cdot \mu - NORM.INV(PXX/100;\ \mu; \mu \cdot \sigma)\ [\text{MWh/a}] \quad (5.5)$$

or:

$$X_{PXX} = \mu \cdot (1 - \sigma \cdot NORM.INV(PXX/100; 0:1)\ [\text{GWh/a}] \quad (5.6)$$

Where:
μ MWh/a: is the arithmetical mean, which is the base case yield MWh/a

σ %: in the standard deviation $\sigma = \sqrt{\Sigma U^2}\ [\%]$ (U: uncertainties)

PXX: Probability e.g. P90 = 0.9 customized

Example 5-4: Exceedance probability

Given:
Base case yield of wind farm: μ=294 GWh/a
Uncertainties:
Wind data 15%; power curve 7%; wake effect 2%
Sought: Exceedance probability P50 and P90
Results:

Standard deviation: $\sigma = \sqrt{15^2 + 7^2 + 2^2} = 16.7\ \%$

P50=0.5: $X_{P50} = 2 \cdot 294 - NORM.INV(P50/100; 294; 294 \cdot 0.167) = 294\,[\text{GWh/a}]$

P90=0.9: $X_{P90} = 2 \cdot 294 - NORM.INV(P90/100; 294; 294 \cdot 0.167) = 231\,[\text{GWh/a}]$

5.2.6 Economics of wind power

Research work on manufacturers´ budget offers [40] has shown that capital cost (CAPEX) of wind turbines depend on the type, hub height, and rotor diameter. Budget prices of wind turbines must state besides the power output also the parameters mentioned above.

Capital cost estimates for wind power plants and wind farms are usually broken down in two groups: Cost of the wind turbines (100%) and costs for associated facilities and works as a percentage related to the costs of the wind turbine. Example 5-5 shows the capital cost estimates for two onshore wind farms with 50 MW and 100 MW installed capacity in Central Europe. Due to modular set up there is virtually little economy of scale. Economics of scale are related to: access roads to the site – see Annex 25 transport of a wind rotor blade, grid connection in particular OHL, equipment mobilization. The difference of the specific cost of the two options is within the estimation uncertainty (± 20%).

Example 5-6 shows a simplified integrated model for the calculation of the electricity generation costs of two wind farms with installed capacity 60 MW and 120 MW. The calculation is conducted in real terms applying the annuity method and is shown in the example in shorted form; the complete calculation is available on the author's Website. The calculation can be conducted with or without escalation rates for O&M costs.

Example 5-5: CAPEX estimate for onshore wind farms in Central Europe

Item		Unit	Wind farm Capacity 60 MW	120 MW
Capacity				
Power output, each Wind turbine		MW	2.40	3.00
Number of Wind turbines		-	25	40
Wind farm capacity		MW	60.0	120.0
WT delivery and assembly		€ /kW	1,350	1,250
Capital costs, 2013, ± 20%				
WT delivery and assembly	100.0%	1,000 €	81,000	150,000
Civil works	3.0%	1,000 €	2,430	4,500
Foundations	4.0%	1,000 €	3,240	6,000
Internal electrical wiring	5.0%	1,000 €	4,050	7,500
Grid connection, 110 kV	5.5%	1,000 €	4,460	8,250
Project development, engineering	5.0%	1,000 €	4,050	7,500
Contingencies	5.0%	1,000 €	4,050	7,500
Total		**1,000 €**	**103,280**	**191,250**
Specific cost		**€ /kW**	**1,720**	**1,590**

Example 5-6: Electricity generation costs, of wind farms

Item	Unit	Wind farm Capacity 60 MW	120 MW
Technical, Operational parameters			
Average wind speed, 30 m above ground	m / s	5.5	5.5
Electrical capacity of each WT	kW	2,400	3,000
Number of wind turbines	Stck.	25	40
Energy yield of each WT	MWh/a	8,590	10,385
Total gross energy yield of the wind farm	MW	60.0	120.0
Energy losses of the wind farm	%	14%	13%
Electricity production, net	MWh/a	184,690	361,399
Full load hours	h/a	3,078	3,012
Technical, economic parameters			
Life time	a	20	20
Construction time	a	1.50	2.00
Inflation	%	2.0%	2.0%
Discount rate in real terms (WACC) *)	%	4.58%	4.58%
CAPEX estimate, 2013 prices, ±20%	**1,000 €**	**103,280**	**191,250**
Operating Costs	**1,000 €**	**5,354**	**10,131**
Maintenance contract	1,000 €	1,847	3,614
Management/technical surveillance	1,000 €	1,343	2,486
Insurance	1,000 €	516	956
Reserves for decommissioning	1,000 €	826	1,530
Costs of personnel	1,000 €	175	280
Leasing costs for site	1,000 €	646	1,265
Annualized CAPEX	**1,000 €**	7,642	14,151
Total annual costs	**1,000 €**	**12,996**	**24,282**
Specific electricity generation cost	**€ / MWh**	**70.36**	**67.19**

*) inflation adjusted, in real terms

5.2.7 Case study – Cashflow & IRR Analysis of a wind farm

In **Case Study 11.6** an integrated techno-economic model for the financial analysis of a wind farm project is presented. The financial analysis comprises a cashflow and IRR model along with the required input and auxiliary calculations. The model includes the following spreadsheets:

1. Summary of results
2. Energy yield calculation
3. Technical-economic inputs
4. WACC
5. CAPEX estimate
6. Cashflow analysis on year by year basis
7. Internal rate of return analysis on year by year basis
8. Graphs: annual cost structure, projection of income and cashflow

A hardcopy of the model is depicted in chapter 11 "Case Studies" and a softcopy is available on the author's Website. Although quite complex, it is clearly arranged in a coherent manner so that the interested reader can follow the approach and retrace the calculations. Nevertheless, some brief explanations is given below.

Starting point of the model is the *energy yield calculation*. This is conducted first for a single wind turbine and projected for the entire wind farm, for exceedance probability P50 as the base case, and for P90 that is required by creditors. The spreadsheet includes also a "yes" or "no" key for conducting the cashflow and IRR analysis for either yield probabilities.

The spreadsheets 3 to 5 include input and auxiliary calculations and are self-explanatory. For practicing the model, inputs can be altered by the user as far as they are not project relevant.

The *cashflow model* is actually a projection of the project's financial performance during its 20 year lifetime. The first four columns include inputs for the actual calculations most of which are self-explanatory and are taken from the previous spreadsheets. The most important are:

- The *electricity production* remains constant during the entire lifetime either for exceedance probability P50 or P90 as chosen in the spreadsheet "yield calculation"
- The *electricity price* is not a feed-in tariff. It is determined with the "goal seek" function of MS-Excel to obtain an *IRR on equity after tax* as stated in the WACC based on an energy yield according to exceedance probability P50
- All payment and revenue series are subject to escalation usually slightly higher than the inflation rate

- The final outcome of the cashflow analysis is the annual *cashflow before amortization* and the *free cashflow* after repayment of the loan. The free cash flow is the amount available for dividend payments for the investors and for provision of reserves for general project risks
- Another important result is the debt service ratio. (DSR). This shall be higher than 1.1 if possible from the 1st or 2nd year of operation.
- Three graphics at the end of the model illustrate the projection of cost structure, income and cashflow

The IRR model is linked with the cashflow model and uses payment and cash inflow and cash outflow series of the cashflow model. It returns as result the IRR on the investment, on the equity before and after tax. Table 5-3 below depicts the main results of the financial analysis.

Table 5-3: Financial analysis of a Windfarm project, Summary of results

Item	Unit	Value
Key technical parameters		
Number of Wind turbines	-	40
Power output of each WT	kW	3,000
Installed capacity wind farm	MW	120
CAPEX	mln €	188.7
of which loan, 15 year maturity	mln €	151.0
Results for Exceedance Probability	**P50**	
Sales of electricityn, net	GWh/a	364
Electricity price, 1st year *)	€/MWh	69.30
Internal Rate of Return		
on investment	-	8.1%
on equity before tax	-	12.2%
on equity after tax	-	10.0%
Cashflow analysis, 1st year results		
Revenues	mln €	25.8
-Operating expenses	mln €	9.1
Operating income (EBIDA)	mln €	16.6
- Depreciation, interest on loans, corporate tax	mln €	16.9
Net income	mln €	- 0.3
+ Depreciation	mln €	9.4
Cashflow bevor principal repayment	mln €	9.2
- repayment of loan	mln €	10.06
Free cashflow	**mln €**	**- 0.89**
Debt coverage ratio (DSR), over load life	-	1.42

5.3 Basics of Solar Energy

5.3.1 Solar energy

The sun produces enormous amounts of energy by nuclear fusion, converting hydrogen to helium, at the sun's core. Solar energy is emitted by the sun in form of solar radiation. The average intensity of solar radiation, measured at the entry of the earth's atmosphere (about 8,000 km) is 1,367 W/m^2. This value is known as the *solar constant*. A part of this energy is lost in the outer layers due to reflection and absorption. Depending on the latitude, time of day and weather conditions, up to a maximum of 1,000 W/m^2 reach the earth's surface.

The term *solar radiation,* or *insolation,* refers to the source from where the solar energy comes (sun); the energy density of solar radiation W/m^2 on a given surface is called *irradiance*. The term *irradiation,* or *insolation incident,* refers to the solar energy falling to a surface over a period of time e.g. kWh/m^{2}d or kWh/m^2 a. The different forms of solar irradiation are illustrated in Figure 5-9.

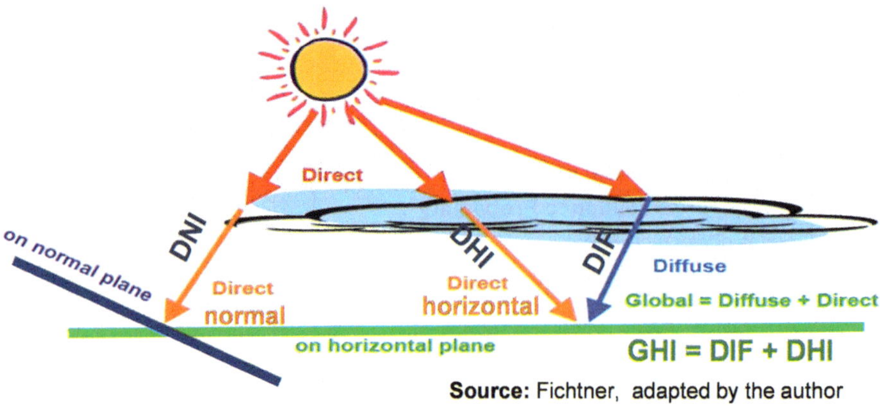

Figure 5-9: Forms of solar irradiation

The part of solar radiation reaching the earth's surface on a horizontal plane consists of two components:

- the *direct horizontal irradiation DHI,* which is received directly from the sun without scattering or reflection
- the *diffuse irradiation DIF,* which is reflected from clouds, water drops and dust particles
- The sum of direct *DHI* and diffuse irradiation *DIF*, measured on a horizontal plane, is called global horizontal irradiation GHI.

The *direct normal irradiation DNI* is measured on a plane normal to the beam of the sun. It is to be distinguished from the direct horizontal irradiation (DHI).

The ratio between *DHI* and *DIF* varies by season, time of the day and latitude. It plays an important role when comparing various technology options. Databases for solar radiation can be found in [41].

Figure 5-10 depicts the global horizontal irradiation (GHI) in two locations with different profiles. Technologies, which use global irradiation, such as photovoltaic, make sense in both locations, because the available resources are sufficient. Kuala Lumpur is apparently a better location as the irradiation remains high throughout the year. Source of data for both figures is NASA atmospheric science Data Center [42].

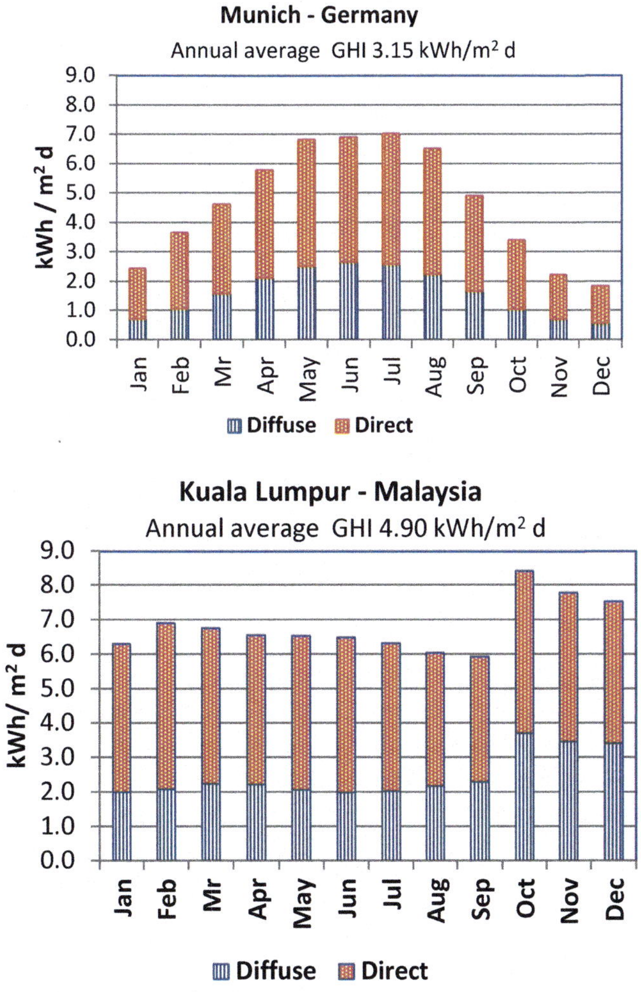

Figure 5-10: Monthly average irradiation on horizontal surface kWh/m^2 a

Generally, there are two main systems of solar technology to generate electricity, namely Photovoltaics (PV) and Concentrating Solar Thermal (CSP). Photovoltaics utilize both direct and diffuse irradiation. Concentrated solar power plants utilize only the direct normal fraction (DNI) of the solar irradiation.

Figure 5-11 depicts the direct normal solar irradiation – DNI – in Munich, Johannesburg and Kuala Lumpur. It becomes evident that Johannesburg is an excellent location for CSP while this technology seems not to make sense for Munich and even for Kuala Lumpur although the latter is near to the equator (see also Figure 5-12 and Figure 5-13) and is a good location for photovoltaics. The reason is apparently the humid climate and a long rain period.

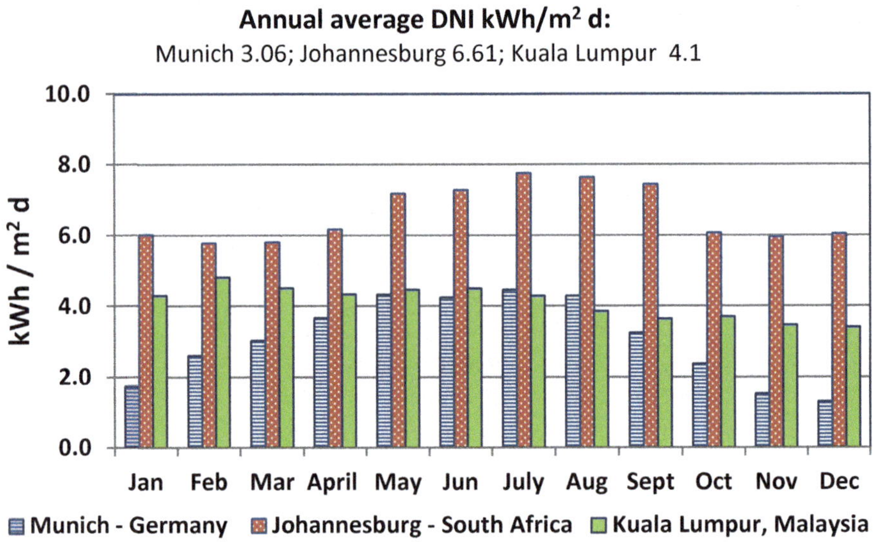

Figure 5-11: Monthly average Direct Normal Irradiation – DNI

Solar irradiation data can be made available by various provider [43], partly free of charge that can be used for a first insight. However, for real projects, long time on-site measurements from specialized companies are indispensable.

Figure 5-12 and Figure 5-13 below depict maps of solar irradiation, primarily for first information purposes. It becomes evident that the best locations for solar energy applications are Africa, South America, Australia and large parts of the USA. European countries around the Mediterranean Sea, notably Spain, Greece, Italy and Turkey, provide sufficient resources for solar energy applications.

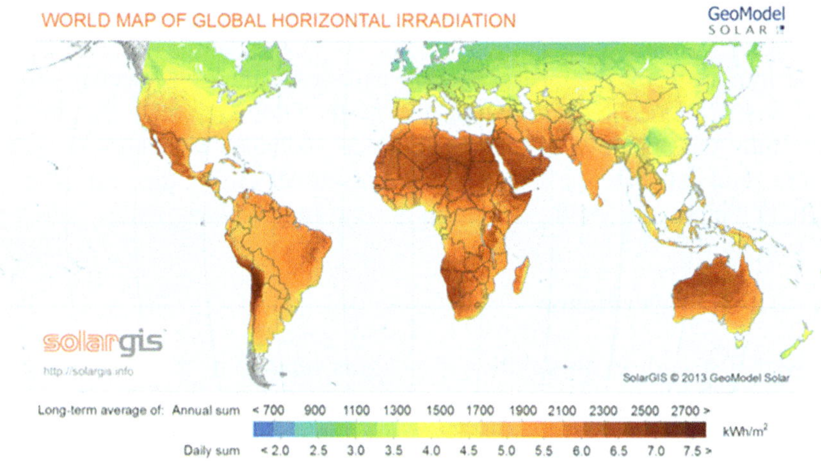

Figure 5-12: World map of Global Horizontal Irradiation

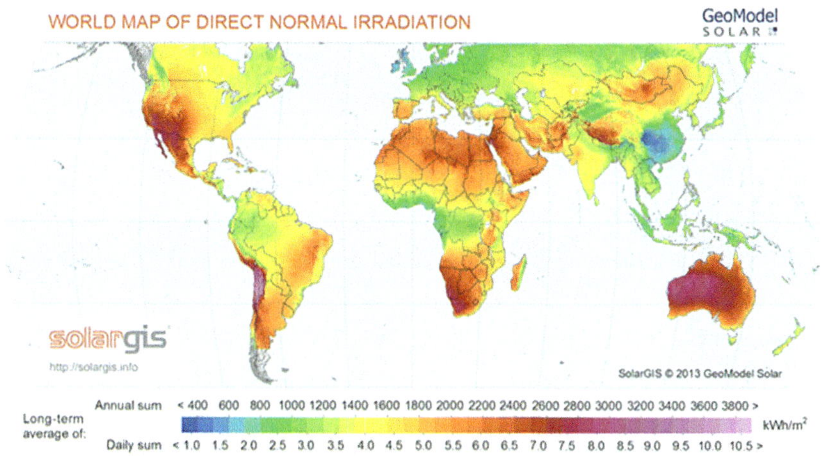

Figure 5-13: World map of Direct Normal Irradiation – DNI

Source: http://solargis.info/doc/maps-for-solar-energy

5.3.2 Sun-Earth geometry

Engineers and consultants, dealing with solar energy, need to have some basic knowledge about the earth-sun relationship to understand how the rotation of the earth around its axis and the motion on its orbit around the sun affects seasonal and daily solar irradiation on the earth. The matter is quite complex and for non-experts difficult to understand. We try to provide the absolutely necessary knowledge as simply as possible. The graphs used are taken from available sources, which are stated in the figures, so we focus on the description and explanation.

Most important for engineers, dealing with solar energy projects, is to define the exact position of the site of the project on the Globus (latitude and longitude), and to study the positions of the sun (altitude, azimuth, solar noon), in order to determine and calculate the available solar resources during the year. The related data are key inputs for yield calculation of solar power plants.

5.3.3 The Earth's rotation

Let's start first with Earth's orbit and position relative to the sun. The earth rotates daily (24 h) around its axis and once a year (365 days) around the sun on an elliptical *orbit* Figure 5-14 [44]. There are three distinct planes: the equator's plane, the plane of the earth's axis and the orbit plane.

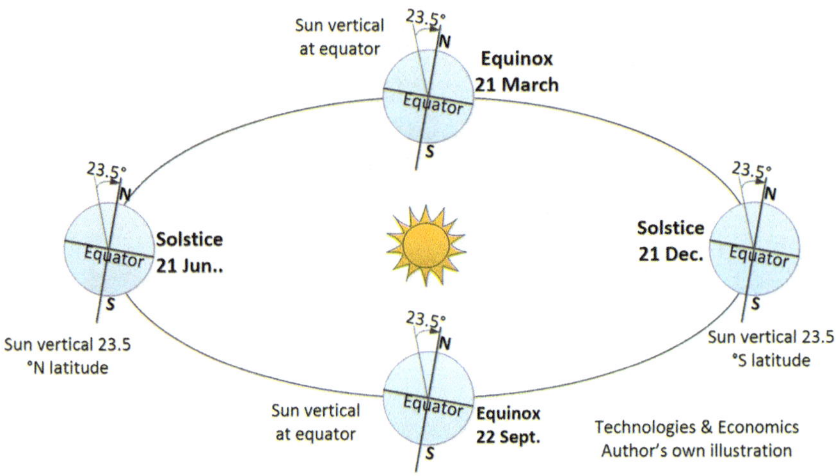

Figure 5-14: Earth-sun positions on the orbit for the northern hemisphere

The earth's axis is perpendicular to the equator's plane and inclined by 23.5° against its orbit around the sun; this inclination is kept the same throughout the year. Thus, during the earth's rotation on its orbit, the incident angle of the sun rays steadily changes its direction, causing a seasonal variation of the insolation of two hemispheres.

There are four distinct points regarding the earth's position to the sun on the orbit: During the two *equinoxes* in March and September, the earth lies in plane to the orbit and its axis plane is parallel to the major axis of the elliptical orbit; the length of the day is equal to the length of the night, and both hemispheres receive 12 hours sunshine.

During the *summer solstice* in June, the earth axis makes an angle of +23.5° with the major axis of the orbit's ellipse and the northern hemisphere is more exposed to the sun's rays than the southern; the area around the north polar circle is insolated 24 hours a day. In contrary, during the *winter solstice* in December (angle -23.5°), the southern hemisphere is more exposed to the sun and the southern pole area receives sunshine throughout the day.

In the southern hemisphere, the summer solstice occurs in December and the winter solstice in June.

5.3.4 Definition of the plants' site

The location of the solar plant's site on the Globus is determined by its coordinates, latitude and longitude – Figure 5-15.

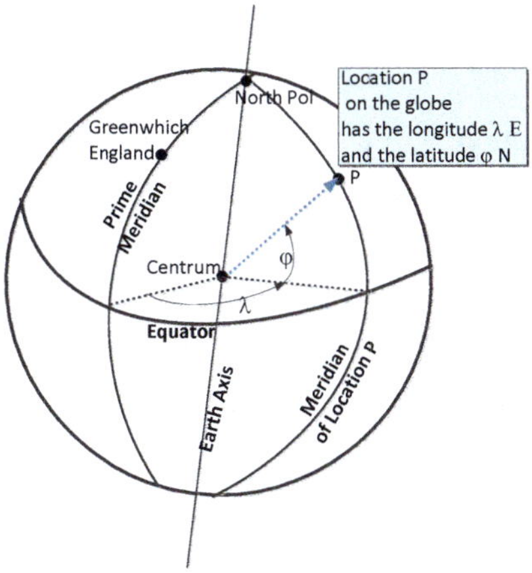

The *latitude* of a location *P* on the earth's surface is the angle $\varphi°$ formed between the line joining point *P* with the center of the earth and the equatorial plane. The earth's equator has the latitude 0. We distinguish between latitude north "$\varphi°$ N" and latitude south "$\varphi°$ S". The semi-circles joining the two poles are called meridians.

Figure 5-15: Angles for definition of site location

The longitude is the angle $\lambda°$ formed between the *prime meridian* with longitude 0, passing through Greenwich in England, and the meridian passing through the point *P*. We distinguish between longitude east and longitude west.

For example: the geographical coordinates of Munich in Germany are longitude 11.58° E and latitude 48.13° N; those of Riyadh in Saudi Arabia are longitude 46.43° E and latitude 24.46° N.

The sun obtains its highest point, called zenith, for all locations along a meridian at the same time of the day; this is defined *as solar noon*. When this occurs, the sun rays are directed perpendicular to the given meridian.

5.3.5 Angles defining the position of the sun

The angles, defining the position of the sun for solar energy applications, are the sun's altitude, sun declination and solar azimuth. A very good website with animations explaining the angles is http://www.pveducation.org/.

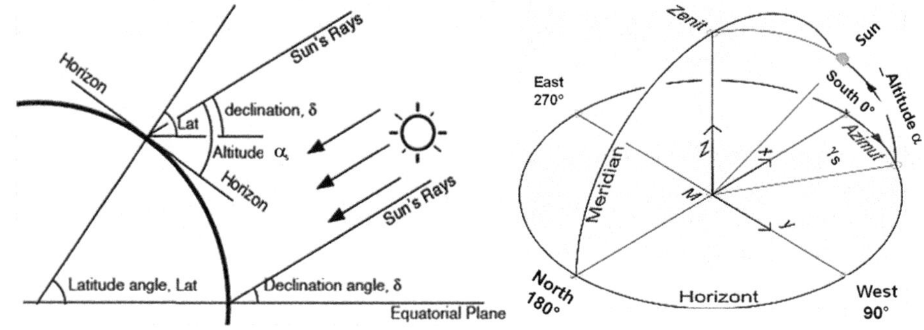

http://www.cibsejournal.com/cpd/2013-07/images/5.jpg https://de.wikipedia.org/wiki/Azimut

Figure 5-16: Angels defining the sun's position

The *sun's altitude or elevation* (left figure) is defined as the angle α_s formed between the direction of the sun's rays and the horizontal plane at the point P. As shown in Figure 5-16, the elevation angle is defined by the equations:

Northern hemisphere: $$a_s = 90° - \varphi + \delta \quad (5.7)$$

Southern hemisphere: $$a_s = 90° + \varphi - \delta \quad (5.8)$$

On the summer solstice, the sun is directly over the tropic circle (tropic of Cancer respectively tropic of Capricorn), the angle is $\varphi=\delta$ and the altitude angle is $\alpha_s=90°$. The altitude angle becomes even >90° for locations between the tropic circle and the equator.

The *Sun declination* is defined as the angle δ between the direction of the sun's rays and the parallel to the equatorial plane or the angle by which the earth axis is inclined against the sun (see Figure 5-14). The angle is taken as positively oriented when the sun rays pass the northern hemisphere. The solar declination varies throughout the year due to the earth's rotation round the sun between -23.5° during the winter solstice, 0° during the two equinoxes and +23.5° during the summer solstice.

Table 5-4: Sun declination

Date	Declination δ in Hemisphere	
	Northern	Southern
21 June	+23.5°	-23.5°
23. Sep	0°	0°
21 December	-23.5°	+23.5°
21 March	0°	0°

Solar azimuth γ_s (Figure 5-16, right) is the horizontally projected angle between the sun's rays and the due-south direction for the northern hemisphere and against the due-north direction for the southern hemisphere. The sun reaches its highest elevation during the solar noon when the sun's rays are parallel to the south direction or at azimuth 0°. During sunrise and sunset, the solar azimuth is -90° and +90° (there is also measurement S=0°, W=90° and E=270°).

Solar plants in the northern hemisphere are orientated north-to-south, their direction azimuth is due zero. Solar plants in the southern hemisphere are orientated south-to-north where again the direction azimuth is zero.

Relative air mass ***AM*** is defined as the path length of the sun's rays through the atmosphere relative to the minimal path length when the sun is at zenith. The rays' path length changes due to the rotation of the earth round the sun.

$$AM = \frac{1}{\sin(\alpha_s)} \tag{5.9}$$

- AM=0, outside of atmosphere, vacuum
- AM=1, for α_s=90°, tropic circle of cancer
- AM= 1.5, α_s=41.8°, average for Central Europe
- AM= 2.0, α_s=30°

The *incidence angle modifier* – IAM – corresponds to the decreased irradiance effectively reaching the solar collector due to cosine effect reflections compared to irradiance under normal incidence. For plain PV panels it is usually considered in the modules definition of losses. For concentrating solar collectors it is calculated from the cosine of sun's rays to collector axis together with collector specific derate with collector specific derate factors.

5.4 Solar Power Plants with Photovoltaic Technology

5.4.1 The photovoltaic process

Photovoltaic is the process of the direct conversion of solar irradiation into electrical energy by means of solar cells. The solar cells comprise two adjoining semiconductor layers that are equipped with separate metal contacts and have each been doped, thus creating an "n" layer (n = negative) with a surplus of electrons, and below that, a "p" layer (p = positive) with electron deficiency [45]. Due to the difference in concentration, the electrons flow from n- into the p-area, creating an electrical field.

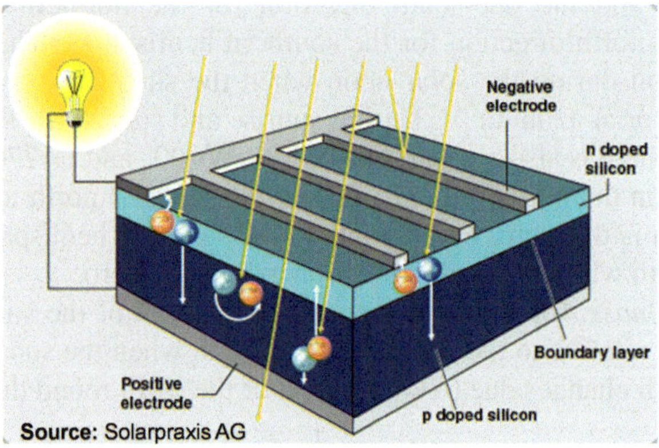

Figure 5-17: Functional principle of photovoltaic cell

The conversion process is based on the photovoltaic effect. Simply described, sunlight consists of energy particles called photons. When a photon hits the upper n-layer of a solar cell, it transfers its energy to an atom within the cell and knocks one of its electrons free. The freed electron follows the electrical field and travels to the p-layer in an electrical circuit. When an electrical load is connected, the power circuit is closed; the electrons flow across the electrical load to the solar cell's rear contact and then back to the space charge zone.

5.4.2 Configuration of PV systems

PV cell technologies are distinguished amongst others by purpose of use, manufacturing process, type of PV material used and electrical efficiency. *Crystalline silicon* and *thin film* cells are the main technologies and have been widely used for commercial and utility-scale PV plants. Silicon cells are usually available in sizes 10x10 cm, 12.5x12.5 cm or 15x15cm. A silicon cell of 100 cm^2 obtains a power output of about 2 W, at irradiance 1000 W/m^2.

PV cells are assembled into a panel or module to increase the power output. The modules are connected in series to reach the specific voltage requirements and then in parallel to a so-called *array* to reach the specific current requirements. The direct current (DC) electricity, generated by the array, is then delivered to an *inverter* which converts it to alternating current (AC) for the electricity grid.

A typical PV system contains several key components, such as modules inverter, mounting structures and balance of system (BOS) systems (e.g. cables, combiner boxes, metering equipment etc.). The structure, holding the modules, is referred to as the *mounting system.* PV modules are oriented to the sun with different mounting concepts in order to achieve higher yields. They may vary from *fixed* structures to *tracking* mounting systems. Generally, ground-mounted systems can be categorized as fixed tilt, single-axis tracked and dual-axis tracked. Utility scale PV plants are currently being built in the scale up to 250 MW.

5.4.3 Performance parameters of photovoltaic systems

The *nominal power (P_N)* of a PV module is stated in watt peak (W_P) or kilowatts peak (kW_P) which is the power of solar modules under the standard test conditions (STC) of 1,000 W/m^2 solar irradiance, air mass of AM 1.5 and 25°C PV module/cell temperature. The *efficiency* (η) of a PV module is a measure of the percentage of solar energy converted into electricity and measured under STC in order to make PV modules comparable. The following Table 5-5 gives an overview of the efficiency of different PV technologies.

Table 5-5: Efficiencies of PV cells and modules

Material	Cell Efficiency %	Module Efficiency %
Crystalline Silicon Cells		
Mono-crystalline	16% to 22%	14% to 20%
Poly-crystalline	14% to 18%	12% to 16%
Thin Film Cells		
Amorphoua Silicon	8% to 10 %	7% to 9%
Cadmium Telluride – CdTe	10% to 17%	11% to 14%
Copper Indium – CIS, CIGS	11% to 14%	10% to 13%

The power *(P)* of a PV module mainly depends on the irradiance (W/m^2) and its operating temperature. The following Figure 5-18 depicts the performance of a PV module at different irradiance levels. The current of the modules increases with rising irradiance while the voltage remains practically constant up to the maximum power point (MPP). This is the point at which the module obtains its maximum power output at the respective irradiance level.

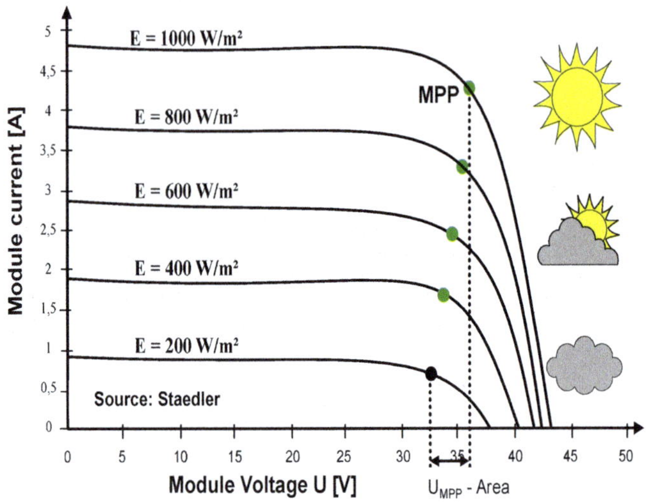

Figure 5-18: Current-voltage diagram of a PV module at 25°C

The power output of a PV module decreases with rising cell temperature. This influence of the temperature on PV module performance is expressed by the *temperature coefficient* defined at the STC temperature 25°C (100%). The temperature coefficient of crystalline technology is in average -0.4%/K; for thin film technologies it varies between -0.2%/K to -0.25/K. Example 5-7 demonstrates the temperature effect on the output of PV modules.
The power output of PV-modules decreases over their lifetime due to *degradation*. Manufacturers of PV modules usually give a guarantee of 90% of the nominal power after the first 10 years and 80% after 25 years lifetime. Other manufacturers have started to give also linear guarantee e.g. 0.7%/a.

Example 5-7: Power output of PV modules vs. operating temperature

Item	Unit	Values		
Cell temperature	°C	25	60	80
Crystalline cells				
Temperature Coefficient	%/K	-0.4		
Power output	kW	100	86	78
Thin film cells				
Temperature Coefficient	%/K	-0.25		
Power output	kW	100	91	86

5.4.4 Techno-economic assessment

5.4.4.1 *Energy yield calculation*

Starting points for an economic assessment of solar power plants are a sound technical concept of the plant and a calculation of the energy yield for a reference year. There are several tools available on the market for modelling, design, performance simulation and economic evaluation of PV plants. Main inputs are among others the technical and performance parameters of the selected PV panels, the site coordinates and the solar irradiation for a reference year.

In the following Example 5-8 a simplified model for the layout and calculation of the annual yield of PV-parks for different locations is presented.

Example 5-8: Technical layout and yield calculation for selected sites

Item	Unit	Germany Munich	Greece Athens	Malysia Kuala Lumpur	Australia Sydney	S. Africa Johannesburg
Technical parameters						
Module area (mono-crystalline)	m^2	489,476				
PV Module efficiency η_M	-	20.43%				
Annual performance ratio *PR*	-	81.50%				
Azimuth *)	degrees	0				
Tracking	-	vertical axis tracking, optimal tilt				
Site latitude	degrees N	48.5	38.5	3.5	-33.5	-26.5
Longitude	degrees E	11.5	23.5	101.5	151.5	28.5
Optimal tilt angle, annual average ***)	-	37.9°	32.3°	14.6°	30.2°	25.9°
Electricity Production						
Nominal capacity STC **)	MW_p	100				
Horizontal global irradiation ***)	kWh /m^2 a	1,149	1,565	1,788	1,620	2,018
Global irradiation, optimal tilted panel ***)	kWh /m^2 a	1,321	1,748	1,821	1,862	2,306
Annual yield	MWh / a	107,661	142,462	148,411	151,753	187,939
Specific yield (1000 W/m^2, 25°C, AM =1.5))	kWh /kW_p	1,077	1,425	1,484	1,518	1,879

*) Northern hemisphere against due south, Southern hemisphere against due north

**) Irradiation 1000 W/m^2; 25°C; AM=1.5)

***) NASA satellite derived meteorological and solar energy parameters

The following formulas are used for the calculation of the nominal capacity and the annual yield:

$$P_N = A \cdot p_{norm} \cdot \eta_m \quad \left[\mathrm{kW_p}\right] \tag{5.10}$$

$$Y = A \cdot \eta_m \cdot PR \cdot H_G \quad [\mathrm{kWh/a}] \tag{5.11}$$

Where:

P_N:	Nominal capacity at STC
A:	Module Area m^2
p_{norm}:	Global irradiance 1 kW/m^2
η_m:	Module efficiency % of solar irradiation
Y:	Annual yield kWh/a
PR:	Annual performance ratio
H_G:	Global irradiation on tilted plain

The performance ratio *PR* is the most characteristic value to evaluate the performance of the plant. It considers all losses of the plant. Some degradation of power production during the lifetime of the plant is also considered in the calculation. The specific yield (kWh/kW$_P$) corresponds to the annual utilization time h/a.

5.4.4.2 ***Electricity generation cost***

Prices for photovoltaic plants have been drastically decreasing in recent years, due to mass production of the main components, increasing energy efficiency of the systems components and rapid growth of the market.

Price trends for photovoltaic systems are provided, e.g., by the LBNL [46] and NREL [47] for the market in the United States[21]. Overnight costs for plants expected to be installed in 2014, are given with 1.92 US$/kWp.

In Example 5-9, a simplified model for the calculation of the levelized electricity generation costs is presented, based on the technical parameters and annual yield of the plants in Example 5-8. The calculation is given in short form; the complete model is available on the author's website.

It becomes evident that the major cost component is by far the capital cost. The CAPEX and all the other financial constraints are assumed to be the same for all plants. Under these assumptions, the LEC of the plant in the best location are almost half as high compared to those of the plant in southern Germany (Munich).

Utility size photovoltaic power plants with capacities of several hundred MW are already in operation in several countries. The largest PV power plant in Europe is currently the **Cestas** near Bordaux with capacity of 300 MWp. The plant covers about 250 hectares and developed for a CAPEX of about 360 million €, Operation start in 2015.

[21] PV Pricing Trends: NREL- National Energy Laboratory, USA, LBNL – Lawrence Berkeley National Laboratory, USA

Example 5-9: Levelized electricity costs of PV-plants

Item	Unit	Germany Munich	Greece Athens	Malysia Kuala Lumpur	Australia Sydney	S. Africa Johannes burg
		100 MWe				
Energy production						
Nominal capacity STC	MWp	100				
Annual yield	MWh / a	107,661	142,462	148,411	151,753	187,939
Financial constraints						
Life time	a	25				
Equity share	%	20.0%				
Inflation	% / a	2.0%				
Discount rate, nominal	% / a	7.5%				
Discount rate, real terms	% / a	5.4%				
O&M Cost	% / a	0.50%				
Site lease	ct / kWh	0.20	0.20	0.20	0.20	0.20
Insurance	% / a	0.75%	0.75%	0.75%	0.75%	0.75%
CAPEX, US$ 2014, ±20%	**Mio. US$**	**215.0**	**215.0**	**215.0**	**215.0**	**215.0**
Specific CAPEX	USD / KWp	2,150	2,150	2,150	2,150	2,150
Annual levelized costs, on real terms						
Annualized CAPEX	1000 US$ / a	15,809	15,809	15,809	15,809	15,809
O&M Costs	1000 US$ / a	1,075	1,075	1,075	1,075	1,075
Lease	1000 US$ / a	215	285	297	304	376
Insurance	1000 US$ / a	1,612	1,612	1,612	1,612	1,612
Total	**1000 US$ / a**	**18,712**	**18,781**	**18,793**	**18,800**	**18,872**
LEC on real terms	**ct / kWh**	**17.38**	**13.18**	**12.66**	**12.39**	**10.04**

5.5 Solar Power Plants with Parabolic Trough Technology

5.5.1 Technology description

The Parabolic trough is the most mature of all Concentrate Solar Power (CSP) technologies. A total installed capacity of more than 3,000 MW is reported worldwide in 2014. The three Andasol solar power plants in Spain have electrical capacity of 50 MW each. Currently, power plant output of 250 MW is considered to be state-of-the-art. The major advantage of (CSP) power plants, compared to photovoltaics, is their ability to store solar heat in thermal energy storage systems (TES).

In contrast to photovoltaics, Concentrated Solar Power (CSP) plants utilize the direct irradiation fraction (DNI) of the sunlight only. Sites with solar irradiation higher than 2,100 kWh/m^2a are considered as suitable, while best sites may ob-

tain up to 3,000 kWh/m^2 a. (Andasol in Spain 2,136 kWh/m^{2}a, Mojave Desert in California 2,700 kWh/m^{2}a, Atacama Desert, Chile 3,000 kWh/m^{2}a).
Solar parabolic trough power plants consist of solar fields with many parabolic trough collectors, a heat transfer fluid piping system, solar steam generator(s), a Rankine steam turbine/generator set, a thermal storage and/or fossil-fired backup systems – Figure 5-19.

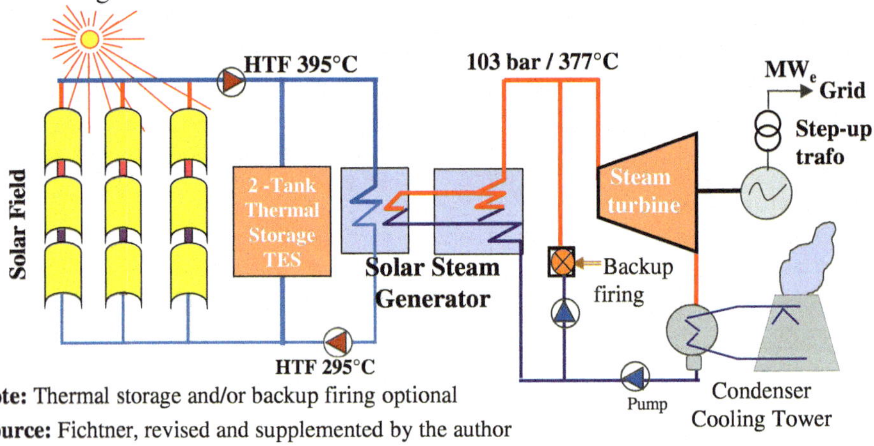

Note: Thermal storage and/or backup firing optional
Source: Fichtner, revised and supplemented by the author

Figure 5-19: Simplified heat flow diagram of parabolic trough solar PP

Solar field thermal efficiencies up to 50% and electrical plant efficiencies up to 18% as annual average, both referred to the solar irradiation (kWh/(m^2 a), are achievable in good sites.

5.5.2 Main system components

The *solar field* is modular in nature and comprises several subfields with many parallel rows of parabolic trough solar collectors. Each *solar collector* has linear parabolic-shaped mirrors – Figure 5-20 – which concentrate the sun's direct beam radiation on a linear *absorber pipe* located at the focus of the parabola. The collectors track the sun from east to west to ensure that the sun is continuously focused on the linear absorber.

The collectors are arranged in *loops*, usually each loop comprising four collectors in two parallel rows. The two ends of the loops are connected to the supply and return line of the piping system of the *heat transfer fluid* (HTF).

The heat *transfer fluid (HTF)* is in most cases thermooil. It can be utilized for temperatures between 12°C and 400°C. The HTF system consists of the piping and pump systems, the pressure holding system (>11 bar to prevent evaporation of HTF), ulllage system and freeze protection system (to maintain HTF temperatures >12°C at any point of the system). The HTF is heated up in the collectors to

393°C and flows through the piping system of the solar field to the steam generator(s) of a conventional steam cycle to generate high pressure steam.

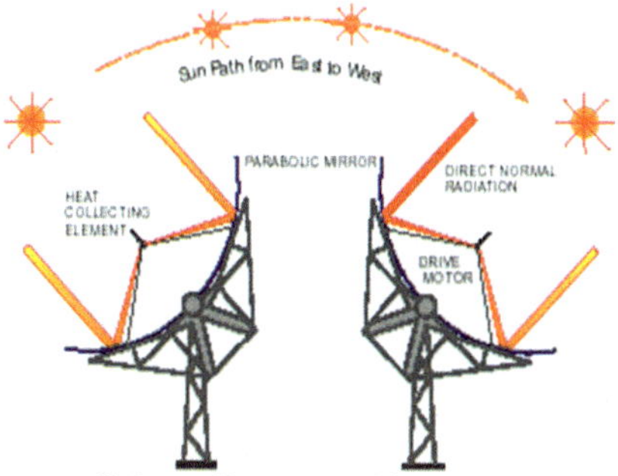

Solar collector tracking system
Source: Concentrated Solar Power

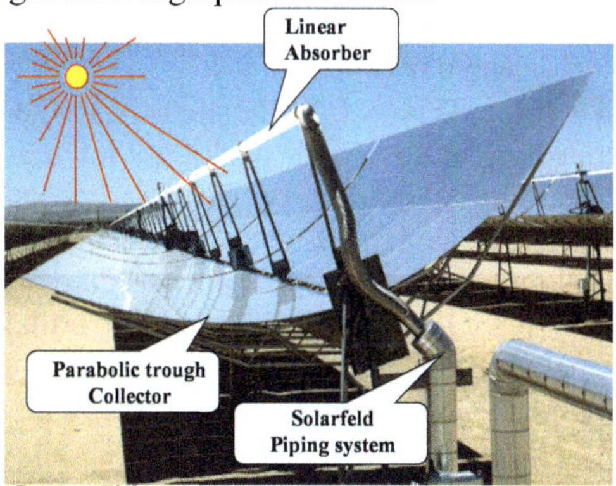

Source: Fichtner

Figure 5-20: Parabolic trough collector and single-axis tracking system

The *thermal storage system (TES)* consists of two tanks containing molten salt – Figure 5-21. Cold molten salt, returning from the solar steam generator, is stored first in the cold storage tank. During excess solar heat production hot HTF, coming from the solar field with 393°C, is pumped through HTF/salt heat exchangers and transfers its heat to the molten salt of the cold tank and is charged into the hot tank. For discharging, the system simply operates in the reverse mode.

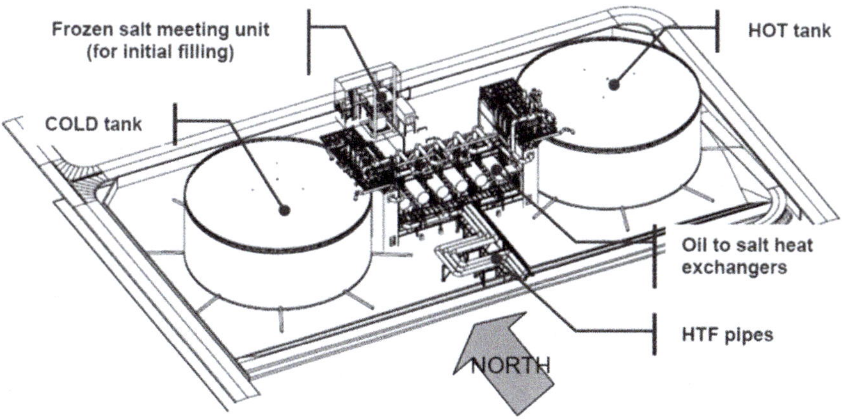

Figure 5-21: Thermal storage system – TES

The stored energy can be utilized/dispatched when additional power is required (e.g. during overcast or in night time period). Plants with thermal storage capacities of 8 to 12 hours achieve capacity factors of 50% to 60% and are capable for intermediate load dispatching. However, the solar field must be adequately oversized in order to be able to charge the thermal storage with excess heat production during day time.

When the TES is full, the solar heat supply is interrupted by tracking the solar collectors in *defocused* position known as *dumping*. The ratio between the solar field's thermal output to the actually required thermal output to drive the power cycle at its nameplate capacity is called *solar multiple*.

The *Solar Steam Power Plant* – Figure 5-22 – consists of the solar steam generator(s) (SSG), usually in several parallel connected units, the steam turbine-generator set, the water-steam piping system and the condenser. The live steam conditions are usually 103bar/377°C with reheat.

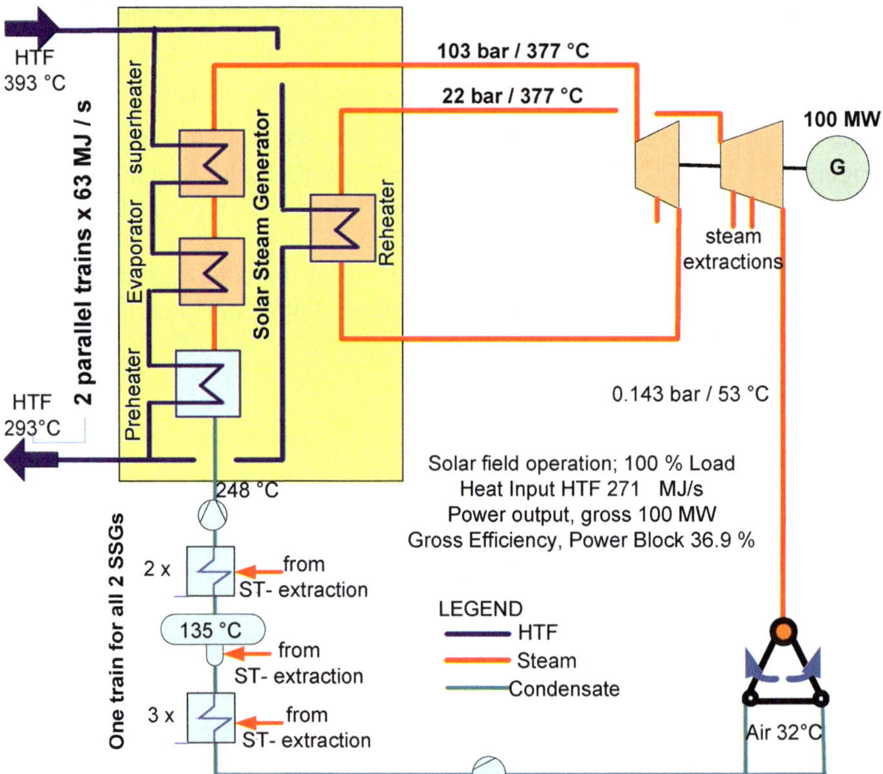

Source: Technologies & Economics, Author's own illustration, Cycle calculation with K

Figure 5-22: Simplified heat flow diagram of a solar power plant

Table 5-6: Power plant performance

Item	Unit	Solar field direct	Thermal storage
Power output	MW	100	85
Heat input	MJ / s	271	234
Electrical efficiency	-	36.90%	36.32%
Live steam	bar /°C	103 / 377	88.6 / 359
Reheat steam	bar /°C	22 / 377	19 / 359

Note: Cycle calculation with KPRO®, Fichtner

5.5.3 Techno-economic assessment

Starting points for a techno-economic assessment of solar power plants is a sound technical concept of the plant and a calculation of the energy yield for a reference year. There are sophisticated software tools available on the market for modelling, and operational performance simulation of parabolic trough plants. They calculate electricity yield for each hour of a reference year based, on the solar irradiation data for the site and several other technical parameters.

Concentrated solar power plant projects are highly capital intensive. Their electricity generation costs mainly consist of capital costs and fixed O&M costs. The main cost component is the size of the solar field and its associated systems.

Case Study 11.7 presents an integrated techno-economic model for a solar power plant project with parabolic trough technology. The model includes three options of a 100 MW solar plant namely: solar only without thermal storage (TES) and with TES of 8 and of 12 hours discharge time. A hardcopy of the model is depicted in chapter 11 "Case Studies" and a softcopy is available on the author's website. The model includes the following spreadsheets.

1. Summary of results
2. Input solar field
3. Input power block
4. Solar field thermal capacity calculation
5. Solar field size calculation
6. CAPEX estimate
7. OPEX estimate
8. WACC
9. LEC calculation
10. Graphs

Although quite complex, the model is clearly arranged in a coherent manner so that the interested reader can follow the approach and retrace the calculations. Nevertheless, some brief explanations are given below.

In the *first spreadsheets* the summary of main results is presented and also depicted in the following Table 5-7, in order to give an impression of the outcomes of the model. All the other spreadsheets are only briefly described and explained in the following.

Table 5-7: Case Study of Parabolic trough PP, Summary of results

Item	Unit	100 MW		
		TES 0 h	TES 8 h	TES 12 h
Power balance				
Power output, solar field operation	MW		100	
Power output, TES operation	MW		85	
Number of collectors	-	655	1,400	1,773
Solar field heat production	MJ / s	271	579	734
Thermal Storage	hours	-	8	12
Solar multiple	-	1.0	2.1	2.7
Energy balance				
Annual irradiation DNI	kWh /(m^2 a)		2,400	
Solar heat to power block	GWh_t / a	591	1,263	1,599
Net electricity production	GWh_e /a	213	456	578
Financial parameters				
Discount rate, in real terms (WACC)	-		4.6%	
Project lifetime	years		25	
Capital expenditures, US$2014, ± 20%	mln US$	402	766	937
Electricity generation costs, in real terms, 2014				
Annual generation cost	mln US$ / a	39.2	71.0	87.1
of which capital cost		70%	73%	73%
Levelized electricity cost	**US$ / MWh**	**214**	**181**	**175**

In *spreadsheets* No 2 and No 3 inputs of some key design and performance parameters of the solar field and power block are shown, which are relevant for the assessment of the annual energy yield and the electricity generation costs. A simplified heat flow diagram of the power block is also shown in Figure 5-22. The power block has the same rated power output of 100 MW for all three options. However, the power output for storage operation drops from 100 MW to 85 MW because the temperature of the HTF, coming from TES to the solar steam generators (SSG), is lower due to the double heat exchange between HTF and molten salt during the charging and discharging process (see Table 5-6).

The model allows calculation of LEC alternatively "with" or "without" escalation of the OPEX. The user is required to enter "yes" or "no" in spreadsheet No 2 "input solar field".

In *spreadsheet No 4* the thermal capacity of the solar field is assessed, based on the irradiation data of the site, the power plant output and storage capacities. The main inputs for the solar field design and performance are the solar irradiance (W/m^2) and geographical coordinates of the site. The site of the plant is assumed to be located in North Africa. The peak optical efficiency of the selected collector type (SKAL-ET 150) is 80% at standard test conditions. However, its effective efficiency at the particular site is determined by the *incident angle modifier (IAM)* and the latitude of the site (see definitions 5.3.4 and 5.3.5). The number of collectors is determined under consideration of following aspects: The aperture area of the solar field for the option without thermal storage is designed to deliver the solar heat, required to drive the power plant at name plate output (solar multiple =1). For the options with thermal energy storage the solar field must be oversized in order to supply additional solar heat in parallel for charging the thermal storage. A charging time of 6 hours is assumed for determining the thermal capacity of the solar field. The main outcomes of the solar field design are the number of collectors and loops and the solar thermal capacity of the solar field for the three options.

In *spreadsheet No 5* the size of the solar field is calculated. This includes the area requirements for collectors, roads between the subfields, area for the power block and supporting facilities.

In *spreadsheet No 6* the CAPEX estimate based on the design parameters of the plant is presented, broken down into the main components. The estimate is derived from bids for real projects. It is evident that the system is capital intensive (see Summary Table 5-7). For comparison, the specific CAPEX for a utility size steam power plant of 600 MW is about 2000 US$/kW, for a CCGT power plant about 900 US$/kW. Furthermore, the solar field and the thermal storage are of modular design; hence, there is very little economy of scale with increasing size of the plant. Some economy of scale can be expected for the HTF system and the power block which are not of modular design.

In the same manner the OPEX estimate is presented in *spreadsheet No 7*. The fixed operating expenses (OPEX) are mainly maintenance costs, which are estimated as a percentage referred to the EPC costs, cost for personnel and insurances. The variable OPEX includes water costs for cleaning the mirrors, costs for refilling HTF to offset HTF-losses and electricity costs during downtime of the plant.

The calculation of the levelized electricity generation costs is conducted in *spreadsheet No 8*. The methodology of the calculations is practically the same as that of the complex tools, however, using average annual date for solar irradiation. Some of the key parameters of the plant performance are derived from actual calculations for projects conducted with the software tool SolPro[22] [48]. The calculation includes also the energy balance based on the annual direct irradiation at the site (kWh/m^2 a). The collector and solar field efficiency over one year is considerably lower compared to design conditions (input from SolPro). Furthermore, the own electricity consumption of the plant for pumps and other consumers is high, about 14% of the annual gross electricity production. (For comparison: coal fired steam power plant about 7%, CCGT power plant about 2.5%). The electricity generation costs of the options with thermal storage are considerably lower owing to the higher capacity factor. Some degradation of the power production during the operation time can also be considered in the calculation (input "yes" for escalation).

5.6 Integrated Solar Combined Cycle Power Plant

5.6.1 Technology description

In a conventional combined cycle power plant, the hot exhaust gases of the gas turbine are used in the heat recovery steam generator (HRSG) to produce steam which is used in the steam turbine generator set for power production. In an inte grated solar combined cycle (ISCC) power plant, additional steam is produced by a solar steam generator, from solar heat coming from a parabolic trough solar field. Thus, during daytime, the electricity production is increased.

The general ISCC power plant configuration is illustrated in Figure 5-23 . The solar energy, collected by the parabolic trough solar field, is transferred to the circulating HTF, which releases the heat into the solar steam generators (SSG). The power plant configuration[23] comprises 2 gas turbines, 2 heat recovery steam generators down-stream of the gas turbine exhaust, two solar steam generators and a steam turbine generator.

The solar heat is delivered as saturated steam entering the superheated of the HRSG with 315°C and is subsequently superheated to a higher temperature of 526°C along with the steam coming from the evaporator of the HRSG. The solar share referred to the total power output of the plant, is limited up to a maximum of about 15% to 17% due to technical constraints regarding the capacity of the superheater. The electrical cycle efficiency under design conditions is considerably higher compared to conventional CCGT plants.

[22] Software tool for design and performance calculations of solar parabolic trough power plants, Fichtner Consulting, Stuttgart, Germany

[23] The cycle configuration is derived from a real project

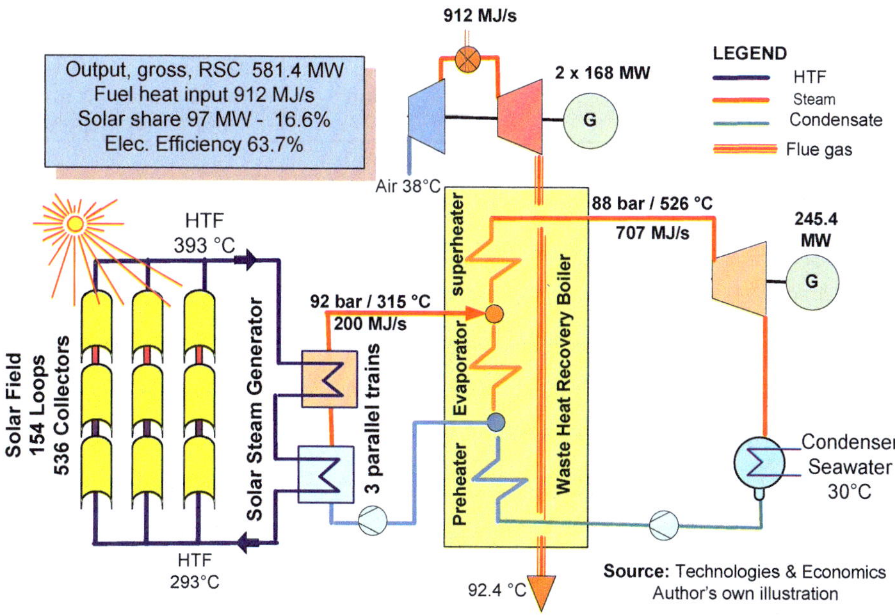

Figure 5-23: Simplified heat flow diagram ISCC-plant

5.6.2 Techno-economic assessment

The concept of the ISCC is flexible in operation as the power output is coincident with the load of the grid. During the daily on-peak time the output is higher, due to the solar heat input, and decreases during the off-peak time in the night. However, this advantage comes at high capital expenditures for the solar part. The specific CAPEX of the conventional part, without solar, is about 900 US$/kW$_e$, while the corresponding for the solar part is about 3,000 US$/kW$_e$.

Example 5-10 demonstrates in a simplified techno-economic model a comparison of ISCC versus conventional CCGT power plant. Some parameters (power output, efficiencies) are based on calculations for a real project.

Based on the levelized cost, the ISCC plant is less attractive. However, it is pointed out that the solar electricity is produced during on-peak hours during day time when more expensive fuels are used for power generation and electricity tariffs are usually higher. Furthermore, ISCC plants, installed recently, benefited from financial support from international donor agencies as, e.g., the World Bank (Egypt, Morocco).

Example 5-10: Performance & cost calculation model CCGT vs. ISCC

Item		Unit	CCGT solar share 0%	ISCC solar share 16%
Solar field				
Number of collectors			-	634
Aperture area	817 m2/Col	1000 m^2	-	518
Solar heat	414 kJ/s Col	MJ /s	-	262
Annual irradiation at site		kWh / a	-	2,400
Annual solar heat to power block *)	46%	MWh / a	-	572
Power & energy balance				
Rated net power output, gross		**MW_e**	**484.4**	**581.5**
of which solar **)	37%	MW_e	-	97
Rated efficiency, gross, day time	912 MWt	-	53.1%	63.8%
Utilization time full load, CCGT part		-	5,500	5,500
Reference electricity generation		GWh_e / a	2,664	2,876
Net electricity production, power plant		GWh_e / a	2,544	2,746
of which, solar electricity **)	37%	GWh_e / a	-	212
Annual average electrical efficiency		-	52%	56%
Fuel consumption		GWh_t / a	5,123	5,123
Financial constraints				
Lifetime power plant		a	25	25
Discount rate on real terms (WACC)		%	6.5%	6.5%
Gas price ref. to LHV		$\text{US\$ / MWh}_t$	25.00	25.00
Non-fuel OPEX		-	5.0%	5.50%
CAPEX, US$ 2014; ± 20%		**mln US$**	**436.0**	**726.9**
Specific CAPEX total plant		$\text{US\$ / kW}_e$	900	1,250
Specific CAPEX solar part only		$\text{US\$ / kW}_e$	-	2,996
Electricity generation costs in real terms		**mln US$ / a**	**185,625**	**227,658**
Annualized CAPEX		"000" US$ / a	35,741	59,592
OPEX		"000" US$ / a	21,798	39,979
Fuel cost		"000" US$ / a	128,087	128,087
Specific electricity generation cost		**$\text{US\$ / MWh}_e$**	**73.0**	**82.9**
Fixed cost		$\text{US\$ / MWh}_e$	22.61	36.26
Fuel cost		$\text{US\$ / MWh}_e$	50.34	46.64

*) solar field thermal efficiency taken from SolPro

*) steam cycle efficiency calculated with KPRO

5.7 Solar Plants with Solar Tower Technology

5.7.1 Technology description

In solar tower power plants, a field of large two-axis tracking mirrors, called heliostats, moving independently of one another, are used to concentrate sunlight onto a central receiver mounted at the top of a tower – Figure 5-24. The field of heliostats may either surround the tower or be spread out on the shadow side of the tower.

Solar tower systems are commonly operated with molten nitrate salt as heat transfer fluid. The salt is heated up at the receiver and pumped first directly to the "hot tank" of the thermal storage system. From there it comes to the solar steam generator unit and transfers its heat content to the water-steam circuit of the steam cycle. The cooled down HTF is then led to the "cold tank" of the thermal storage system and back to the receiver.

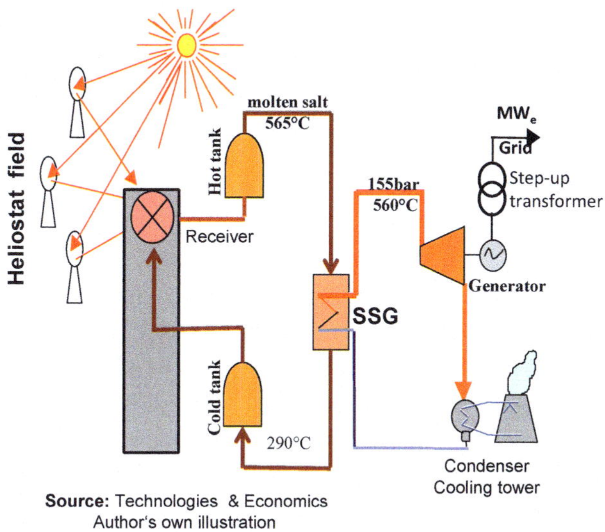

Figure 5-24: Schematic of solar tower power plant

The solar tower concept is technically and from the thermodynamic point of view superior compared to parabolic trough technology: For heat transfer and storage the same fluid is used that flows in a common closed circuit. Due to the high concentration ratios of the sun rays, higher fluid temperatures and hence higher efficiencies are obtainable compared to parabolic trough technology – Figure 5-25. The temperature difference between the hot and cold fluid of 275 K (565°C to 290°C) compared to 100 K (393°C to 293°C) of the parabolic trough system allows smaller storage tanks.

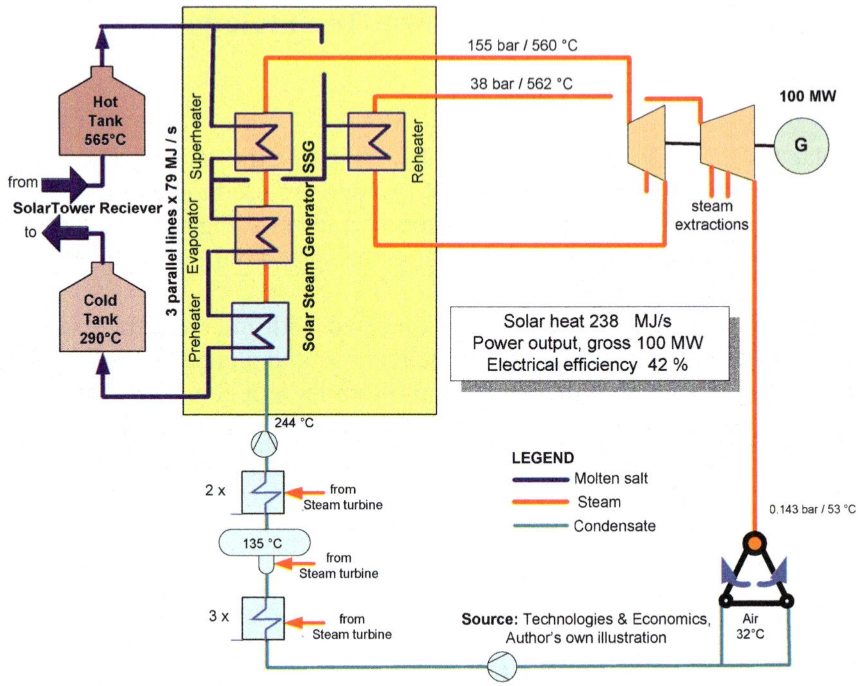

Figure 5-25: Simplified heat flow diagram, Solar Tower power plant

5.7.2 Techno-economic assessment

The solar tower technology in utility scale is less mature compared to parabolic trough technology. However, several plants have been field-tested around the world in the last decades, demonstrating the technical and economic feasibility of the solar tower concept. Utility scale plants are under construction and recently also in operation (e.g. Ivanpah Solar Power, California, USA, 392 MW; Crescent Dunes, Nevada, USA, 110 MW; Ashalim Plot B, Israel, 121 MW).

Modelling of layout and performance simulation of Solar Tower plants is complex and challenging as each heliostat is focusing the tower from a different angle and distance and needs to be tracked individually to deliver its optimal performance. Also the height of the tower increases with the size of the solar field. The simulation must be done for each hour of the year and the control hardware for tracing must be programmed accordingly. In Germany the software tool "Visual HFLCAL (Heliostat Field Layout Calculation)" of the German Aerospace Center (DLR) [49] is usually applied.

In **Case Study 11.8** a simplified concept is used for modelling the plant, assessing the plant performance and calculate of the electricity generation costs of a solar tower power plant with a rated power output of 100 MW and three different thermal storage options.

The methodology of the calculations is in conformity with the complex tools, however, using average annual data for solar irradiation and performance. Some of the key parameters of the plant configuration and performance are derived from actual calculations for project studies conducted with the HFLCAL software tool.

In Table 5-8 the key layout and performance at design conditions are depicted. Table 5-9 presents the energy balance and the generation cost calculation. Both tables are presented in shorted form; the complete model is shown **Case Study 11.8** on the author's website.

Table 5-8: Case Study – Key layout and performance of solar tower PP

Item		Unit	100 MW		
			TES 9 h	**TES 12 h**	**TES 15 h**
Solar Field					
Latitude		degrees	28		
Design Reference DNI		W/m^2	900		
Design Reference DNI			Solar noon, Equinox		
Solar efficiency (DNI to heat)			80.0%		
Number of Heliostats		-	7,158	8,978	11,074
Aperture area, total	121 m2	$1000\ m^2$	866	1,086	1,340
Tower height		m	280	315	320
Solar field heat output, design point		MJ /s	624	782	965
Receiver thermal power	72.5%	MJ /s	452	567	699
Solar heat to power block		MJ / s	238	238	238
Solar Multiple		-	1.90	2.38	2.94
Thermal Storage		MWh_t	2,143	2,857	3,571
Technical Parameters, Power Block cycle					
Rated power output, field operation		MW	100	100	100
Live steam parameters		bar / °C	155 / 550		
Electrical efficiency, gross		%	42.0%	42.0%	42.0%
Condenser cooling		-	ACC, air 32°C		

The plant site is assumed to be in North Africa. The power block is designed for a power output of 100 MW. The model composes three options with nine, twelve and fifteen hour thermal storage capacity; a heat flow diagram of the power block cycle is also shown in Figure 5-25. The number of heliostats and the aperture area are chosen to meet the required solar heat demand for the power block and to charge the thermal storage during daytime under design conditions. The height of the tower is aligned to the size of the solar field of the three options.

Note, however, that the electrical cycle's efficiency is considerably higher compared to the same of the cycle with parabolic trough technology (Figure 5-22, 36.8%) due to the higher steam conditions. Furthermore the cycle output is the same in solar field operation and in TES-operation because there are no losses due to heat exchange.

Table 5-9 presents the annual energy balance along with the calculation of the electricity generation costs for the Solar Tower plant in the Case Study. It is evident that options with large TES with regard to the generation cost are more favorable.

Table 5-9: Case Study –Techno-economics, Solar Tower, Results

Item	Einheit Unit	100 MW		
		TES 9 h	TES 12 h	TES 15 h
Technical parameters				
Rated power output	MW		100	
Site latitude	grd		28	
Number of heliostats	-	7,158	8,978	11,074
Solar irradiation	kWh / m^2 a		2,400	
Net electricity production	GWh /a	379.6	476.2	587.3
Financial parammeters				
Discount rate in real terms	-		4.6%	
Project lifetime	a			25
CAPEX, US$ 2014, ±20	**mln US$**	**784**	**933**	**1,094**
Electricity generation costs in real terms				
Annual costs	mln US$ /a	66.9	79.4	92.8
of which capital cost		79.7%	79.9%	80.1%
Levelized electricity cost	**US$ / MWh**	**176.3**	**166.7**	**158.1**

5.8 Solar Power Plants with Fresnel Technology

The linear Fresnel reflectors concentrate sunlight by long plane mirror strips up to 1000 m long called linear Fresnel reflectors (LFR), which are grouped to a mirror field close to the ground. The sunlight is then focused onto a linear fixed absorber located above this mirror field – Figure 5-26. With this concept no rotating parts are required for the HTF system as for parabolic trough plants. Each mirror strip is driven by an individual electric motor and rotates around the north-south axis following the position of the sun. LFC plants are normally operated with water/steam as heat transfer fluid. Evaporation takes place directly in the fixed absorber tubes; a separate solar steam generator unit is not required. The fluid is normally passed once-through. Most commercial plants generate saturated steam only. Concepts with superheated steam are also expected and partly realized.

Figure 5-26: Fresnel collector system (AREVA)

Figure 5-27: Solar field of linear Fresnel power plant

Fresnel trough power plants consist of the Linear Fresnel Reflectors, a heat transfer fluid system, a steam generation system, a Rankine steam turbine/generator cycle and optional thermal storage and/or fossil-fired backup systems. The main difference between the parabolic trough technology and the Fresnel technology is the reflector system, the heat transfer fluid and the direct evaporation.
The Fresnel trough technology is seen as a lower cost alternative to parabolic trough technology for the production of solar steam for power production. The main advantages, compared to parabolic trough technology, are seen to be:

- Inexpensive planar mirrors and simple tracking system
- Direct steam generation within the absorber tubes
- No separate steam generator necessary
- Fixed absorber tubes with no need for flexible high pressure joints
- One absorber tube with no need for thermal expansion bellows
- Lower maintenance requirements

5.9 Aspects of Integration of Renewable Power

The rapidly increasing penetration of power generation from renewable energy in power systems creates challenges for the security of supply of electrical networks. The main issues are provision of adequate backup power, reliable frequency and voltage control.

Table 5-10 below highlights some aspects resulting from the penetration of renewable electricity in the power system, taking Germany as an example.

Table 5-10: Installed capacities vs. electricity production in Germany

Item		2005	2014	
		stand	stand	share
Capacity	**GW**	**136.9**	**202.5**	**100.0%**
Wind	GW	18.4	39.2	19.4%
PV	GW	2.1	38.2	18.9%
Total production	**TWh**	**622.6**	**627.8**	**100.0%**
Wind	TWh	27.2	57.4	9.1%
PV	TWh	1.3	38.5	6.1%
Capacity factor	TWh	0.52	0.35	-
Full capacity hours	h /a	4,548	3,100	-

While the total electricity production remained almost the same in the 10year period, installed capacity has substantially increased and the capacity factor decreased accordingly. The increase of installed capacity is solely attributed to renewable energies. The shares of renewable power are much higher than those of their share in electricity production. The reason is the lower capacity factors of renewable power generation, resulting from the fluctuations of renewable power from zero to maximum output.

Power output of PV directly depends on the insolation and is available during daytime only. Worth mentioning as an extreme case is the solar eclipse in 2015; PV power output became temporarily zero during the eclipse and rose to maximum seconds after the eclipse. The challenge for the operators was to balance the rapid changes of output by rump-up and bringing down conventional power within seconds.

Wind power output is proportional to the third power of the wind speed. Even small changes of the wind speed induce a big change of the power output. For instance, change of wind speed from 6.5 m/s to 5.2 m/s results to a halving of power output.

In general, we can draw the conclusion that renewable energy without storage capability requires almost 100 percent backup power.

A major challenge in the integration of renewable energy recourses in power generation is to reliably balance power output and load of the system in order to maintain frequency and voltage stability. Hence, besides quantity also the quality of backup power is a most important issue. The intermittency of solar and wind power requires the availability of conventional power capacities with quick response to fluctuation of demand and fast startup ability from zero to full output.

Traditionally, utilities have preferred technologies as large steam power plants, usually fired by coal or liquid fuels for base load, and combined cycle gas turbine power plants for intermediate and cycling load and simple cycle gas turbines for peak load. These technologies are at an advanced state-of-the-art and ensure a safe and cost efficient electricity supply. However, after a substantial power generation is supplied from renewable resources, these systems cannot provide the required quality backup power.

Hence, the existing power generation systems need to be complemented by quickly dispatchable, dynamic capacity with the capability of handling frequent fast starts, stops and load ramps. In this context, aeroderivative gas turbines and multiunit internal combustion engine (ICE) power generation technologies provide all these advantages and will become an indispensable part of large scale power systems.

5.10 Promotion Schemes or Renewable Energy

For the first time ever, Germany launched the *Electricity Feed-In Act* (Stromeinspeisegesetz[24]) in 1990, for the promotion of electricity produced from renewable energy. According to the act, utilities were obliged to purchase electricity from renewable sources and to remunerate with statutorily set-out tariffs. The tariffs amounted, depending on the source, between 75 to 90 percent of the average end-user price. The objective was to stipulate market opening and technical development of renewable energies, and to protect them from direct competition with the established conventional fossil-fueled power generation technologies in their emergence and development phase.

[24]) Law for feed-in electricity from renewable energies in the public grid – acronym: Stromeinspeisegesetz

After liberalization of the energy markets the electricity tariffs temporary fell drastically due to the competitive environment. Tariffs linked to the average end-user price, as in the 1990 act, were not cost covering for renewable energy sources any more. On the other hand, the commitment from the Kyoto Protocol to reduce CO_2 emissions required an accelerated expansion of renewable energy in the power sector. Thus, the *Renewable Energy Act* – known as *EEG* (Erneuerbare-Energien-Gesetz – acronym EEG) was introduced and replaced the 1990 act.

The new, so-called, *feed-in tariffs* of the EEG were fixed and binding for 20 years after commissioning of new plants. The EEG included also an annual digression clause of tariff for new plants, taking into consideration the development of the technologies,. The act has been amended several times; the latest is from 2014.

In 2001, the Directive on Renewable Energies (EC, 2001) was enacted aimed at increasing the share of electricity from renewable energies from 14% in 1997 to 22% in 2010.

Meanwhile, the overall principle of the promotion scheme has been taken over from many other countries in their energy legislation, and renewable energy receives government support worldwide. Currently, there are two promotion schemes in application:

- Renewable Portfolio Standard regulation (acronym RPS)
- The Feed-in Tariff regulation (acronym FIT)

RPS regulation is a quantity-based approach. It obliges electricity utilities to install an amount of power generation from renewable energy. This is a practical approach for achieving share targets of renewable electricity, especially in countries with regulated end-user tariffs. Commonly, governments make construction permits for new power generation plants dependent on the installation of a share of renewable energy usually referred to the installed capacity (see also Table 5-10 for clarification).

FIT schemes oblige electricity utilities to purchase electricity generated from renewable energy and to remunerate with the feed-in tariff, which in most cases is relatively high and cost covering.

6 Cogeneration of Power and Heat

6.1 Introduction to the Cogeneration Cycle

The following figure depicts simplified schematics of the three most common steam cycles, along with indicative energy efficiency standards:

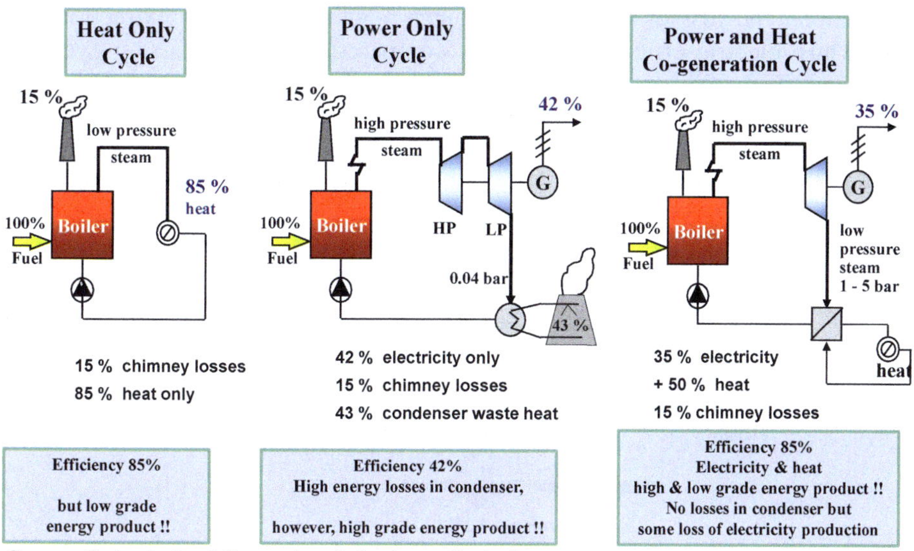

Source: Technologies & Economics, Author's own illustrations

Figure 6-1: Overview of steam Cycles

The oldest and basic form of energy conversion is the heat generation in a *heat only boiler (HOB)*. In the boiler, the chemical energy of the fuel is converted into thermal energy in the form of low pressure steam or hot water. The efficiency of modern heat only boilers is quite high, between 85% up to 95%, however, the product heat is a low grade form of energy; it can only be used for heating purposes in an industrial process or for space heating.

In a *power only cycle* (condensing power generation cycle) the boiler of a power plant generates high pressure and high temperature steam which is led to a steam turbine where its pressure and temperature are converted by expansion to

P. Konstantin and M. Konstantin, *The Power Supply Industry*,
https://doi.org/10.1007/978-3-319-72305-1_6

mechanical energy. The turbine drives the generator where the mechanical energy is turned into electrical energy. Electricity is the energy form with the highest grade, as it can be converted into any other form of energy. The disadvantage of this cycle is that almost half of the fuels' chemical energy is contained in the turbine exhaust steam with temperature nearly equal to the ambient temperature and therefore too low to be utilized as usable energy. It is therefore dissipated into the environment in the condenser. This results in relatively modest cycle efficiency in the range of 42% to 46%.

Cogeneration cycle is a combination of both cycles, utilizing the fuel input for the production of heat and electricity in a combined heat and power (CHP) plant. The steam is extracted from the steam turbine at a higher pressure and temperature so it can be utilized as usable energy for industrial processes or for space heating. On the other hand, the steam extraction induces some loss of electricity production. The loss of electricity becomes bigger with higher extraction pressure. Thus, electric efficiency is lower, compared to the condensing cycle, but about 85% of the fuel energy can be utilized as usable energy in form of heat and electricity.

The advantage of cogeneration with regard to energy efficiency is also demonstrated in the following figure.

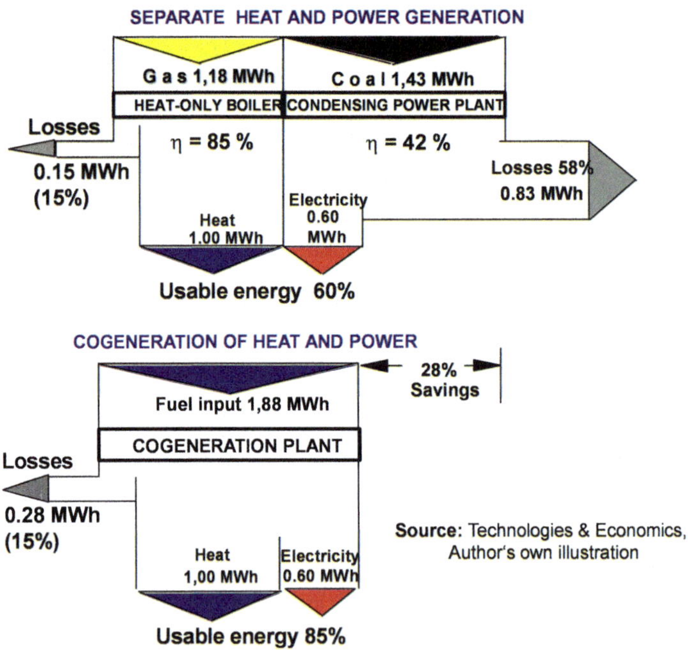

Figure 6-2: Efficiency of useful energy generation in comparison

6.2 Types of Combined Heat and Power Plants

The development of cogeneration started with the steam process at the beginning of the 20th century. Meanwhile, several types of Combined Heat and Power Plants (CHP) have been developed and are in operation. These are:
Engine CHPs, gas turbine with waste heat recovery steam generator CHPs (GT-HRSG), steam turbine CHPs and Combined Cycle Gas Turbine (CCGT) CHPs. Brief technical descriptions are outlined in the following.

6.2.1 Combustion Engine CHPs

The main application of *combustion engine CHPs* – Figure 6-3 – are municipal small scale district heating systems. The engine drives the generator and heat is recovered from the cooling water and the hot exhaust gas of the engine in heat exchangers (HEX). The produced heat is hot water of about 90°C.

Combustion engine CHPs are delivered as completely prefabricated, turn-key units. Engines are available for several types of fuel as natural gas, biogases, sewage gas, mine gas, diesel etc. Most engine CHP plants consist of 3 to 5 units with unit capacity between 250 kW to 1000 kW. However, there are small units on the market, starting from 7 kW for energy supply in larger apartment blocks or commercial buildings.

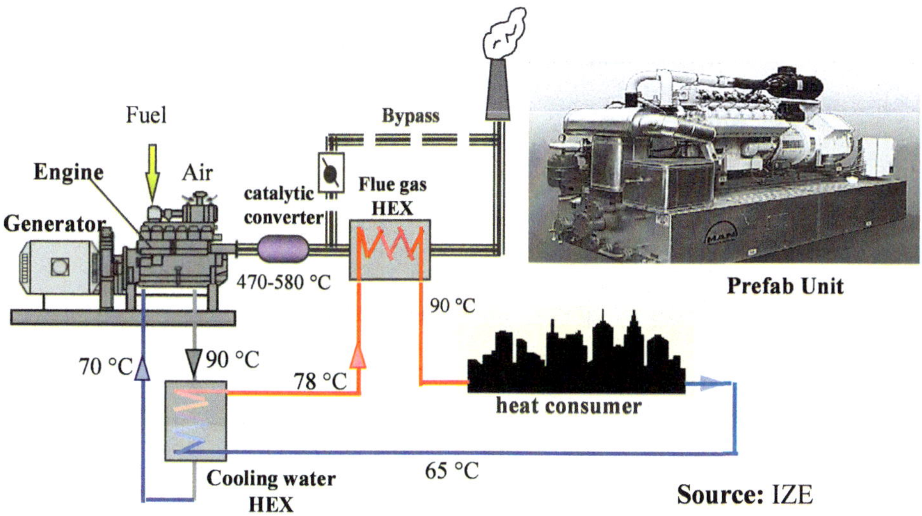

Figure 6-3: Combustion engine CHP unit

In normal arrangement (without bypass, see below), heat and power output are firmly coupled with each other. This means, the power output directly depends on the heat output. Low or no heat demand results in low or no electricity production.

In order to make operation more flexible, the units may be equipped with a bypass-duct to the flue gas heat exchanger (HEX). Usually such plants are heat-led operated. But in the case of short peaks on the power side, during low heat demand, they can be partly or fully shifted in bypass operation to meet the power peak (see Figure 6-4).

6.2.2 Gas turbine CHPs

Gas turbine CHPs – Figure 6-4 – are mainly used for cogeneration of electricity and low pressure steam (<12 bar) for production processes and space heating in factories. The steam is recovered from the hot exhaust gases in a Heat Recovery Steam Generator (HRSG). The capacity of the gas turbines ranges, in most cases, between 10 to 25 MWe. The most common fuel is natural gas.

They may be equipped with a bypass to the HRSG for more flexible operation in times of low or no heat demand. The HRSG may be equipped with a supplementary firing in order to meet short heat peaks (Figure 6-4).

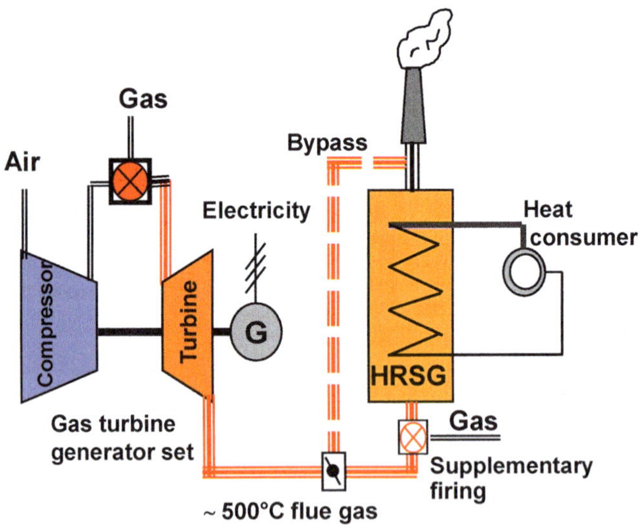

Source: Technologies & Economics, Author's own illustration

Figure 6-4: Gas turbine CHP

> **Important note:** In contrast to heat extraction from steam turbines (see next section), the generated heat of engine or gas turbine CHPs is *real waste heat*. If not recovered, it must be dissipated in the environment. It does not cause any electricity loss; the loss of electricity generation is zero!

6.2.3 Steam turbine CHPs

There are two different arrangements of steam turbine CHPs. The schematic in Figure 6-5 shows a plant with a backpressure steam turbine and a plant equipped with an extraction-condensing steam turbine.

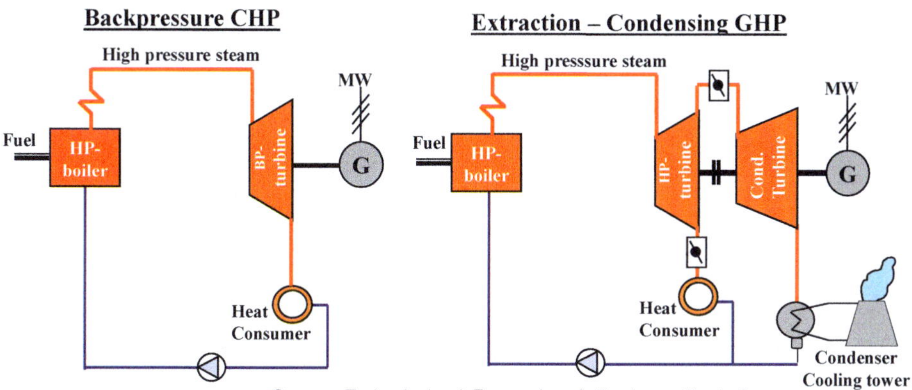

Source: Technologies & Economics, Author's own illustration

Figure 6-5: Steam turbine CHPs

A *backpressure steam turbine* discharges the steam into a pressurized piping system to be used for process heating or for the supply to district heating systems. The steam has a pressure considerably higher than the atmospheric pressure, depending on the discharge pressure and the temperature requirements of the heat consumer. Cogenerated electricity and heat are firmly coupled with each other.

Large steam turbine CHPs are usually equipped with *extraction condensing steam turbines*. A part of the live steam is extracted from the turbine to be utilized for heating purposes, while another part is led to the condensing turbine. Thus electricity is generated in parallel in cogeneration mode as well as in condensing mode. The steam flows to the extraction and to the condenser can be controlled by the dampers following the actual demand.

In most cases steam turbines have more than one extraction in different pressure levels. Main application of the steam turbine CHP is cogeneration of steam and power for industrial processes and for district heating networks.

6.2.4 Combined cycle gas turbine CHPs

The most energy efficient developments are *Combined Cycle Gas Turbine (CCGT) CHPs* – Figure 6-6. The hot exhaust gases of the gas turbine(s) generate high pressure steam in the HRSG which is lead to the steam turbine. Power is produced in the gas turbine generator as well as in the steam turbine generator.

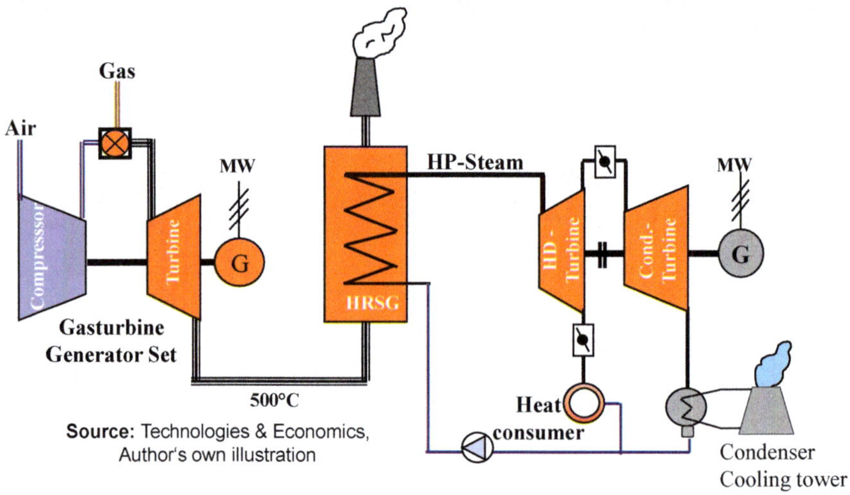

Source: Technologies & Economics, Author's own illustration

Figure 6-6: Combined cycle gas turbine CHP

Process steam is generated from the exhaust of the backpressure steam turbine. Large CCGT CHPs are equipped with extraction-condensing steam turbines for more flexible operation.

6.3 Performance Parameters of Cogeneration

6.3.1 Key performance parameters

The following performance parameters are referred to a pure cogeneration of heat and power. This means, both products, electricity and heat, are fully produced in cogeneration and there is no electricity generation in bypass or in condensing mode operation.

The CHP plant's capacity is defined by the following parameters. They are calculated in thermodynamic cycle simulations with appropriate software tools (steam and CCGT CHPs) or provided by the manufacturers of the plants (engine or GT CHPs). These are:

P_{cogen}: Rated power output, cogenerated electricity kW_e
$\dot{Q}_t$: Rated thermal output, cogenerated heat kJ/s, KW_t
Q_t: Heat generation during operation period (year) KJ/a, kWht/a,
$\dot{Q}_{f\_cogen}$: Rated fuel energy input kJ/s, kW_t
Q_{f_cogen}: Fuel energy consumption operation period kJ/a: kWh_t/a

Note: Characters with a dot above designate energy flow $\dot{Q}$ [kJ/s]. Characters without dot above Q [kWh, kWh/a] designate energy amounts produced in a period of time, usually year (a).

a) The ISO unit for heat is J/s (or KJ/s), which is numerically equal to W_t or (kW_t). For practical reasons we use the unit kW_t , e.g., in power and energy balances
b) The subscript: "$_e$" stands for electricity, "$_t$" for thermal and "$_f$" for fuel
c) The subscript "$_{cogen}$" is used to distinguish cogenerated from non-cogenerated electricity, or for fuel input for cogeneration only.

Based on the above parameters, the following performance terms are determined:

The *electricity-to-heat ratio* σ is the electricity generated by the steam from its entry into the turbine up to its extraction. It is defined as the ratio of the cogenerated electricity to the cogenerated heat.

$$\text{El.-to-heat ratio: } \sigma = \frac{P_{cogen}}{\dot{Q}_t} \left[\frac{\mathrm{kW_e}}{\mathrm{KW_t}}\right] \quad \sigma = \frac{W_{cogen}}{Q_t} \left[\frac{\mathrm{kWh_e}}{\mathrm{KWh_t}}\right] \tag{6.1}$$

For power output and electricity generation in cogeneration, we get (see application in Example 6-5, splitting cogen from non-cogen electricity):

$$\text{Output:} \quad P_{cogen} = \sigma \cdot \dot{Q}_t \ [\mathrm{kW_e}] \quad W_{cogen} = \sigma \cdot Q \ [\mathrm{kWh_e/a}] \tag{6.2}$$

The *electrical equivalent* β of steam defines the loss of electricity generation caused by steam extraction. It is the ratio of electricity loss caused by the extracted steam (vs. condensation of the same amount of steam) to the heat content of the extracted steam. It is designated as the *electricity equivalent* β of the extracted steam.

Electrical equivalent: $\beta = \frac{\Delta P_e}{\dot{Q}_t} \left[\frac{\text{kW}_\text{e}}{\text{kW}_\text{t}}\right] \quad \beta = \frac{\Delta W}{Q_t} \left[\frac{\text{kWh}_\text{e}}{\text{kWh}_\text{t}}\right]$ (6.3)

By rearranging the equation we get for power and electricity loss, caused by extracted steam:

$$\Delta P_{cogen} = \beta \cdot \dot{Q}_t \left[\text{kW}_\text{e}\right] \quad W_{cogen} = \beta \cdot Q \left[\text{kWh}_\text{e}/\text{a}\right] \qquad (6.4)$$

Note: Electricity loss occurs if steam is extracted from steam turbines (steam CHPs or CCGT CHPs).
Cogenerated heat from engine CHPs and gas turbine/HRSG CHPs does not cause any electricity loss ($\beta = 0$).

Fuel energy input: $\dot{Q}_{f\_cogen} = \frac{P_{cogen} + \dot{Q}_t}{\eta_{tot}} = \frac{\dot{Q}_t \cdot (\sigma + 1)}{\eta_{tot}} \quad \left[\text{kW}_\text{t}\right]$ (6.5)

Overall or total efficiency "η_{tot}" is the ratio of the co-generated useful energy (electricity and heat) to the fuel energy input for both products:

$$\eta_{tot} = \frac{P_{cogen} + \dot{Q}_t}{\dot{Q}_{f\_cogen}} \times 100 = \frac{\dot{Q}_t \cdot (\sigma + 1)}{\dot{Q}_{f\_cogen}} \times 100 \quad [\%] \qquad (6.6)$$

Important notes: (a) Electrical efficiency of cogeneration cycles – ($\eta_e = P_{cogen}/\dot{Q}_{f\_cogen}$) – makes sense for engine and GT/HRB CHPs only because the cogenerated heat does not cause any loss of electricity.

(b) In contrast, electrical efficiency of cogeneration cycles, such as steam CHPs or CCGT CHPs, is a misleading performance parameter for comparison of cycles; the same cogeneration cycles give different electrical efficiencies if the extraction pressures are different.

(c) Neither the total η_{tot} efficiency is an appropriate criterion for thermodynamic performance of cogeneration cycles. This is a measure of the thermal (i.e. fuel utilization rate) and not of the electrical efficiency. Even badly designed cogeneration plants have a high total efficiency (e.g., as an extreme case, a heat only boiler may have thermal efficiency of 90% to 95%.

(d) Appropriate performance parameters for comparing the performance quality of cogeneration cycles are:

- the equivalent condensing power output and
- the equivalent condensing electrical efficiency.

(e) For this purpose, the cogeneration cycle is converted into an *equivalent condensing cycle* without steam extraction. The formulas are shown below.

Equ. cond. power output: $P_{equ\_cond} = P_{cogen} + \Sigma\beta_i \cdot \dot{Q}_t = \dot{Q}_t \cdot (\sigma + \beta)$ (6.7)

Equivalent cond. efficiency: $\eta_{equ\_cond} = \dfrac{\dot{Q}_t \cdot (\sigma + \beta)}{\dot{Q}_{f\_cogen}}$ (6.8)

Note: For reasons of simplifications, we assume for calculations in the examples, that the performance parameters at rated conditions and operation over the year remain the same (kW_e/kW_t equal kWh_e/kWh_t).

The unit for electricity-to-heat ratio σ and electrical equivalent β is commonly given in kW_e/kW_t. However, often the unit kWh_e/kg or (MWh_e/t) is used, referred to the mass of extracted steam. The electrical equivalent in MW_e/MW_t depends on the temperature and rate of return and of the condensate and may be different for the same amount of extracted steam. This is demonstrated in Example 6-1 below.

Example 6-1: Electrical equivalent MWh_e/MWh_t and MWh_e/t of steam

Item		Unit	Value		
Power output					
In condensing mode		MWh_e	135.7		
In cond.-extraction mode		MWh_e	125.6		
Lost electricity	ΔP	MWh_e	10.1		
Steam extraction:					
Mass flow m	50.00 t/h	kg /s	13.89		
Pressure		bar	12		
Temperature		°C	250		
Enthalpy		kJ/kg	2,936		
Condensate:					
Pressure		bar	1	1	1
Temperature		°C	95	70	0
Enthalpy		kJ/kg	398	293	0
Return rate *)		-	100%	70%	0%
Heat output	Q	MW_t	35.2	37.9	40.8
Electrical equivalent	$\Delta P/Q$	MW_e/MW_t	0.287	0.266	0.248
	$\Delta P/m$	MWh_e/t	0.202	0.202	0.202

*) Condensate may be consumed in industrial processes or lost in piping system or unclean for reuse

The following two examples Example 6-2 and Example 6-3 demonstrate in a simple manner the calculation of the performance parameters electricity-to-heat ratio σ and electrical equivalent β. The values are taken from calculation with cycle simulation tools for real projects.

Example 6-2: Simplified calculation of electricity-to-heat ratio σ

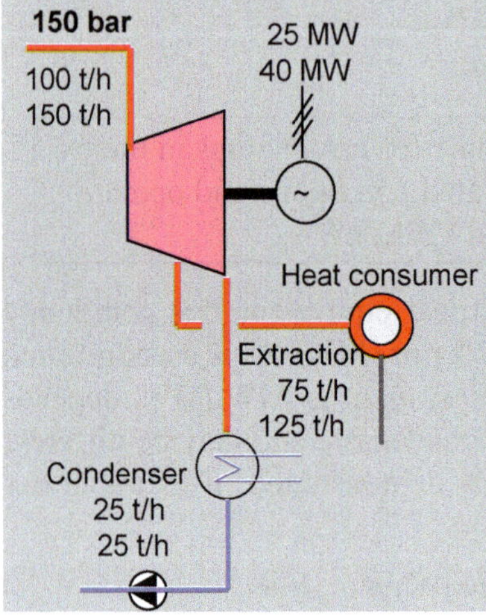

Item	Unit	Load 1	Load 2
Live steam	t /h	100	150
Steam extraction	t /h	75	125
Steam to condenser	t /h	25	25
Power output	MW	25	40

$$\text{Electricity-to-heat ratio: } \sigma = \frac{40-25}{125-75} = 0.30\left[\frac{\text{MWh}}{\text{t}}\right]$$

Annual Production			
Electricity production		MWh /a	50,000
Steam extraction		t /h	100,000
Cogen electricity	σ=0.30	MWh /a	30,000
Cond. Electricity		MWh /a	20,000

Example 6-3: Simplified calculation of the electrical equivalent β

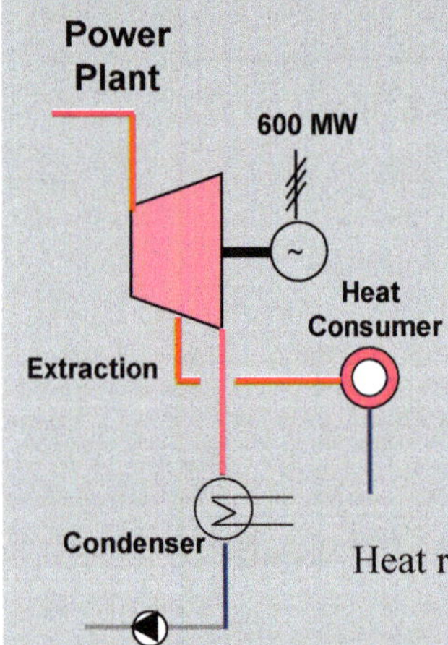

Item	Unit	Operation Mode	
		Power Only	Cond.-extraction
Power output	MW	600	548
Steam extraction	t / h	0	303
Fuel input	MJ / s	1,033	1,033

$$\text{Electrical equivalent: } \beta = \frac{600-548}{303} = 0.172 \left[\frac{\text{MWh}}{\text{t}}\right]$$

$$\text{Heat rate of electricity: } q_e = \frac{1{,}033}{600} = 6.2\left[\frac{\text{MJ}}{\text{MWh}_e}\right]$$

$$\text{Heat rate of steam: } \quad q_t = 0.172\left[\frac{\text{MWh}_e}{\text{t}}\right] \times 6.2\left[\frac{\text{MJ}}{\text{MWh}_e}\right] = 1.064\left[\frac{\text{MJ}}{\text{t}}\right]$$

$$\text{Control}: \dot{Q}_f = 548[\text{MW}] \times 6.2\left[\frac{\text{MJ}}{\text{MWh}}\right] + 303\left[\frac{\text{t}}{\text{h}}\right] \times 1.064\left[\frac{\text{MJ}}{\text{t}}\right] = 3{,}720\left[\frac{\text{MJ}}{\text{h}}\right] \triangleq 1{,}033\left[\frac{\text{MJ}}{\text{s}}\right]$$

6.3.2 Performance parameters of selected CHP plants

Key performance parameters of selected cogeneration plants are show in the following Table 6-1 and in Figure 6-7 to Figure 6-9.

Table 6-1: Performance parameters of selected CHP plants

Type of plant	Rated Power output 1) kW	Heat extraction kW_t	Heat transport medium	Electrical cond. Efficiency η_e	Total efficiency, cogen only η_{tot}	Electricity-to-heat ratio σ kW_e/kW_t	Electrical equivalent β kW_e/kW_t
Combustion engine CHP 2)	250 kW	338	Hot water 90	37.4%	88.0%	0.740	0
	600 kW	654	Hot water 91	42.0%	87.8%	0.917	0
	1,132 kW	1,265	Hot water 92	41.1%	87.0%	0.895	0
Gas turbine CHP 3)	4,427 kW	7,300	Steam 10 bar	30.5%	80.8%	0.606	0
	10,313 kW	15,756	Steam 10 bar	32.5%	82.2%	0.655	0
	24,170 kW	35,875	Steam 10 bar	33.6%	83.5%	0.674	0
Steam power plant Extraction condensing 4) 5)	600 MW	as required	Steam 23 bar	41.2%	82%	0.237	0.454
			Steam 6 bar		83%	0.447	0.258
			Steam 3 bar		83%	0.528	0.232
CCGT power plant 4) 5)	457 MW	as required	Steam 14 bar	51.3%	84.5	1.233	0.315
			Steam 5 bar		84.5	1.180	0.215
			Steam 1 bar 6)		n.a.	n.a.	n.a.

1) at full condensation
2) Source: ASUE BHKW Kenndaten 2011
3) Source: ASUE Gasturbinen Kenndaten 2006
4) Own calculation
5) h_e is the equivalent cond. efficiency, this means without steam extraction
6) low pressure steam directly from HRB

The electrical equivalent and the electricity-to-heat ratio of the cogenerated steam extracted from steam turbines strongly depend on the extraction pressure. Guide values are depicted in Figure 6-7 (Rankine cycle without reheat), Figure 6-8 (Rankine cycle with reheat) and Figure 6-9 (CCGT cycle). In real projects the values must be calculated with appropriate cycle simulation tools.

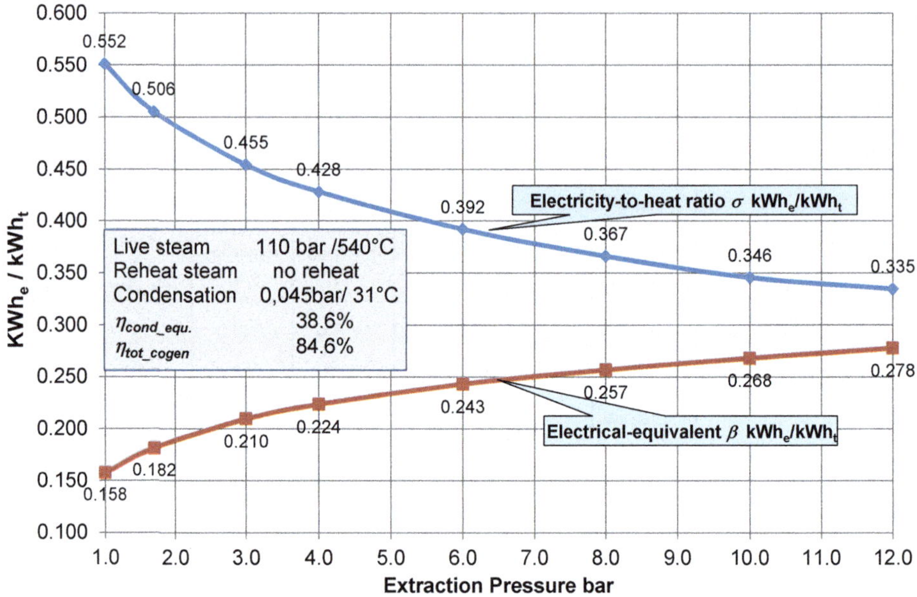

Figure 6-7: Performance parameter of extracted steam CHP, no reheat

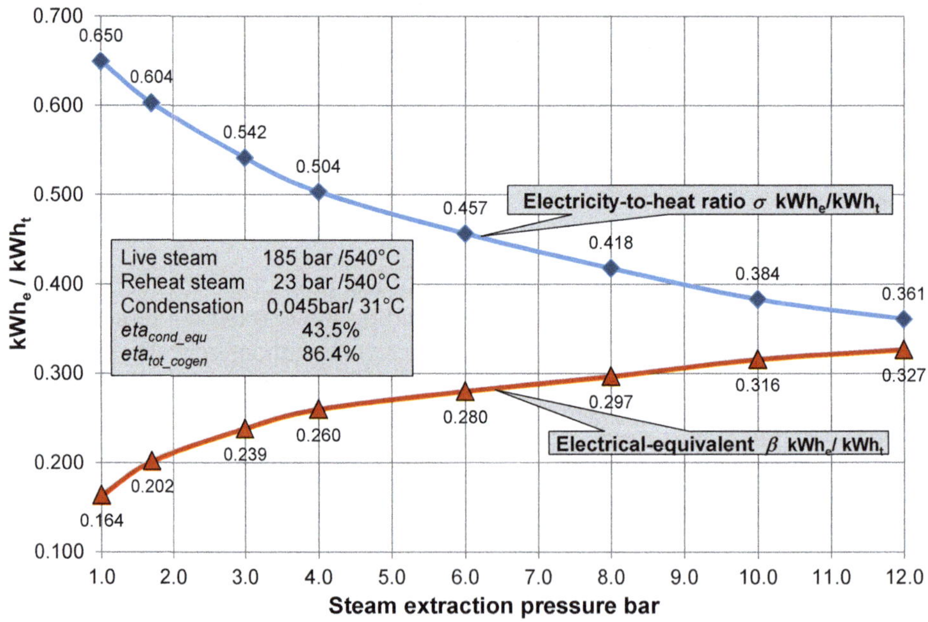

Figure 6-8: Performance parameters of extracted steam, CHP with reheat

The electricity-to-heat ratio of CCGT cogeneration cycles is considerably higher, compared to steam cycles, due to the added power of the gas turbine (in most cases $\sigma > 1$). It is also highly dependent on the specific gas turbine type and cycle conditions; the values depicted in the following figure apply for the given gas turbine and cycle conditions only.

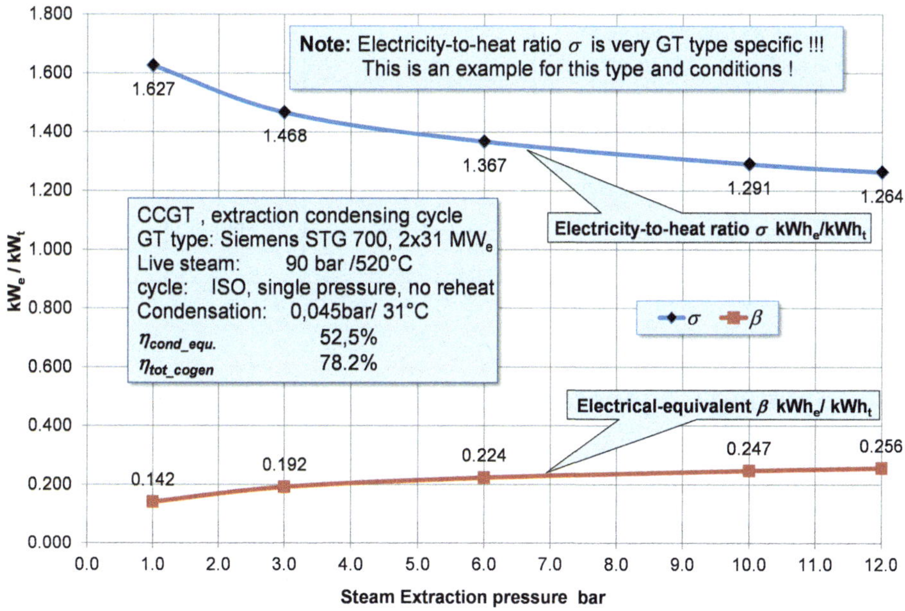

Figure 6-9: Performance parameters of extracted steam from CCGT CHP

Following example demonstrates the application of performance parameters.

Example 6-4: Performance of a steam backpressure CHP
Two industrial steam CHP plants supply cogenerated heat and power for two paper factories. Both turbines have an extraction at 12 bar and backpressure at 6 bar but different live steam parameters.

Calculate the power output, annual electricity and heat production and fuel consumption. Calculate also the equivalent condensing power output and efficiency for assessing the thermodynamic performance quality of the plants. Performance parameters can be taken from Figure 6-7 and Figure 6-8.

Item	Formula	Unit	Case 1	Case 2
Given	type of plant	-	110bar no reheat	185bar reheat
Heat extraction at 6 bar	Q_{6bar}	MW_t	200	200
Heat extraction at 12 bar	Q_{12bar}	MW_t	150	150
Electricity-to-heat ratio, 6 bar steam *)	σ_{6bar}	kWh_e / kWh_t	0.392	0.457
Electrical equivalent, 6 bar steam	β_{6bar}	kWh_e / kWh_t	0.243	0.280
Electricity-to-heat ratio, 12 bar steam *)	σ_{12bar}	kWh_e / kWh_t	0.330	0.361
Electrical equivalent, 6 bar steam	β_{12bar}	kWh_e / kWh_t	0.277	0.327
Total efficiency	η_{tot}	-	84.6%	86.2%
Performance Parameters, in cogeneration, calculated				
Rated cogen electrical output, gross	$P_{cogen} = \Sigma\sigma_i \times Q_i$	MW_e	127.9	145.6
Fuel input	$Q_f=(P_{cogen}+\Sigma Q_i)/\eta_{tot}$	MJ/ s	565	575
Equivalent cond. power output	$P_{cond\_equ}=P_e+\Sigma\beta_i \times Q_i$	MW_e	**218.1**	**250.6**
Equivalent cond. Efficiency	$\eta_{cond\_equ.}=P_{cond\_equ.}+Q_f$	⇨	**38.6%**	**43.6%**

*) taken from electricity-to -heat-ratio figures from text part

6.4 Splitting Cogenerated and Non-cogenerated Electricity

Cogeneration is a highly energy efficient technology. Therefore, the energy policy in many countries supports the development of cogeneration with promotion schemes. In general, legislation for promotion of cogeneration requires that power system operators are obliged to purchase and remunerate cogenerated electricity with a preferred feed-in tariff.

This applies for cogenerated electricity only! However, CHP plants usually generate cogenerated and non-cogenerated electricity in parallel (e.g., in bypass or condensing mode of operation). The operator of the plants must provide evidence on the share of cogenerated electricity to the total electricity production. Therefore it is necessary to split the total electricity production in cogenerated and in non-cogenerated (see Example 6-5).

The German promotion law for cogeneration stipulates that cogeneration plants need to be certified by an accredited consultant. The consultant calculates and certifies the rated cogeneration output of the CHP plant. The operator of the CHP plant is obliged to meter and record the heat and the electricity production and separate the amounts of cogenerated from non-cogenerated electricity and heat. These records need certification as well, usually monthly.

A further requirement of eligibility for promotion is the average total efficiency η_{tot}, during the operation period must not be lower than 70%. In pure cogeneration mode the total efficiency is in the range of 85% and higher. This means, the 70% limit accepts some amount of non-cogenerated electricity to be eligible for promotion; this is necessary to enable some flexible operation during on-peak times for one of the two products.

Example 6-5: Splitting total production in cogen + non-cogen electricity

A gas turbine CHP is operated in 100% cogeneration mode; another of the same type is operated in cogen plus bypass operation. Calculate the total electricity production, the cogenerated electricity only, the total fuel consumption and that contributed to cogenerated electricity only.

Note: The government promotions scheme subsidizes cogen electricity only!

Item	Symbol	Unit	Operation mode: cogen only	Operation mode: mixed cogen+bypass
Given rated output *)				
Gas turbine CHP	-	GT type	Mars 90	Mars 90
Rated electric output ISO	P_e	kW	9,111	9,111
Rated heat output (5 bar steam)	Q_t	kW_t	15,128	15,128
Rated firing capacity	Q_f	kW_t	29,352	29,352
Fuel load operating hours, plant	t_{FLH}	h /a	7,300	7,300
Performance parameters at rated conditions (calculated)				
Electric efficiency **)	$\eta_e = P_e/Q_f$	-	31.0%	31.0%
Total efficiency, cogen only	$\eta_{tot} = (P_e + Q_t)/Q_f$	-	82.6%	82.6%
Electricity-to-heat ratio	$\sigma = P_e/Q_t$	kW_e / KW_t	0.602	0.602
Electricity loss ratio	β	kW_e / KW_t	0	0
Annual performance (calculated)				
Annual elec. production, total **)	$W = P_e \times t_{FLH}$	MWh / a	66,510	66,510
Annual heat production, <u>metered</u>	Q_t	MWh_t / a	110,434	85,000
Cogenerated electricity	$W_{cogen} = \sigma \times Q_t$	MWh / a	66,510	51,192
		-	100%	77%
Fuel consumption, total	Q_f	MWh_t / a	214,270	214,270
contributed to cogeneration	$Q_f = (Pe+Qt)/\eta_{tot}$	MWh / a	214,270	164,921
Total efficiency, mixed mode		-	82.6%	70.7%

*) Source of capacity parameters: ASUE Gasturbinen-Kenndaten

**) the same in cogen and mixed mode operation as there is no electricity loss β=0

6.5 Model for Performance Parameters Relationships

In order to make the calculations of power and energy balances feasible and more transparent, we developed the following theoretical model. The model depicts the energy flow diagrams of a pure cogeneration cycle and of the equivalent power only (condensing) cycle – Figure 6-10. The purpose of the model is to show that there is a firm relationship between the key performance parameters of the two cycles. Thus, they must not be used independently from each other.

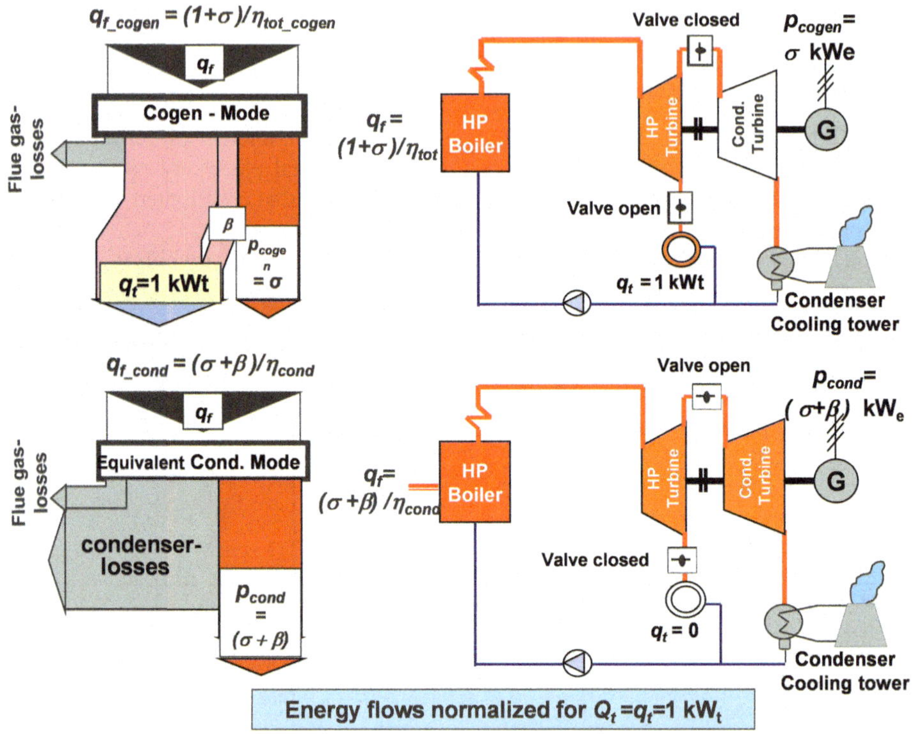

Source: Technologies & Economics, Author's own illustration

Figure 6-10: Energy flow diagrams of cogen and equivalent cond. cycle

The energy flow diagrams and all the key parameters are normalized for 1 kW_t (=1 kJ/s) cogenerated heat and the fuel energy input is the same for both cycles. The following equation applies (see also indications in the diagram):

$$\text{For } \dot{Q}_t = \dot{q}_t = 1, \text{ we get: } \dot{q}_f = \frac{(\sigma+\beta)}{\eta_{cond}} = \frac{(1+\sigma)}{\eta_{tot\_cogen}} \tag{6.9}$$

By solving the equation (6.9), we obtain the following formulas for the interrelationship between the key performance parameters:

$$\sigma = \frac{\eta_{cond} - \beta \cdot \eta_{tot\_cogen}}{\eta_{tot-cogen} - \eta_{cond}} \quad \left[\frac{kW_e}{kW_t}\right] \text{ or } \left[\frac{kWh_e}{kg}\right] \tag{6.10}$$

$$\beta = \frac{\eta_{cond} \cdot (1+\sigma)}{\eta_{tot\_cogen}} - \sigma \quad \left[\frac{kW_e}{kW_t}\right] \text{ or } \left[\frac{kWh_e}{kg}\right] \tag{6.11}$$

$$\eta_{cond} = \eta_{tot\_cogen} \cdot \frac{\sigma + \beta}{1 + \sigma} \quad [-] \tag{6.12}$$

$$\eta_{tot\_cogen} = \eta_{cond} \cdot \frac{1 + \beta}{\sigma + \beta} \quad [-] \tag{6.13}$$

The key performance parameters are strictly linked to each other and the equations above must be fulfilled.

For GT-HRSG CHP plants and for combustion engine CHP plants the power loss of the extracted heat is zero β=0. So from equation (6.9) we get:

$$\frac{\sigma}{\eta_{cond}} = \frac{1 + \sigma}{\eta_{tot\_cogen}} \tag{6.14}$$

Note: For cost allocation between cogeneration products we need to calculate first the electricity generation cost in power only (condensation) mode of operation. This is also necessary for pure cogeneration cycles without condensing part. Therefore there is a need to assess η_{cond} although it is a theoretical value.

6.6 Modelling a Cogen Cycle into Cond. Equivalent Cycle

As already shown in section 6.2, CHP plants are in most cases designed for mixed operation mode producing cogenerated and non-cogenerated electricity in parallel. This is required for technical reasons and also for shifting between the two modes whenever peaks occur for one of the two products.

The specific electricity generation cost of *power only* power plants is simply calculated by dividing the total generation costs by the produced amount of electricity during a reference period. The same approach, applied to an extraction-condensing CHP, would deliver overstated specific cost. This is because the loss of electricity production caused by heat extractions results to a lower electrical efficiency in cogeneration mode of operation. Therefore, the electricity generation cost must be first calculated for an equivalent condensing mode of operation and, on this basis, the cost allocation to cogeneration and non-cogenerated products can be done.

In order to calculate the *actual specific cost* of an extraction-condensing CHP we first convert the cycle into an equivalent condensing cycle by adding the loss of electricity production, caused by the extracted heat, to the rated output of the extraction-condensing cycle. The model is outlined in Figure 6-11 below.

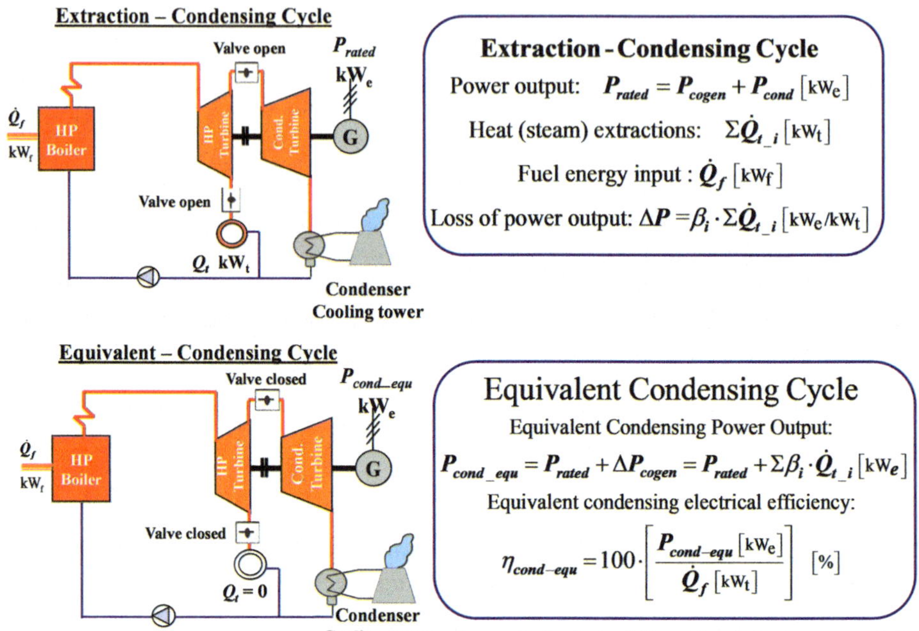

Source: Technologies & Economics, Author's own compilation

Figure 6-11: Converting extraction-cond. cycle to equivalent cond. cycle

Some explanations are stated below:

The *rated output* P_{rated} is usually higher than the power output of the plant operating in pure cogeneration mode. *Heat* Q_t is generated in cogeneration mode only. The performance parameters σ, β, η_{tot} for the cogeneration cycle and η_{cond} for the equivalent condensing cycle are calculated with appropriate thermodynamic cycle simulation software tools for each of the two cycles separately. Indicative figures for studies can be taken from Figure 6-7 and Figure 6-8.

For gas turbine or engine CHPs is β=0, σ, η_{tot} and $\eta_{cond.}$ can be taken or easily calculated from the nameplate of the machine. Initiative figures for selected plants are shown in Table 6-1.

For establishing power and energy balances for the reference period of operation, being usually one year, the following equations apply:

Heat production, cogen: $$\Sigma Q_{t\_i} \quad \left[\mathrm{MWh_t/a}\right] \tag{6.15}$$

Electricity production: $$W_{mixed} \quad \left[\frac{\mathrm{MWh_e}}{\mathrm{a}}\right] \tag{6.16}$$

Cogen electricity: $W_{cogen} = \Sigma\sigma_i \cdot Q_{t\_i}$ $[\mathrm{MWh_e / a}]$ (6.17)

Non-cogen electricity: $W_{cond} = W_{mixed} - \Sigma\sigma_i \cdot Q_{t\_i}$ $[\mathrm{MWh_e / a}]$ (6.18)

Fuel: $Q_f = \dfrac{W_{cogen} + Q_t}{\eta_{total}} + \dfrac{W_{cond.}}{\eta_{cond}} = \dfrac{Q_t \cdot (\sigma + 1)}{\eta_{total}} + \dfrac{W_{cond}}{\eta_{cond}}$ $[\mathrm{MWh_f / a}]$ (6.19)

The conversion of a extraction-condensing cycle into an equivalent condensing cycle is demonstrated in the following Example 6-6. It is recommended to follow the calculation steps in the softcopy on the author's website, because they are quite complex.

Important Note: Cycle simulation software tools are presented in **Case study 11.11 and 11.13**. They are available as softcopy in MS-Excel on the author's website. The models enable the calculation of performance parameters for condensing and cogeneration cycles. For their application the user needs to upload the software FluidEXL which is also available on the website. Furthermore, the use needs some experience and routine in cycle calculations.

Therefore, it is recommended to practice first the simple models presented in **Case Studies 11.1 and 11.2**.

Example 6-6: Conversion of extraction-cond. to equivalent cond. cycle

An extraction-condensing CHP is operated at constant electrical output during working time throughout the year (7500h/a) and supplies process steam of 12 bar and 6 bar to a paper factory. The full capacity hours for steam are different (6500 and 4500 h/a respectively).

Convert the CHP cycle in an equivalent condensing cycle, conduct power and energy balance, and calculate cogenerated and non-cogenerated electricity and the respective fuel consumption for both.

Item		Formula	Unit	Value
Rated power and heat output, given				
Electrical output, gross		P_{rated}	MW_e	87.6
Heat extraction at 6 bar	65 t / h	Q_{12bar}	MW_t	46.0
Heat extraction at 12 bar	100 t / h	Q_{6bar}	MW_t	67.3
Fuel input		Q_f	MW_t	305.4
Performance parameters, given				
Electricity-to-heat ratio, 12 bar steam		σ_{12bar}	kWh_e / kWh_t	0.335
Electrical equivalent, 12 bar steam		$\beta_{12\ bar}$	kWh_e / kWh_t	0.278
Electricity-to-heat ratio, 6 bar steam		σ_{6bar}	kWh_e / kWh_t	0.392
Electrical equivalent, 6 bar steam		$\beta_{6\ bar}$	kWh_e / kWh_t	0.243
Power balance				
Cogen power output		$P_{cogen} = \Sigma\, \sigma_i \times Q_i$	MW_e	41.8
Non-cogen power output		$P_{non-cogen}$	MW_e	45.8
Power loss		$\Delta P = \Sigma \beta_i \times Q_i$	MW_e	29.1
Equivalent condensing power		$P_{cond\text{-}equ} = P_m + \Delta P$	MW_e	116.7
Equivalent condensing efficiency		η_{cond_equ}	-	38.2%
Total efficiency cogeneration		η_{tot_cogen}	-	83.6%
Fuel input for non-cogen electricity		$Q_{f\_non\text{-}cogen} = P_{non\text{-}cogen} / \eta_{cond\_equ}$		119.8
Fuel input for cogen Heat and power		Q_{f_cogen}	MW_t	185.6
Annual Energy Balance				
Electricity generation, total	t=7,500 h/a	$W_{tot} = P_{mrated} \times t$	MWh / a	657,000
of which in cogeneration		$W_{cogem} = \Sigma\sigma \times Q_t$	MWh / a	218,882
Heat generation 12 bar	6,500 h/a	$Q_{t,\ 12\ bar}$	MWh / a	299,000
Heat generation 6 bar	4,500 h/a	$Q_{t,6\ bar}$	MWh / a	302,850
Fuel consumption		$Q_f = \frac{W_{cogen} + \Sigma Q_t}{\eta_{tot\_cogen}} + \frac{W_{cond.}}{\eta_{cond\_equ}}$ $[MWh_t / a]$	MWh/a	2,128,118

7 Cost Allocation to Cogeneration Products

7.1 Overview of Allocation Methods

In general, cogeneration is the thermodynamic cycle of electricity and heat production in a common generation facility, namely a combined heat and power (CHP) plant. Each of the two cogeneration products, electricity and heat, may consist of different kinds of sub products with regard to their production costs or their thermodynamic grade. A CHP plant may generate cogenerated electricity and non-cogenerated electricity, e.g., in bypass or in condensing operation. Heat may be extracted from a steam turbine at different pressure levels. Thus, in many cases the cogeneration products will be more than two.

One important task in investment appraisal for CHP plants is also the proper allocation of production costs to the different cogeneration products, which will be further allocated to other production goods or commodities of a manufacturing process. There are two main groups of methods as shown in the figure below.

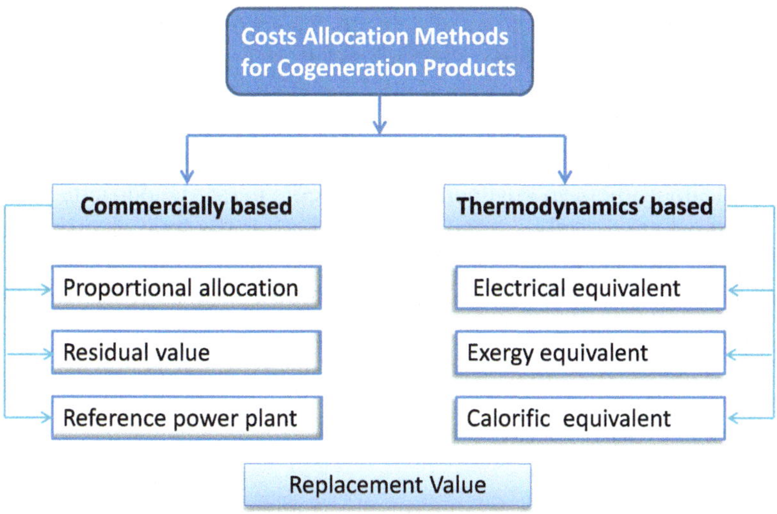

Figure 7-1: Cost allocation methods for cogeneration products

P. Konstantin and M. Konstantin, *The Power Supply Industry*,
https://doi.org/10.1007/978-3-319-72305-1_7

The main difference of the two groups is that commercially based methods do not distinguish production costs differences, e.g., for different steam extraction pressures or between cogenerated and condensing electricity. In contrary, thermodynamics' methods consider this kind of costs differences.

Besides costs, the methods can also be applied for the allocation of fuel or of emissions to the cogeneration products.

> The methods cannot be arbitrarily applied. In general, for each individual application only a single method is the most appropriate, depending on the purpose and the economic philosophy that lies behind.

7.2 Proportional Allocation

Base case: Heat is produced in a captive, heat only boiler plant (HOB) and electricity purchased from the grid. The costs are charged to the production goods as incur proportional to the consumption.

Cogeneration: In the case of *cogeneration* in a captive CHP plant, the costs are allocated in the same proportion to the cogeneration products heat & power, as incurred in the base case, and subsequently charged in the same manner to the production goods. In so far, the cost allocation is self-explanatory.

Philosophy: The cogeneration benefit is equally shared by cogeneration products. Cogeneration may provide some relative reduction of production costs of goods which have heat or/and power demand. The proportional method is applied for engine CHP or gas turbine CHP plants. It is not applicable for steam- or CCGT CHPs with steam extraction at different pressures.

Pros and Cons of the method: The method seems to be just; however, it does neither provide any significant economic advantage to production goods nor an incentive for investment in cogeneration. The cogenerated electricity in a captive CHP, is in most cases, only a relatively small part of the total electricity demand of the factory; thus the cost advantage is marginal referred to the total electricity consumption. Neither does the heat benefit significantly as it has to share with the cogenerated electricity.

7.3 The Residual Value Method

The residual value method is very commonly applied in municipal utilities, small scale distribution companies or factories that operate captive engine or gas turbine CHP plants. The approach is shown below:

Approach:

Total production costs for cogeneration of heat & power

./. Minus a *credit* for avoided costs for one of the two products

Residual costs are allocated to the second product

Philosophy

The cogeneration benefit is fully allocated to the second product

The costs of the first product are kept unchanged

The method finds application mainly for small or medium scale engine- or gas turbine/HRB CHP plants. The credit may be either an *electricity credit* for avoided costs for electricity purchase from the grid or a *heat credit* for avoided heat production costs in a captive heat only boiler (HOB) plant. The former is mainly practiced by municipal utilities, the latter mainly by industrial factories.

Application in municipal utilities: In general, the method is commonly applied for cogeneration of heat and power in CHP plants that supply heat to small and medium size district heating networks. Municipal utilities usually purchase electricity from the public grid and redistribute it to their consumers. Some part of their electricity demand may be produced in own cogeneration plants. District heating can be competitive with decentralized boiler plants on the consumer side only if the high costs of the heat distribution networks can be compensated by lower heat generation costs. This can be achieved by cogeneration if the benefit is fully allocated to the heat.

Hence, the avoided costs of purchased electricity from the grid are deducted as *electricity credit* from the total production costs of the cogeneration, and the residual costs are allocated to the heat which becomes less costly. On the other side, the electricity consumers do not have any disadvantage, as they are charged the same costs.

Note: The avoided costs for electricity purchase are deducted as a credit from the total production costs. This requires that quantity and load profile of the cogenerated electricity is equal with that of the avoided power purchase. In other words, the credit must be composed of the avoided costs for capacity (kW) plus for energy (kWh) purchase.

Application in industrial plants: Operators of industrial CHP plants follow a different philosophy: The heat must be produced, in any case in a captive heat only boiler plant because it cannot be supplied by an external heat network while electricity can be purchased from the grid. Hence, the avoided heat production costs in the captive boiler plant are deducted as a *heat credit* from the total production costs of the cogeneration and the residual costs are allocated to the electricity. The cogeneration benefit is fully allocated to the electricity, which may become less costly.

However, the avoided costs (credit) are usually only fuel costs. This is because heat supply is an indispensable commodity for the production. Therefore, industrial companies always maintain standby boilers for safety reasons and their capital cost cannot be considered in the credit.

Pros and cons of the Method: The residual value method is purely commercially based. Production cost differences caused by different extraction pressure of the steam are not considered.

Another major shortcoming is that the residual value is very sensitive and may deliver extremely overstated or understated residual costs, depending on the ratio between fuel price and electricity price. Electricity is usually the product with the higher costs. If, e.g., the electricity prices from the grid become temporarily high, the heat cost may be too low or even negative. In contrary, if electricity prices become low, the heat cost will be too high. Both scenarios are a real possibility in liberalized markets. This is more likely to happen for CHP plants with a high electricity-to-heat ratio σ (see Table 6-1: Performance parameters of selected CHP plants).

In general, cogeneration is a cost effective option in base load application only and needs a high utilization time of the installed capacity (full capacity hours).

Following Example 7-1 demonstrates the application of the residual value method in a spreadsheet as the base case, following a sensitivity analysis and breakeven point analysis.

It is evident for the base case calculation that the costs for heat production in cogeneration are considerably lower compared to heat production in a HOB. However, this advantage may be reversed if the relation between the main parameters, influencing the costs, is changed as shown in the sensitivity analysis.

CHP plants are commonly equipped with stand-by HOBs. These are operated for covering peak loads and serve as back-up in the case of a forced or planned outage of the cogeneration plant. The following example demonstrates when shifting from cogeneration to HOB heat production becomes more favorable.

The break-even point shows that the threshold of production cost is obtained at a quite higher gas price compared to the production cost of the base case (see sensitivity analysis). This is because at rising gas prices both the production cost in cogeneration and in heat only boiler increase.

Example 7-1: Residual value allocation with electricity credit, base case

The example demonstrates the cost allocation to cogeneration products for a small scale gas fired GT -CHP plant, the cost-sensitivity of the cogenerated heat against the main parameters, utilization time, gas price and electricity purchase price (credit).

Item		Unit	Value
Power balance			
Electrical output, net	P_e	kW_e	4,985
Thermal output	P_Q	kW_t	7,510
Fuel Input	η=88.0%	kW_f	14,199
Annual energy balance			
Full load hours		h / a	7,000
Electricity generation *)		MWh_e / a	34,895
Heat generation	Q_t	MWh_t / a	52,570
Fuel consumption	Q_f	MWh_f / a	99,392
Financial constraints			
Gas price in LHV		€ / MWh_f	25.00
Electricity generation price, flat tariff, 220 kV grid		€ / MWh_e	50.00
Use of system fees **)			
Capacity price		€ / (kW_e*a)	55.00
Energy price		€ / MWh_e	8.00
Annual production costs, total *C*			
Fixed costs ***)	C_{fixed}	th. € / a	880.0
Variable cost (Fuel costs)	C_{fuel}	th. € / a	2,484.8
Subtotal		**th. € / a**	**3,364.8**
Electricity credit *CC* (avoided purchase cost)			
Costs of electricity from grid	CC_{gen}	th. € / a	1,744.8
Use of system costs:			
capacity costs	CC_{fixed}	th. € / a	274.2
energy costs	CC_{energy}	th. € / a	279.2
Electricity credit, total		**th. € / a**	**2,298.1**
Electricity credit, specific		€ / MWh_e	*65.86*
Residual costs of heat		**th. € / a**	**1,066.7**
Capacity cost	$(C_{fixed}\text{-}Ccfixed)/P_Q$	€ / (kW_t*a)	*80.67*
Energy cost	$(C_{fuel}\text{-}CC_{gen}\text{-}CC_{energy})/Q_t$	€ / MWh_t	*8.77*
composite heat cost		**€ / MWh_t**	**20.29**

Generation cost of HOB **)**	η=85.0%	**€ / MWh_t**	**29.41**

*) full cogeneration, no bypass operation
**) cost for transmission & distribution
***) annualized CAPEX + O&M costs)
****) for comparison only

Example 7-1 Continuation, Sensitivity analysis

Full load hours		Gas price		Electricity price	
h /a	€ / MWh$_t$	€ / MWh$_t$	€ / MWh$_t$	€ / MWh$_e$	€ / MWh$_t$
base case	**20.29**	base case	**20.29**	base case	**20.29**
4000	28.93	15	1.38	45	23.61
5000	24.90	20	10.84	50	20.29
6000	22.21	25	20.29	55	16.97
7000	20.29	30	29.74	60	13.65
8000	18.85	35	39.20	65	10.33

Heat generation cost of a HOB (only fuel cost)	€ /MWht	29.41

Example 7-1, Continuation, Break-even point heat cost Cogen vs. HOB

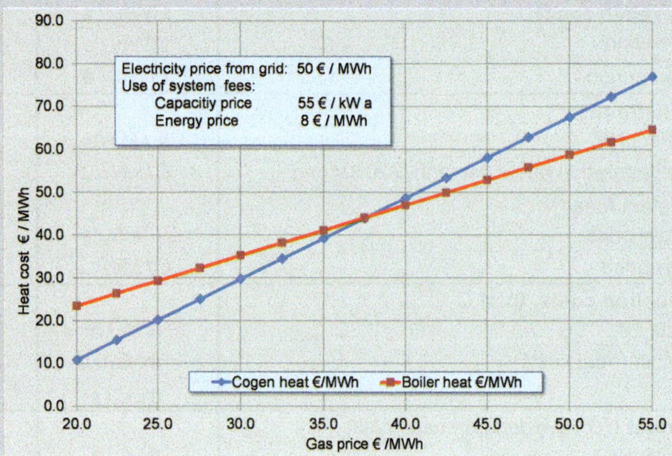

Example 7-1, Continuation, break-even point heat cost Cogen vs. HOB

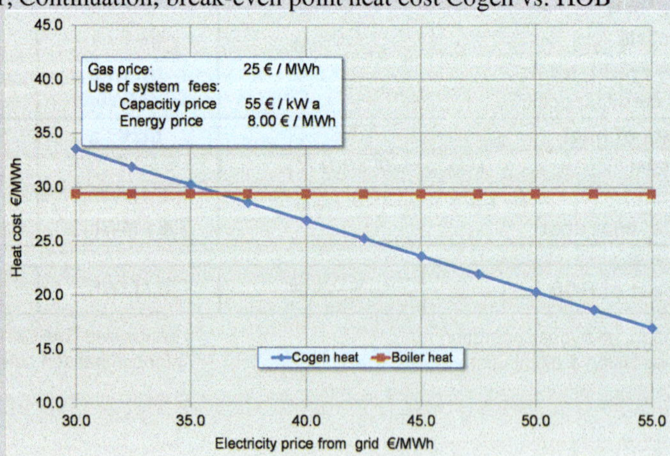

7.4 The Residual Value with Reference Power Plant

Cost allocation based on the production costs of a reference utility power plant is a special case of the residual value method. It is applied in the same manner as the residual value method; the only difference is that the electricity credit is calculated based on the electricity generation costs of a Reference Power plant instead of power purchase from the grid.

The method's main applications are dual purpose power plants, which deliver cogenerated steam to desalination plants (MSF or MED plants) and are equipped with backpressure steam turbines.

The method is simple in application; its disadvantage is that different pressure levels of steam extraction cannot be charged with different costs. For dual purpose power plants this is not a real disadvantage as steam extraction occurs mainly at 2.7 bar and only a marginal quantity at 18 bar.

Approach

Total generation costs of the dual purpose power plant for power and process steam

./. Minus a credit for the same electricity based on the production costs of the Reference Power Plant

= Residual costs are allocated to the extracted steam

Dual purpose power plants are usually large, utility size power plants with similar thermodynamic process parameters (steam turbine or CCGT). They are called dual *purpose because* their primary objective is the supply of cogenerated steam to seawater desalination plants and their secondary product is cogenerated electricity. They produce electricity, however, with a significantly lower electrical efficiency due to the extraction of steam at a relatively high pressure (about 2.7 bar and a small amount at 18 bar) compared to the condensation pressure (about 0.05 bar).

The reference power plant method finds application for such kind of plants if they are equipped with backpressure steam turbines without any technical means to generate electricity in condensing mode. There are also extraction-condensing dual purpose plants; for those, the electricity equivalent method is more appropriate.

The *Reference Power Plant* must be of the same type (Steam or CCGT), operating in full condensing mode, firing the same fuel and commonly used for power generation for the public grid by the power utility.

In order to define the size of the reference power plant, first the equivalent condensing power output of the dual purpose power plant must be estimated. The capacity of the reference power plant of the grid must be equal or the most closest. The equivalent condensing power output of the dual purpose plant can be estimated with the following equation. For a rough estimate, the electrical equivalent β can be taken from Figure 6-7 or Figure 6-8:

$$P_{cond\_equ.} = P_{e\_DP} + \beta \cdot \dot{Q}_t \quad [MW] \tag{7.1}$$

In the following Example 7-2 the technical parameters of typical new dual purpose power plants and corresponding reference utility power plants from public grid are shown.

Example 7-2: Typical dual purpose reference PPs

Item		Unit	Steam Rankine Cycle HFO	CCGT Natural gas
Dual purpose plant				
Live steam		bar / °C	140 /535	124/555
Type of steam turbine		-	back pressure	
Rated power output, gross		MW_e	294.0	302.6
HP Steam for desal	18.0 bar	MW_t	9.5	8.5
LP steam for desal	2.7 bar	MW_t	668	227
Efficiency: electrical/total		-	27.5% / 90%	49.7% / 87%
Equivalent cond. Power *)	beta=0.23	MW_e	450	357
Corresponding reference power plant, from public grid				
Live steam		bar / °C	160/535/535RH	124/555
Type of steam turbine		-	Condensing	
Rated power output, gross		MW_e	600	350
Electrical efficiency, gross		.	41.4%	57.3%

Source: Energy Efficiency Study, Saudi Aramco-Fichtner, 2010

*) average for 18bar and 2.7bar steam, β=0.23

In the continuation of, next Example 7-3, the calculation of the corresponding (residual) steam costs of both dual purpose power plants, broken down in capacity and energy costs, are demonstrated.

Example 7-3: Steam cost of a dual purpose PP, based on reference PP

Item		Unit	Steam Rankine Cycle HFO	CCGT Natural gas
Reference Power Plant (RPP), Electricity Generation Cost [1)]				
Spec. Capacity cost		US$ / kWa	199.2	102.6
Spec. Energy cost		US$ / kWh$_e$	17.7	21.5
Composite cost	7920 h/a	US$ / MWh$_e$	42.83	34.42
Dual Purpose PP, Power and Energy balance				
Electrical output, gross	HFO	MW$_e$	294	303
Electrical output, net		MW$_e$	279	293
Steam output (18 bar + 2.7 bar)		MW$_t$	678	236
Electricity production	7920 h/a	GWh$_e$ / a	2,210	2,321
Steam production for desal	7920 h/a	GWh$_t$ / a	5,366	1,865
Annual costs dual purpose PP [1)]				
Fixed Costs, dual purpose		mln US$ / a	89.0	35.3
minus credit for fixed elec. costs; RPP		mln US$ / a	55.6	30.1
Residual capacity costs for steam		**mln US$ / a**	**33.4**	**5.2**
Variable costs, dual purpose		mln US$ / a	59.8	58.5
minus credit for variable elec. costs; RPP		mln US$ / a	39.1	49.8
Residual energy costs for steam		**mln US$ / a**	**20.7**	**8.7**
Residual costs dual purpose PP, total		**mln US$ / a**	**54.1**	**13.9**
Specific steam cost for desalination plant				
Capacity cost		US$ / kW$_t$ a	49.3	22.2
Energy cost		US$ / MWh$_t$	3.86	4.66
Composite cost	7920 h/a	**US$ / MWh$_t$**	**10.09**	**7.47**

Source of data: Saudi Aramco Energy Efficiency Study [1)] Detailed costs calculation in separate files

7.5 The Electrical Equivalent Method

7.5.1 Principle and application forms

The electrical equivalent method is based on the fact that steam extracted from a steam turbine causes a loss of electricity production. This is referred to as the *electrical equivalent* of the extracted heat and is denoted with the symbol β [kWh$_e$/kWh$_t$] or [kWhe/kg steam]. Hence, the steam is charged the cost of the equivalent electricity production. The same principal is applied for the calculation of the heat rate and the emissions' rate of the extracted steam. We can express this mathematically with the following equations:

$$\text{Heat generation costs: } c_h = \beta \times c_e \quad \left[\frac{\text{CU}}{\text{kWh}_t}\right] \tag{7.2}$$

$$\text{Heat rate: } \dot{q}_s = \frac{\beta}{\eta_e} = \beta \times \dot{q}_e \quad \left[\frac{\text{kWh}_f}{\text{kWh}_t}\right] \tag{7.3}$$

$$\text{Heat emission rate: } e_s = \beta \times e_e \quad \left[\frac{\text{kg}_{CO2}}{\text{kWh}_t}\right] \tag{7.4}$$

Where:

c_e: Electricity generation cost in condensing mode of operation (CU/ kWh_e)
c_h : Generation cost of the extracted heat
β : Electrical equivalent of heat (kWh_e / kWh_t)
η_e : Electrical efficiency in condensation mode
$\dot{q}_e$: Fuel rate of electricity in condensing mode (kWh_t / kWh_e)
$\dot{q}_s$: Fuel rate of the extracted heat (steam) kWh_f/kWh_t
e_e : Specific emissions of condensing electricity (kg_{CO2} / kWh_e)
e_s : Specific emissions of extracted steam (kg_{CO2} / kWh_t)

The electrical equivalents shall be calculated with thermodynamic cycle simulation programs as Fichtner's KPRO® . For preliminary calculations in studies it can be taken from Figure 6-7 and Figure 6-8.

With regard to the application of the electrical equivalent method, we have to distinguish between two different cases:

- Application in connection with *actual condensing power plants*. The main duty of these plants is to generate electricity in condensing mode of operation with a relatively small heat extraction. These are large utility size power plants.
- Application in connection with *actual combined heat and power (CHP) plants*. These plants are designed for operation in extraction-condensing mode. They produce primarily cogenerated heat and, in parallel, additional to the coupled cogenerated electricity, also some electricity in condensing mode.

7.5.2 Application to actual condensing power plants

Actual condensing power plants are those designed for electricity generation in full condensing mode of operation. The steam turbine can absorb the entire live steam production of the boiler in condensing mode of operation but steam extractions are technically feasible and often practiced. Electricity remains, however, by far the main product.

Typically a 600 MW_e utility size steam power plant is designed for heat extraction of up to 300 MW_t. Its output in condensing mode will be 600 MWe, in full extraction-condensing mode 555 MW_e and 300MW_t. Such kind of plants supply base load heat to large district heating networks and can easily be shifted to full condensing mode of operation during on-peak time of the electrical grid.

The application of the method is quite simple, provided that the electricity generation costs in condensing mode of operation have been calculated and are known. For the calculation of the heat cost the electrical equivalents of the steam extractions "β_i" and the electricity generation costs in full condensing mode of operation "c_e" are needed.

Example 7-4: Cost of extracted steam at different pressure levels

The example demonstrates the calculation of the cost for several steam extractions from a 600 MW utility power plant. In the upper table it is assumed that steam is being extracted at different pressure levels but with the same equivalent operating hours (6000 h/a). Hence, heat of higher steam extractions is more costly. In the lower table the pressure level of the extraction is the same but the operating hours are different, which results in higher heat costs for lower operating hours.

Heat Extraction from a 600 MW power plant						
	Electricity capacity cost c_C		€ / (kW_e a)	198.00		
	Electricity energy cost c_e		€ / MWh_e	30.00		
Steam Extraction			Specific heat cost			
Steam-pressure	Steam Temperature	Electrical equivalent	Capacity cost	Energy cost	Equivalent full capacity hours	Composite cost
p	t_s	β	β x c_C	β x c_e	hours	
bar	°C	kWh_e / kWh_t	€ / (kW_t a)	€ / MWh_t	h / a	€ / MWh_t
12.0	218	0.251	49.70	7.53	6,000	15.81
6.0	189	0.220	43.56	6.60	6,000	13.86
1.7	145	0.164	32.47	4.92	6,000	10.33

Note: Performance parameters from Cycle calculation

Steam Extraction			Specific heat cost			
Steam-pressure	Steam Temperature	Electrical equivalent	Capacity cost	Energy cost	Equivalent full capacity hours	Composite cost
p	t_s	β	β x c_C	β x c_e	hours	
bar	°C	kW_e / kW_t	€ / (kW_t a)	€ / MWh_t	h / a	€ / MWh_t
6.0	189	0.220	43.56	6.60	6,000	13.86
6.0	189	0.220	43.56	6.60	3,500	19.05

Example 7-5: Heat supply cost of district heating

A district heating network is supplied with base load heat from a utility size power plant. The steam is extracted in two pressure stages of 3.2bar and 0.9bar. The heat condensers (exchangers) are arranged in series. The 0.9bar extraction is operated throughout the year and generates heat of 90 °C. The 3.2bar extraction is operated only during the on-peak period in winter in order to arise the heat temperature to 130°C. The electricity generation costs are given as well as the electrical equivalents of the extracted steam. Calculate the heat costs broken down in capacity and energy costs.

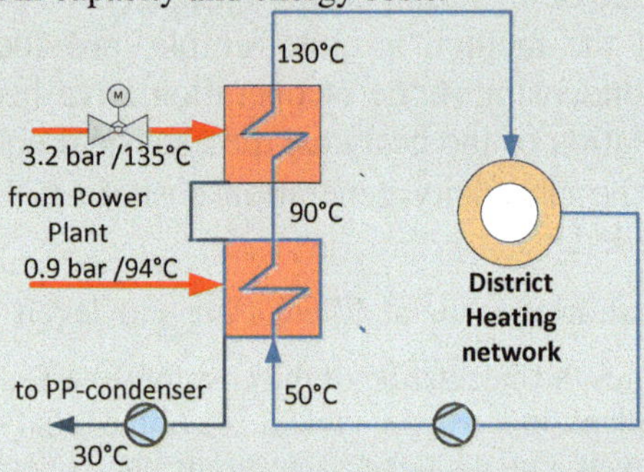

Item			Formula	Unit	Value
Given					
Capacity cost, electricity			cc_e	€ / (kW*a)	198.00
Energy cost, electricity			ce_e	€ / MWh	30.00
Heat supply, peak load			Q_{max}	MW	75
3.2 bar / 0.9 bar			-	-	50%
Annual energy supply	5500 h/a		W_{DH}	MWh / a	412,500
of which 3.2 bar heat condenser			r_{3bar}	-	15%
of which 0.9 bar heat condenser			$r_{1.2bar}$	-	85%
Electrical equivalent, 3.2 bar steam			β_{3bar}	kWh_{el} / kWh_{th}	0.218
Electrical equivalent, 0.9 bar steam			β_{1bar}	kWh_{el} / kWh_{th}	0.115
Cost of District Heat					
Electrical equivalent, average of the two extractions					
for capacity cost	50%	50%	β_c	kW_{el} / kW_{th}	0.167
for energy cost	15%	85%	β_E	kWh_{el} / kWh_{th}	0.130
Specific capacity cost			$c_{C,DH} = \beta_C \times cc_e$	€ / (kW*a)	32.97
Specific energy cost			$c_{E,DH} = \beta_E \times ce_e$	€ / MWh	3.91
Annual cost			$C_{DH} = CC_{,DH} + C_{A,FW}$	**th. € /a**	**4,087**
of which capacity costs			$C_{,DH} = cc_{DH} \times Q_{max}$	th. € /a	2,473
of which energy costs			$C_{EDH} = c_{E,DH} \times W_{FW}$	th. € /a	1,614
Composite cost			c_{DH}	**€ / MWh**	**9.91**

Example 7-6: Calculation of heat rate and emissions of extracted steam

In the example the calculation of the heat rate and carbon emissions to the extracted steam with the electrical equivalent method is shown.

Technical Parameters	Symbol	Unit	Value
Efficiency, PP, net	η_e	-	42%
Electrical equivalent, 6 bar process steam	β	kWh_{el} / kWh_t	0.220
CO_2 fuel emission factor, hard coal	e_f	kg / MWh_t	342
Allocation			
Heat rate, electricity, net	$q_e = 1/\eta_e$	kWh_t / kWh_e	2.38
Heat rate, process steam, net	$q_s = q_e \times \beta$	kWh_t / kWh_t	0.52
Emission factor electricity	$e_e = e_f \times q_e$	kg / MWh_e	814
Emission factor process steam	$e_s = e_e \times \beta$	kg / MWh_t	179

7.5.3 Application for extraction-condensing CHP plants

In the former section, the electricity generation cost could be calculated as the plant was primarily a utility scale condensing power plant. Heat extraction was feasible, and we could calculate the heat cost simply by multiplying the electricity generation cost with the electrical equivalent of the extracted steam.

Steam CHP plants or CCGT CHP plants are designed for continuous operation in extraction-condensing mode, while heat is the primary product. They produce heat and electricity in cogeneration and non-cogenerated electricity in parallel. For a proper costs allocation to the different products, it is necessary that the electricity generation cost must be first calculated in *equivalent condensing mode* of operation. For this purpose, we need to apply a quite complex technical-financial model that includes several modules with thermodynamic and financial calculations.

The model's structure is presented in Figure 7-2; it comprises seven modules including the main algorithms for the calculations. A brief description of the modules is given after the figure.

Note: The electrical equivalent method is the most appropriate for costs allocation to cogeneration products for large scale CHP plants. Some knowledge in thermodynamics' is a mandatory prerequisite. Thermodynamic cycle calculations to define performance parameters are essential. Its application is primary a task for engineers with some background in economics rather than for economists.

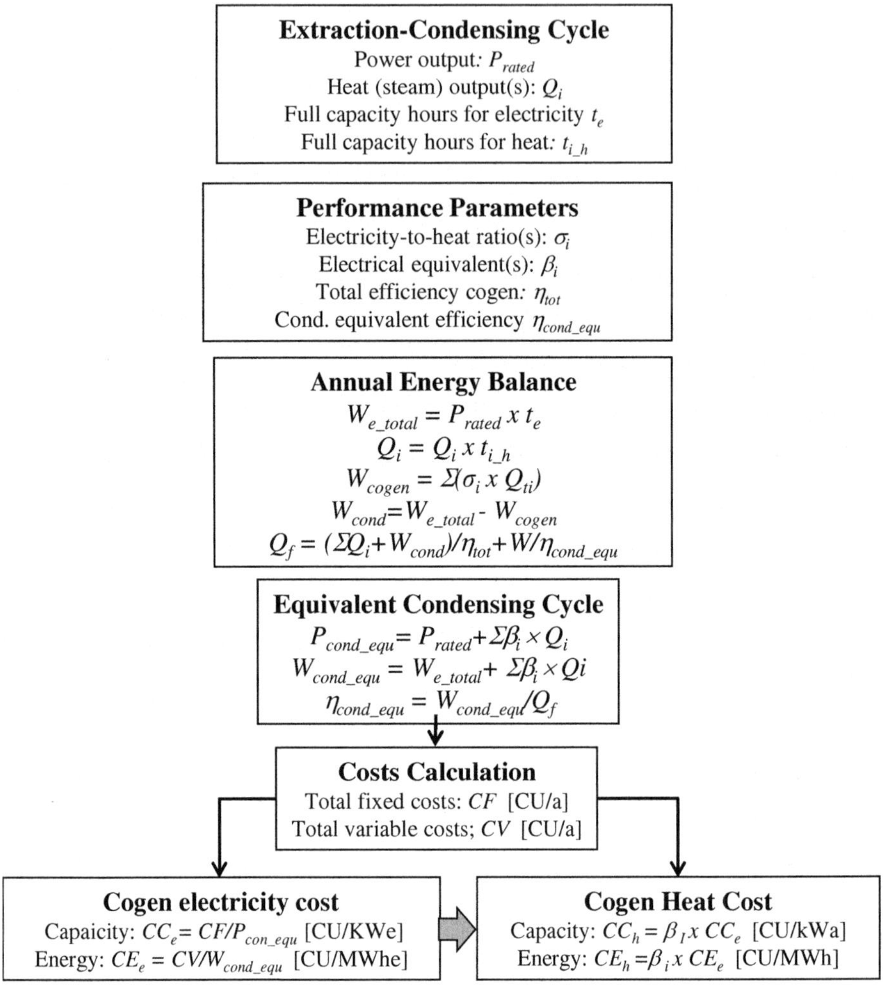

Figure 7-2: Model for the costs allocation based on electrical equivalent

Extraction-condensing cycle: The output at rated conditions in condensing-extraction cycle and the performance parameters are calculated with thermodynamic cycle simulation software tools. For preliminary calculations in studies, the required performance parameters can be taken from Figure 6-7 and Figure 6-8. Cycle simulation software tools in MS-Excel, available on the author's website, can also be used. However, some background and routine by the user in cycle simulation is indispensable. For real projects professional software tools, such as Fichtner's KPRO®, or equivalent shall be used.

The *annual energy balance* is conducted with the rated output(s) for power P_e (MW$_e$) and the different heat extractions Q_i (MW$_t$) and their expected equivalent full capacity hours (t_{FCH}). The annual electricity production is broken down in cogenerated and non-cogenerated (cond.) electricity. Fuel consumption is calculated for cogenerated and non-cogenerated electricity separately.

Equivalent condensing cycle: The conversion of the extraction-condensing cycle into an *equivalent condensing cycle* (power only) is done by adding the lost power generation caused by the extracted heat to the rated electricity generation (see also modelling description in section 6.6).

Costs Calculation: First the annual generation costs for actual operation under rated conditions are calculated; they are broken down into fixed and variable costs.

The specific *electricity generation* costs are calculated from the total annual costs and the amount of equivalent condensing power and electricity. It is important to split the costs in capacity and in energy costs because the equivalent operating hours of the products may be different.

The *heat generation costs* are calculated by multiplying the respective electrical equivalents of the extracted heat with the specific electricity cost.

The model, presented in Figure 7-2, is applied in **Case Study 11.9** for cost allocation of an extraction-condensing CHP plant. The spreadsheets of the case study correspond with the modules described in the model. In addition, the MS-Excel software tool for simulation of the cycle and calculation of its performance parameters is included. The results of the case study are shown in short in the following table. The complete model is presented in the case study section as hard copy and is also available as soft copy on the author's website.

Table 7-1: Case Study, Cost allocation with electrical equivalent, Results

Item		Formulas	Unit	Value
Electricity generation cost				
Composite cost	*7,500 h/a*	$ce=1000*cc_e/t_{FCH}+cv_e$	€ / MWh	61.03
Capacity cost		cc_e	€ / (kWa)	238.68
Energy cost		cv_e	€ / MWh	29.20
Heat generation cost, 12 bar steam				
Composite cost	*6,500 h/a*	$c_{H12}=1000*cc_{12}/t_{FCH}+c_{v12}$	€ / MWh	19.06
Capacity cost	*beta=0.289*	$cc_{12}=\beta_{12bar}\times cc_e$	€ / (kWa)	69.00
Energy cost		$cv12=\beta_{12bar}\times cv_e$	€ / MWh	8.44
Heat generation cost, 6 bar steam				
Composite cost	*4,500 h/a*	$c_{H6}=1000*cc_6/t_{FCH}+cv_6$	€ / MWh	20.96
Capacity cost	*beta=0.255*	$cc_6=\beta_{6bar}\times cc_e$	€ / (kWa)	60.83
Energy cost		$cv6=\beta_{6bar}\times cv_e$	€ / MWh	7.44

7.6 The Exergy Method

7.6.1 What is exergy

Exergy is defined as the maximum amount of mechanical work that can be obtained from the energy content of an energy carrier, under ideal conditions, in a reversible process, using the environment as the lower heat reservoir.

Usually, we have to deal in thermodynamics with gaseous or liquid fluids which contain thermal energy. The formula for the calculation of the exergy of a fluid in a steady flow is as follows:

$$e = h - h_a - T_a \times (s - s_a) \quad \left[\frac{\text{kJ}}{\text{kg}}\right] \tag{7.5}$$

h : Enthalpy of the energy carrier (kJ /kg)

h_a: Enthalpy of the ambience (kJ/kg)

T_a: Thermodynamic temperature of the ambience (K)

s: Entropy of the energy carrier (kJ/kg K)

s_a: Entropy of the ambience (kJ/kg K)

Note: For the calculation of the thermodynamic properties vapor tables or better appropriate software tools are needed. In this book the software tool FluidEXL [50] is used.

Although the unit for energy and exergy is the same (kJ/kg or kWh/kg), there is a fundamental difference between these two properties:

- In a closed thermodynamic system energy is converted during a process into different energy forms but the sum of energies remains constant (1st law of thermodynamics)
- in contrary, exergy is destructed during the process flow and becomes zero after the process has reached environmental level (2nd law of thermodynamics).

Usually the lowest energy level (temperature and pressure) of a process is defined as the zero exergy level. The exergy of the ambience is zero. Electricity, in contrast, contains hundred percent exergy as it can be nearly completely converted into any other form of energy.

In this context, exergy can be defined as the percentage of the energy content of an energy carrier that can be converted into electricity.

The following example gives some indication of the energy and exergy content for different energy carriers.

Example 7-7: Energy and exergy content of energy carriers

Energy carrier		bar	°C	Energy content	Exergy	
				kJ / kg	kJ / kg	%
Electricity		n.a.	n.a.	100%	n.a.	100%
Steam levels of steam turbines	live steam ultra super critical	245	620	3,557	1,702	48%
	live steam sub critical	180	540	3,390	1,555	46%
	extraction	20	250	2,903	1,018	35%
	extraction	6	159	2,756	810	29%
	cond. turbine discharge	0.045	31	2,641	2	0.1%
Condenser cooling Water		1	15	63	(0)	0%
Note: Exergy is the part of the embodied energy in an energy carrier that can be converted into mechanical energy in an ideal reversible process						

It is obvious that the exergy of the discharge steam of a condensing turbine (0.045 bar /31°C) still contains a high amount of energy that, however, is useless for power generation because its exergy is only 0.1%. The exergy of the condenser cooling water is zero. This is the zero exergy level of the steam cycle.

Our objective in this section is to use exergy for cost allocation to cogeneration products, especially for cogeneration cycles with a steam turbine (steam CHP or CCGT CHP plants). The extraction from the steam turbine usually occurs at pressure levels between 1bar to 12bar; small amounts may be extracted at higher pressure levels, for instance, motive steam for desalination processes at 18 bar. In general, however, cogeneration does not provide any energetic advantage at higher extraction pressures.

Real conversion processes are non-reversible, and there is always some loss of exergy. Only a certain part of the available exergy can be converted into mechanical energy, respectively electricity. We know from the previous section 7.5 that the electricity, extracted steam could produce in a real cycle up to the condensing pressure, if it was not extracted, is denoted as the electrical equivalent β [kWh_e/kWh_t or kWh_e/t]. From this we can deduce:

> The share of exergy that can be converted into electricity in a steam cycle is equal to the electrical equivalent β [kWh_e/kWh_t] of the extracted steam.

Hence, the *exergy loss* can be determined as the difference between the exergy of the extracted steam and its electrical equivalent. Based on this, we get the ratios, depicted in Figure 7-3, which are derived from a typical real cogeneration cycle.

$$\text{Exergy loss: } \Delta e = e - \beta \quad [\text{kWhe/kg}] \quad \text{ratio: } \Delta e = \frac{\Delta e}{e} \cdot 100\,[\%] \tag{7.6}$$

$$\text{Exergy conversion ratio:} \qquad r_{ex} = \frac{\beta}{e} \quad [-] \tag{7.7}$$

So we can assign an electrical equivalent for the exergy of the extracted steam as:

$$\text{Electrical equivalent of exergy: } \zeta = \frac{r_{ex} \times e}{q_h} \quad \left[\frac{\text{kWh}_e}{\text{kWh}_t}\right] \tag{7.8}$$

Where, q_h is the heat content of the extracted heat.

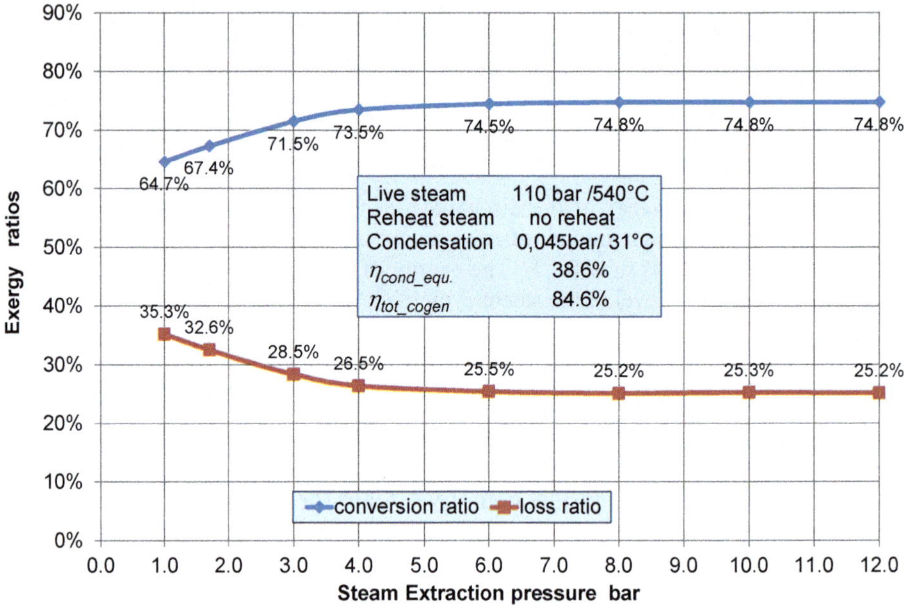

Figure 7-3: Exergy loss and utilization rate

The ratios seem to remain quite stable over a large range of extraction pressures. For practical applications, we can assume that the ratios in Figure 7-3 are typical for cogeneration cycles with an acceptable accuracy margin. This has the advantage that we can proceed with cost allocation without the need a complete cycle simulation or an extensive and complex exergy balance.

7.6.2 **Cost allocation based on exergy, simplified model**

Based on the equation and ratios presented in the previous section, we proposed the following practical approach for cost allocation based on exergy.

The exergy method can be applied in a similar way as the electrical equivalent method. If the generation cost of electricity "c_e" is given, or taken from a reference power plant, the cost of the heat of different steam extractions are simply calculated by multiplying with the electrical equivalent of the exergy of the individual extractions.

$$\text{Heat cost of extracted steam}: \quad c_h = \zeta \cdot c_e = \frac{r_{ex} \times e}{q_h} \times c_e \quad \left[\frac{\text{CU}}{\text{kWh}}\right] \qquad (7.9)$$

The electrical equivalent of electrical energy is per definition = 1!

The proposed method is demonstrated in the following example.

Example 7-8: Heat generation costs calculation with exergy

An industrial Rankine cycle CHP plant with a power output of about 110 MW in extraction-condensing mode has three steam extractions. The electricity generation costs, broken down in capacity and energy cost, are taken from **Case study 11.9**. The specific steam costs shall be calculated with the exergy method. The full load hours are required for the calculation of the composite cost only and are different for power output and for extractions.

Cost Allocation Based on Exergy

Item	Pressure 1)	Temperature 1)	Heat content of extracted steam 3)	Specific exergy 2)	Exergy conversion ratio	Exergy converted	Electrical equivalent
	bar	°C	kJ/kg	kJ/kg	ζ	kJ / kg	kWh$_e$/kWh$_t$
Steam	12.0	244	2,545	963	74.8%	720	0.283
	6.0	178	2,424	826	74.5%	615	0.254
	3.0	134	2,349	712	71.5%	509	0.217

1) from cycle calculation
2) zero exergy level, condenser cooling water, 1 bar — 15 °C
3) process steam condensate return 100%, 90°C — 377 kJ/kg

Item	Pressure	Electrical equivalent	Capacity cost	Energy cost	Composite cost 5)
	bar	kWh$_e$/kWh$_t$	€ /kW a	€ /MWh	€ /MWh
Electricity, given 4)		1.000	243.89	29.86	62.38
Steam	12.0	0.283	69.01	8.45	19.07
	6.0	0.254	61.89	7.58	21.33
	3.0	0.217	52.87	6.47	18.22

4) electricity costs taken from Case Study, cost allocation extraction-condensing CHP
5) referred to the stated full load hours only

The exergy method delivers almost identical cost for the extracted heat as the electrical equivalent method (compare results of case study, in Table 7-1).

7.6.3 Cost allocation model based on exergy balance

The advantage of the exergy method is that it is generally applicable. The electrical equivalent method is strictly applicable only for CHP plants equipped with a steam turbine that is capable to generate electricity in condensing mode of operation. This is not a requirement of the exergy method because all the energy streams, electricity and heat, are evaluated based on their exergy which is equivalent to electrical work that could be produced from their energy content.

Within a refinery or a chemical complex for instance, there are different energy streams contained in different fluids (gases, steam, product streams etc.) with different pressures and temperatures; their energetic grade and their value in monetary terms can be evaluated and determined based on exergy. However, we have to conduct exergy balances and apply relatively complex calculation models.

In the following we shall demonstrate the application of the exergy method for cost allocation for CHP plants with backpressure steam turbines (Steam or CCGT CHPs).

Basic approach for costs allocation

The exergy of each cogeneration product, the total exergy (kJ or KWh) and the exergy share of each cogeneration product (%) are calculated.

The total production costs CU/a are allocated to the different cogeneration products (electricity and heat) based on their exergy shares (%) related to the total exergy (100 %) of all cogeneration products.

The allocation model is, of course, more complex consisting of several calculation steps – Figure 7-4. The positive about it is that for our purposes we do not need to conduct a complete exergy balance for the whole thermodynamic cycle. For our purpose, it is sufficient to consider the exergies of the co-generation products only.

Pros and Cons

The exergy method is appropriate for cost allocation for CHP plants and complex energy systems (e.g. refineries, chemical complexes with different energy streams), especially if there is no possibility for generation of electricity in condensing mode that can be used as the reference system. The results of the costs allocation are comparable to those of the electricity equivalent method. A disadvantage may be to explain the theoretical basis of exergy to financial managers without background in thermodynamics. The calculation effort is also complex and time consuming.

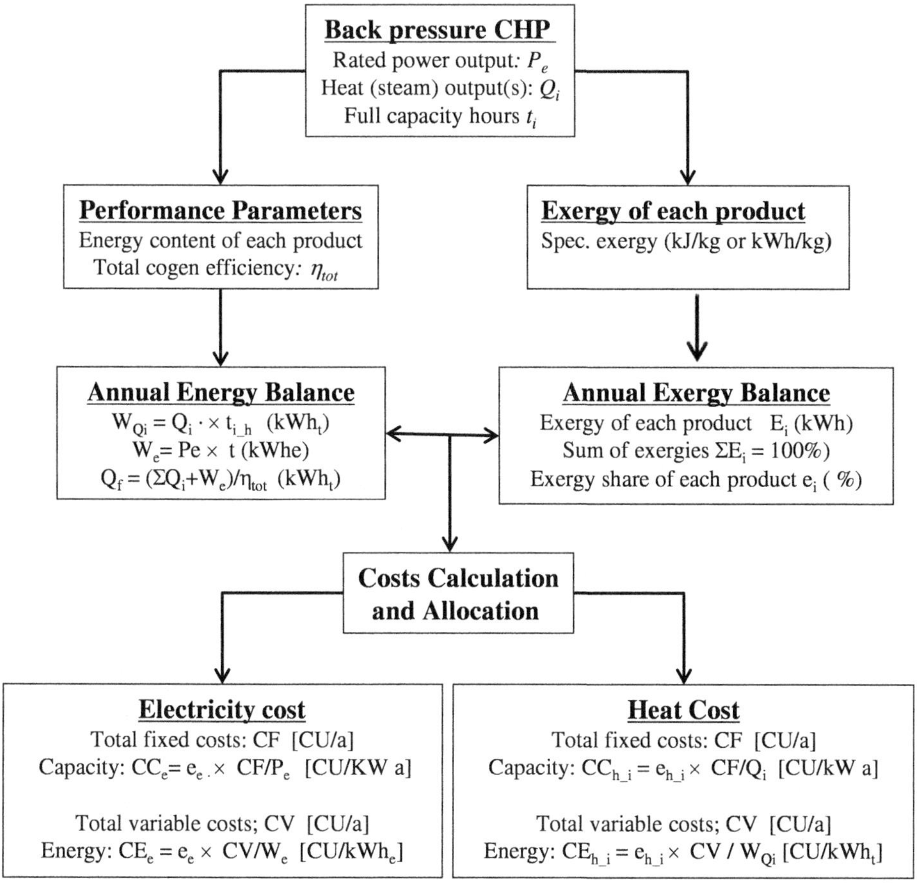

Figure 7-4: Cost allocation model based on exergy

The method is demonstrated in the **Case Study 11.10** for a CCCGT CHP plant with backpressure steam turbine. The case study includes:

- A simplified energy flow diagram (see below)
- A cycle simulation depicted in a heat flow diagram with the main results of the simulation
- A spreadsheet with the following calculation modules:
    - Power and energy balance
    - Calculation of the specific exergy of the extracted steam
    - Calculation of the total exergy streams
    - Energy-Exergy balance sheet
    - A rough calculation of the total cost
    - The cost allocation to the different cogeneration products

Case Study 11.10: Cost allocation based on exergy balance

The CCGT heat and power plant– see flow diagram below– is equipped with a backpressure steam turbine with 6 bar steam discharge pressure and a steam extraction at 12 bars. The extracted steam is supplied into industrial processes. Furthermore, 3 bar steam is generated in the low pressure part of the HRSG. This steam cannot be used for power generation in the steam turbine because its pressure is lower than the backpressure; therefore it is directly supplied to a heat consumer. The task is to calculate and allocate the costs to the cogeneration products. The equivalent operating hours and the required performance parameters are stated in the tables.

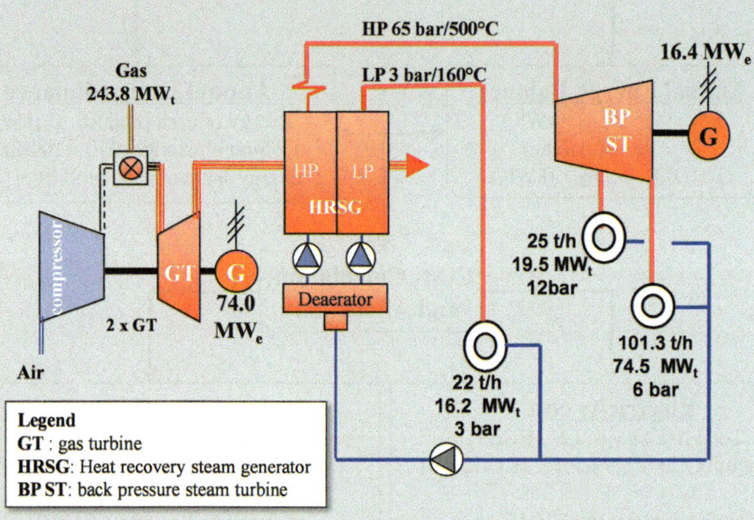

Note: The calculation is quite complex therefore interested readers are advised to follow the calculation steps in the case study, section 11.10.

The following two tables depict the energy- exergy balance and the cost allocation spreadsheet only.

Table 7-2: Case Study – Cost allocation based on exergy balance

Energy - Exergy Balance						
	Output			Energy		
Item	Output MW	Exergy MW	Exergy share %	Annual production MWh /a	Exergy MWh /a	Exergy share %
Electricity, net	88.0	88.0	82.3%	653,673	653,673	83.2%
Steam 12 bar	19.1	3.9	3.7%	95,410	19,598	2.5%
Steam 6 bar	73.8	12.8	11.9%	553,517	95,635	12.2%
Steam 3 bar	15.9	2.2	2.1%	118,898	16,733	2.1%
Total	n.a.	**106.9**	**100.0%**	**1,421,499**	**785,639**	**100.0%**

Costs Allocation					
Item	Capacity costs *)		Energy costs **)		Composite cost
	Fixed costs th. € / a	Specific Capacity Cost € / (kW*a)	Variable Costs th. € / a	Specific Energy Cost € / MWh	€/ MWh
Electricity, net	13,038	143.7	19,936	29.58	48.93
Steam 12 bar	581	30.4	598	6.26	12.35
Steam 6 bar	1,889	25.6	2,917	5.27	8.68
Steam 3 bar	331	20.8	510	4.29	7.07
Total *)**	**15,838**	n.a.	**23,961**	n.a.	n.a.

*) Allocation based on the Exergy-power share

**) Allocation based on the exergy-energy share

7.7 The Calorific Method

The allocation is done based on the energy content of the cogeneration products in relation to the total energy output of all products. The total production costs CU/a are allocated to the different cogeneration products (electricity and heat) based on their energy share (%) to the total energy (100%). The thermodynamic grade of energy is hereby ignored. This means 1 kWh electricity is equal with 1 kWh of thermal energy and the specific costs in CU/MWh are equal for all cogeneration products. The application is conducted in similar way as that of the exergy method.

Pros and Cons

The method does not make sense neither from the point of view of thermodynamics neither in terms of the costs.

However, it can be justifiable for regions where excess hydropower resources are available and still heat generation in HOB plants is required.

The method is demonstrated in the Example 7-9.

Example 7-9: Cost allocation based on the calorific method

Power- and Energy balance					
Item	Power		Full load hours	Energy	
	Output MW	Share %	h/a	Production MWh/a	share %
Power, total, net	87.7	44.1%	7,500	657,750	45.6%
12 bar stean extraction	19.5	9.8%	5,000	97,500	6.8%
6 bar stean, backpressure	75.5	38.0%	7,500	566,250	39.2%
3 bar steam from HRSG	16.2	8.1%	7,500	121,500	8.4%
Total	**198.9**	**100.0%**	-	**1,443,000**	**100.0%**

Costs Allocation					
item	Capacity costs *)		Energy costs **)		Composite
	Total T€ / a	specific € / (kW*a)	Total T€ / a	specific € / MWh	Costs € / MWh
Electricity	4,123	47.01	21,905	33.30	39.57
12 bar stean extraction	917	47.01	3,247	33.30	42.71
6 bar stean, backpressure	3,549	47.01	18,858	33.30	39.57
3 bar steam from HRSG	762	47.01	4,046	33.30	39.57
Total **)	**9,350**	-	**48,057**	-	-

*) allocation based on the power share
**) Allocation based on the energy share
***) Annual costs calculation in separate spreadsheet

7.8 Replacement Value

The Replacement Value Method is often used to assess the value or price of a fuel versus the price of a competitive fuel on the market place. Usually the fixed costs of the fuel, for which the replacement value is sought, are lower and there is some space for added value for the fuel costs. Actually, this is a price setting approach and not a costs allocation method.

The method is commonly applied for the calculation of the replacement value of natural gas for gas purchase agreements [51]. Its main competitor on the power sector in Europe is hard coal in the base load and intermediate load segment. In countries with crude oil resources the competitor may be HFO.

The basic approach is depicted in the following figure. The philosophy behind is that the electricity generation cost CU/MWh shall be the same for both options. In general, the fixed costs of the coal fired PP are higher compared to those for naturel gas. So the difference "Δ" can be balanced with higher fuel cost for the gas option. The difference "Δ" is the so-called gas *premium* which remains constant also in the case of a price increase of hard coal.

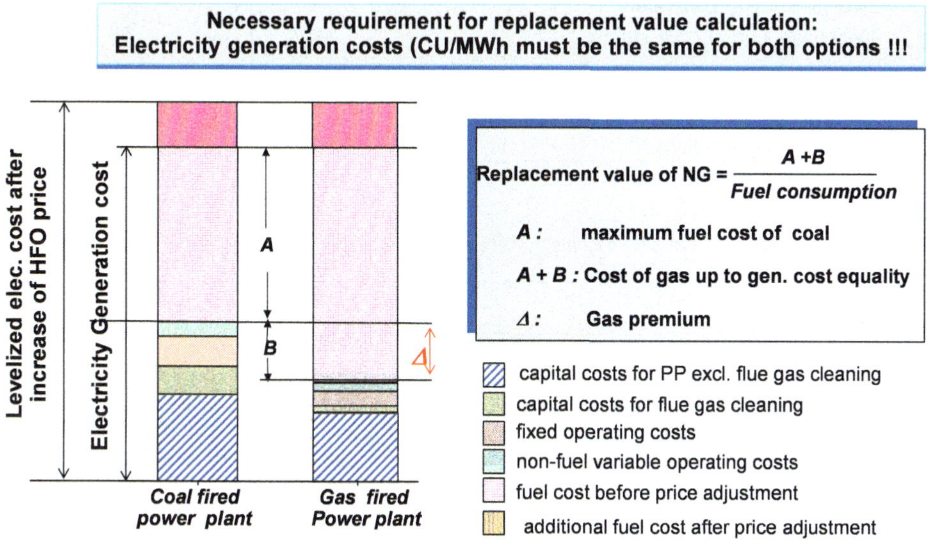

Figure 7-5: Replacement value calculation principle

In context with the costs allocation the method can be used for price setting for cogeneration products after their production cost have been determined with the one of the costs allocation methods.

Example 7-10: Replacement value of heat for process heat supply

A paper factory needs 12 bar process steam that is currently generated in a heat only boiler (HOB) plant within the factory. The base load steam could be supplied from nearby (CHP) cogeneration plant which is operated by the local utility. The existing boiler plant must remain in operation for peak load and backup supply during planned or forced outages of the cogeneration plant.

A steam/condensate pipeline and substations for heat delivery are to be installed. The factory would avoid fuel costs only because the existing boiler plant is still required and cannot be shutdown. Some O&M costs may be saved but the amount is marginal and shall not be considered

The replacement value and the premium of the cogenerated heat vs. avoided fuel costs at the HOB are to be calculated.

Apparently, heat supply from the CHP plant is economically favorable and there is some margin for bargaining as it becomes evident from the calculation in the example.

Item			Unit	Cogen Plant	Status Quo HOB
Power & Energy Balance					
Heat base load, 12 bar steam *)			MW_t	25	
Annual heat consumption *)		7,000 h/a	MWh_t / a	175,000	
Fuel consumption HOBs		eta=85.0%	MWh_t / a	n.a	205,882
Electricity for pumping **)	0.8 MW	7,000 h/a	MWh_e / a	5,600	n.a
CAPEX, total			**mln €**	**8.0**	**0.0**
Pipeline (2 km, DN 300)			mln €	4.0	0.0
Heat substation at the factory			mln €	2.5	0.0
Heat substation at CHP plant			mln €	1.5	0.0
Fuel price for HOB			€ /MWh_t	n.a	25.0
Electricity generation cost, cogen plant, given ***)			€ /MWh_e	70.0	n.a
Heat generation cost CHP gate β		=0.25	€ /MWh_t	17.6	n.a.
Annual costs			th.. € /a	**4,213**	**5,147**
Annualized CAPEX	6 %/a	25 a	th. € /a	626	0
O&M cost	1.5 %/a	-	th. € /a	120	0
Heat generation cost			th. € /a	3,075	5,147
Electricity for condensate pumping			th. € /a	392	0
Spec. heat cost, free factory			**€ / MWh_t**	**24.07**	**29.41**
Replacement value			**€ / MWh_t**	**24.07**	
Cogen heat premium			**€ / MWh_t**	**5.34**	

*) base load heat only **) for condensate return

***) rqrd for heat cost calculation

8 Transmission and Distribution of Power

8.1 Basic Electricity

8.1.1 The nature of electricity

Matter is anything that has a mass and occupies space [52]. The smallest constituents of matter are the atoms. The atoms again are composed by smaller particles: the electrons, protons and neutrons in various combinations. The electrons are the fundamental negative charge of electricity. Electrons revolve around the nucleus of the atom in several concentric shells or orbits. The protons are contained in the nucleus of the atom and are the fundamental positive charge.

In a stable state the negative and the positive charges are in equilibrium. If external energy is applied to a certain material, its electrons gain energy and become unstable and the atom is in excited state. Some of the electrons in the outer shell will become free and leave the atom. The movement of free electrons in a metal conductor generates electric current.

Electric current is the movement or flow of electrons in a conductor. In order to generate current, the electrons must be moved by a potential difference at the ends of the conductor, called the voltage. There are two types of electric current: When electrons' flow in a conductor occurs in one direction only the current is called *direct current – DC.* With *alternating current AC*, the direction of the flow of the electrons is changing back and forth in regular time intervals, *the cycles.* The number of cycles per second is the *frequency* (Unit Hertz, Hz =Cycle/second).

8.1.2 Direct current

The movement of the electrons from one end of the conductor to the other is called electric circuit. A simple direct current DC circuit includes a power source, e.g., a battery, the conductor wires and a load (Resistance), e.g. an electric heater and a switch. Table 8-1 shows the key parameters of electric circuit.

P. Konstantin and M. Konstantin, *The Power Supply Industry*,
https://doi.org/10.1007/978-3-319-72305-1_8

Table 8-1: Key parameters of electric circuits

Physical parameter		Unit	
Name	**Symbol**	**Name**	**Symbol**
Voltage	U	Volt	V
Current	I	Ampere	A
Resistance	R	Ohm	Ω
Power	P = U×I	Watt	W

For a better understanding of physical phenomena analogies with the mechanical equivalent are often helpful as it is demonstrated in Figure 8-1 between a closed water circuit and an electrical circuit.

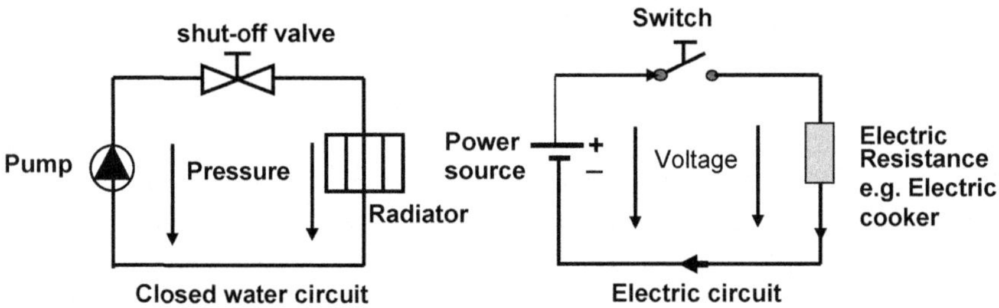

Figure 8-1: Water and electrical circuits' analogies

Note, however, that for illustrations in electrical circuits the flow of the current is indicated from the plus pole to the minus pole although, in reality the electrons' flow occurs from minus to plus polarity.

8.1.3 Basic laws of electrical circuits

8.1.3.1 *The Ohm's Law*

According to the *Ohm's Law* the basic parameters of an electric circuit namely Voltage U, Resistance R and current I stand in a fixed relationship to one another. Based on this other related laws are stated below:

The Ohm's Law: $$U = R \cdot I \quad [\mathrm{V}] \tag{8.1}$$

Electrical power: $$P = U \cdot I = R \cdot I^2 \quad [\mathrm{W}] \tag{8.2}$$

Resistances in series: $$R_T = R_1 + R_2 + \ldots + R_n = \sum_{i=1}^{n} R_i \quad [\Omega] \tag{8.3}$$

Resistances in parallel: $$1/R_t = 1/R_1 + 1/R_2 + \ldots + 1/R_n = \sum_{i=1}^{n} \frac{1}{R_i} \tag{8.4}$$

Example 8-1: Ohmic loads in series and in parallel

Two Ohmic loads of 200 Ω and 250 Ω are connected in a circuit with a voltage of 230V, first in series then in parallel. Find the total power used in both cases:

Item	Unit	Series	Parallel
Given:			
Voltage	V	230	
Load 1	Ω	200	
Load 2	Ω	250	
Results:			
Total resistance	Ω	450	111
Current	A	0.51	2.07
Power	W	118	476

Circuits with loads connected in parallel consume a multiple of power compared to circuits with loads connected in series.

8.1.3.2 *The Kirchhoff's Laws*

The two Kirchhoff's Laws are the "voltage law" and the "current law" also called "node law".

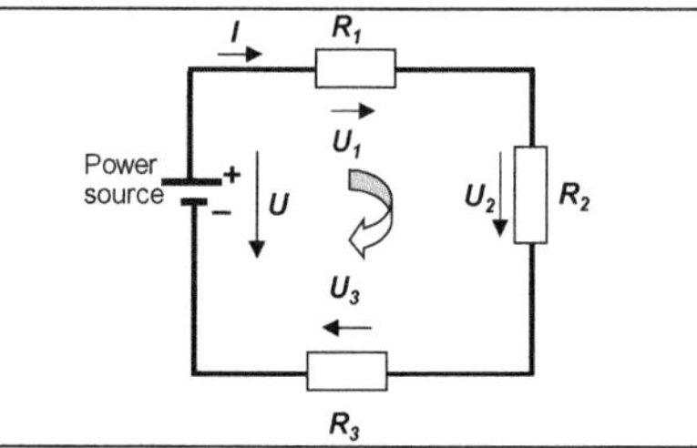

The *Kirchhoff's voltage law* states that the sum of the voltages in a closed circuit is zero.

$$U - U_1 - U_2 - U_3 = 0$$

$$U - I_1 R_1 - I_2 R_2 - I_3 R_3 = 0$$

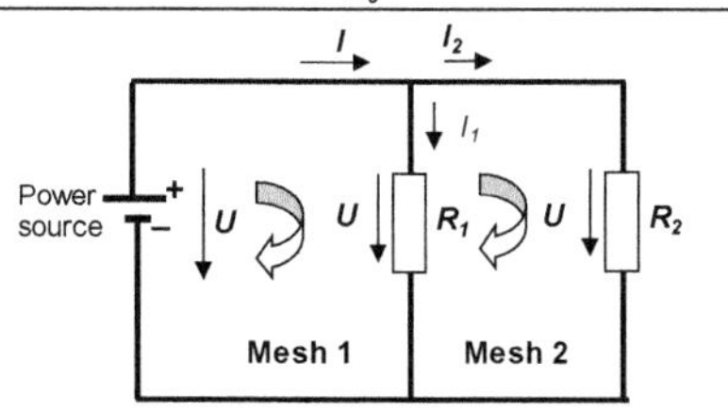

The *Kirchhoff's current law* states that, the sum of the currents at any *node* (point) of the circuit is zero.

$$I - I_1 - I_2 = 0$$

$$I - U / R_1 - U / R_2 = 0$$

Figure 8-2: Illustration of Kirchhoff's Laws

8.1.4 Alternating current

Alternating current (AC) is a form of electric power that changes continuously and periodically its voltage, current values and its polarity. For most technical applications alternating current in sinusoidal form is used whose instantaneous values for voltage and current have the form of a sinus curve. AC is produced by a generator, called also alternator, using the principle of the electromagnetic induction. When a conductor's loop rotates through a magnetic field and cutes magnetic flux lines a voltage is induced across its terminals. This physical phenomenon is called electromagnetic *induction* and is used in alternators to generate AC power.

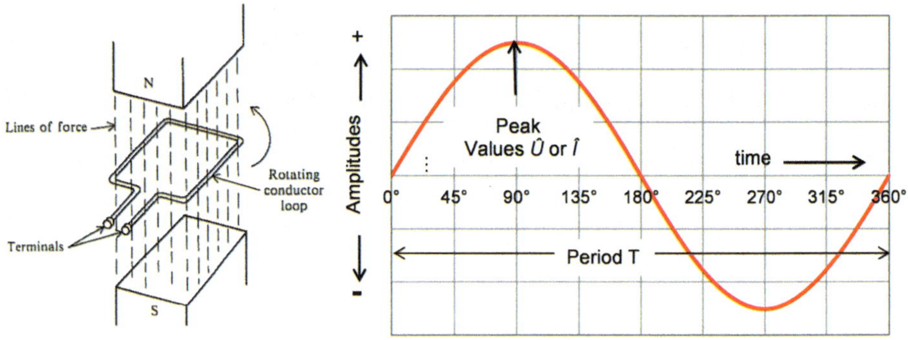

Figure 8-3: AC current in sinus wave form, one cycle

The values of voltage and current steadily change; for calculations the *effective* values, called ***root mean square*** (*rms)* values, are used. The relationship between the peak values and the rms values are:

$$U = U_{rms} = \frac{U_M}{\sqrt{2}} = 0.707 \cdot U_M \tag{8.5}$$

$$I = I_{rms} = \frac{I_M}{\sqrt{2}} = 0.707 \cdot I_M \tag{8.6}$$

$$P = U \cdot I = U_{rms} \cdot I_{rms} \tag{8.7}$$

Where U_M and I_M (M for maximum) are the amplitudes of voltage and current. Instead, $\hat{U}$ or $\hat{I}$ are often used for the amplitude values.

The *effective* or *rms* values correspond to the same direct current values that produce the same physical effect. For instance, an AC *rms* voltage of 230 V will cause a filament lamp to shine as intensive as a 230 V voltage of direct current.

Commonly the letters *U* and *I* without any indication of the subscript *rms* denote *rms* values of AC ($U=U_{rms}$ *and* $I=I_{rms}$). Lower case letters such as *u*, *i* or *p* are used to indicate instantaneous values of AC current.

The amount of time to complete a sinus wave is called *period*; it is indicated in seconds with the symbol *T*. A complete sinus wave is a *cycle*. The number of cycles per second is called *frequency*; it is denoted with the symbol *f* and is given in Hz (Hertz). Frequency and period are reciprocal values.

$$T = \frac{1}{f} \; [\mathrm{sec}] \qquad f = \frac{1}{T} \; [\mathrm{Hz}] \tag{8.8}$$

In Europe the frequency of the power of the public grids is 50 Hz. In USA and Saudi Arabia and some other countries the frequency is 60 Hz.

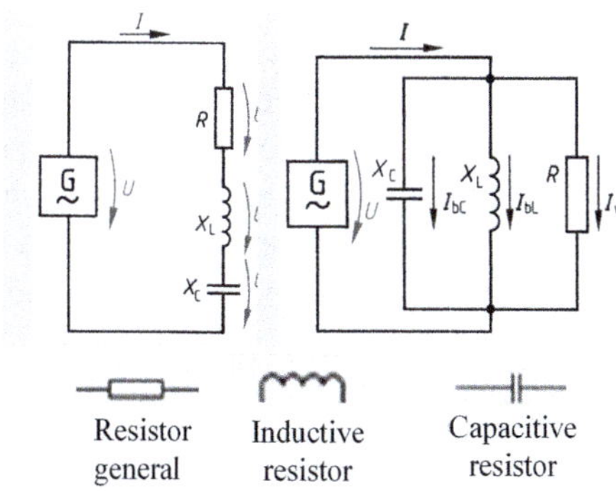

Besides the ohmic resistance AC circuits may include also inductive resistors (inductance) and capacitive resistors (capacitance). Inductive resistance is caused by coils, e.g., in electric motors. Capacitive resistance is caused by condensers. Also the wires of power lines cause these types of resistances.

Figure 8-4: Examples of AC circuits

Figure 8-5 depicts the voltage *u*, current *i* and power *p* curves of AC circuits. In an AC circuit with ohmic resistance only, voltage and current are in phase. As both instantaneous values are either positive or negative their product, which is the power *p*, is always positive. A positive power means that there is an energy flow from the generator to the consumer. The frequency of the power is double as that of the voltage or current.

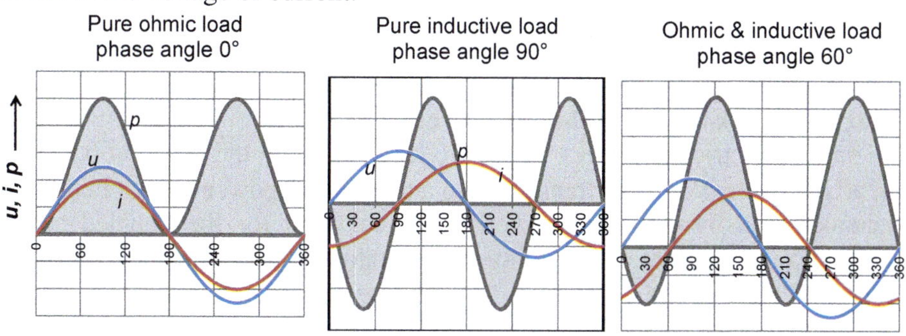

Figure 8-5: Voltage *u*, current *i* and power *p* diagrams in AC circuits

In circuits with pure inductance or capacitance there is a 90° phase difference between voltage and current. Their product results to periodic equal positive and negative power segments which neutralize each other so that the total power becomes zero (middle figure). The figure on the right shows a circuit with 60° phase difference between voltage and current. There are periodic segments with positive and negative power. Negative power segments imply an energy flow from the network to the generator, while positive power segments imply that energy flows from the network to the consumers. The difference between positive and negative power is the real power.

Due to the phase difference between voltage and current, power in AC circuits is produced in two different forms, active power P and reactive power Q. *Active or real power* are effectively the energy flow supplied to the load. Resistors, e.g., convert active power in heat or light. Inductors use *reactive power* for the formation of electro-magnetic fields, e.g., in electric motors and generators. Capacitors use reactive power to form electrostatic fields. Active and reactive power supplied to a load cannot be added together arithmetically, they must be added as vectors geometrically. *Apparent power S* is the geometric sum of active and reactive power – Figure 8-6. The cosine of the phase difference angle φ is called *power factor.*

Following figure depicts the *power triangle* along with the formulas and their units. Similar triangles can be drawn also for voltage, current and resistances.

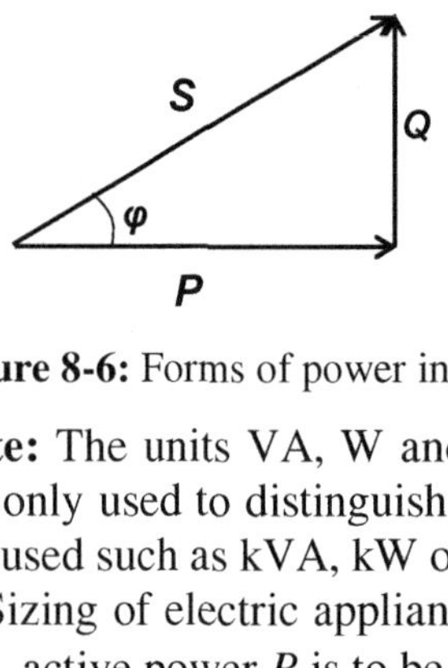

$$S = U \cdot I = \sqrt{P^2 + Q^2} \quad [\mathrm{VA}] \tag{8.9}$$

$$P = S \cdot \cos\varphi = U \cdot I \cdot \cos\varphi \quad [\mathrm{W}] \tag{8.10}$$

$$Q = S \cdot \sin\varphi = U \cdot I \cdot \sin\varphi \quad [\mathrm{Var}] \tag{8.11}$$

$$\text{Power factor}: \cos\varphi = \frac{P}{S} \tag{8.12}$$

Figure 8-6: Forms of power in AC circuits

Note: The units VA, W and Var are numerically equal. The different notations are only used to distinguish the power form. In practice, multiples of these units are used such as kVA, kW or MW and kVar.

Sizing of electric appliances and networks is based on the apparent power. If, e.g., active power P is to be transferred to a load with a power factor of $\cos\varphi=0.5$ generators, transformers and network must be sized for the doubled current $I = P / 0.5$. Therefore the capacity of generators, transformers and motors is given in kVA along with the *cos*φ.

Reactive power compensation: Utilities usually require consumers to maintain a power factor of $\cos\varphi=0.8$ or higher ($0.8<\cos\varphi\leq 1$). Power factor 0.8 corresponds to a reactive-active power ratio of Q/P of 75% – see equation(8.13). If this ratio is exceeded, a *reactive power compensation* system must be installed. This is done by parallel arrangement of suitable capacitors (compensators) to inductive loads of AC circuits, see Figure 8-4. The Q/P ratio can be calculated with the following formula.

Q/P ratio:

$$\frac{Q}{P} = \sqrt{\frac{1}{\cos^2 \varphi} - 1} \qquad (8.13)$$

By analogy to the power there is also active and reactive electrical energy. In power purchase agreements for industrial consumers there is a price for both energies in \$/kWh and \$/kVarh, and they are both metered. While the full amount of kWh active electricity consumed must be paid, the kVarh reactive power is usually delivered free of charge if it does not exceed 50% of the consumed active power. A higher consumption must be paid.

8.1.5 Three-phase alternating current

Three-phase alternating current is a combination of three single phase sinusoid currents with the same frequency, amplitude and rms-values, each of which is out of phase with the other by ⅓period or 120°. It is generated by an alternator, the main parts of which are the stator and the rotor. Three wire windings are arranged in 120° apart to each other around the circumference of the stator. The rotating magnetic field, applied on the rotor, induces alternating current in the wire windings of the stator. The rotor is driven, e.g., by a turbine or an engine.

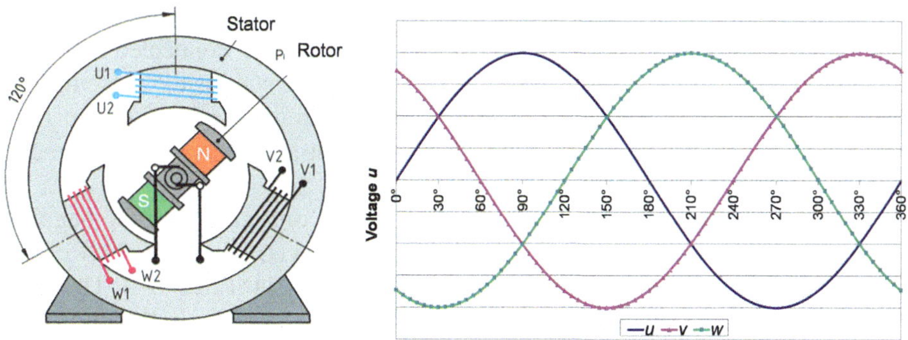

Courtesy: www.europa - lehrmittel.de

Author's own illustration

Figure 8-7: Three phase alternating current system

Either ends of the wire windings of the rotor are denoted with *U1* and *U2*, V1 and *V2* and *W1* and *W2*.

The sum of the instantaneous values of the voltages and currents of a three-phase alternating current system is zero.

Figure 8-8 depicts a simple 3-phase 400/230 V alternating current system including generator, wires and connected loads. The current flows to the loads via the three conductors (wires) L_1, L_2, L_3, called the *phases*. The neutral conductor *N* returns current flows to the alternator. The phase-to-phase voltages are U=400V, *the* phase-to-N voltages are 230V ($400/\sqrt{3}$). The string current is $I_{str}=I/\sqrt{3}$ (relations see Table 8-2).

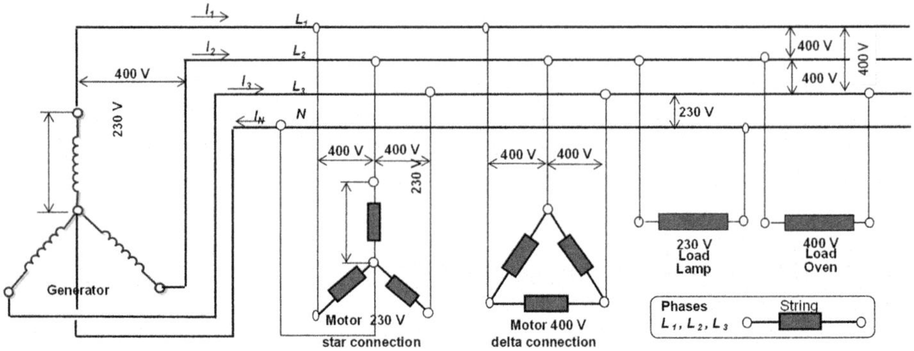

Figure 8-8: Three-phase alternating current system 400/230 V

Lamps are connected to one phase and to *N*-wire. Ovens may be connected between two phases. The *N*-conductor is currentless if the loads of the phases are equal. If they are different (asymmetrical), the *N*-conductor leads current back to the source and balances the asymmetrical loads.

Motors can be connected either in Star (called also Y-) or in Delta connection. Their wire windings are called *strings.*

In a *star* arrangement, the ends of the three strings are connected together at the neutral point. If the strings have the same load, the sum of the currents at the neutral point is zero. The *N*-conductor is then currentless. The voltage of the strings is U_{str}=230 V (or $U/\sqrt{3}$), while phase current and string current are equal ($I=I_{str}$).

In a *delta arrangement*, the end of one string is connected with the beginning of the next. Phase voltage and string voltage are equal ($U=U_{str}=400$ V). Relations between voltage, current and power are summarized in Table 8-2. The power consumption of a load, connected to a 3-phase AC system, is the sum of the power consumption of its strings.

Table 8-2: Voltage, current and power relations of 3-phase AC

	Symbol	Star connection	Delta connection
Voltage			
Phase-to-phase	U	$U=\sqrt{3}\cdot U_{str}$	$U = U_{str}$
String-to-string	U_{str}	$U_{str} = U/\sqrt{3}$	$U_{str} = U$
Current	I	$I = I_{str}$ $I_{str}= I$	$I = \sqrt{3}\cdot I_{str}$ $I_{str}=I/\sqrt{3}$
Power (symmetrical loads)			
Apparent	S	$S = 3\cdot P_{str} = 3\cdot U_{str}\cdot I_{str}$	
		$S=3\cdot U/\sqrt{3}\cdot I=\sqrt{3}\cdot U\cdot I$	$S = 3\cdot U\cdot I/\sqrt{3} = \sqrt{3}.U\cdot I$
Active	P	$P = S\cdot cos\varphi = \sqrt{3}\cdot U\cdot I\cdot cos\varphi$	

The power of a star and delta connections is calculated with the same formula. However, if the same voltage is applied to both, their power consumption will be different, due to the different current values, as shown in the example below.

Example 8-2: Power of 3-phase loads

Two 3-phase loads are connected to a 400 V net in star and delta connection respectively. Calculate the Power input to each load.

Item	Symbol	Unit	Star	Delta
String resistance	R	Ohm	100	
Phase-to-phase voltage	U	V	400	
String voltage	U_{str}	V	230	400
String current	$I_{str} = U_{str}/R$	A	2.30	4.00
Phase current	I	-	$I = I_{str}$	$I = \sqrt{3}\cdot I_{str}$
	I	A	2.30	6.93
Power factor	$\cos\varphi$		1.00	
Power	$P=\sqrt{3}\cdot U\cdot I\cdot cos\varphi$	W	1,593	4,800
Ratio	-	-	1.0	3.0

The power in delta connection is 3-times that of in star connection.

8.2 The Architecture of the Power Transport System

8.2.1 Background

When the first power systems were built at the beginning of industrialization end of the 19th century, a single power plant provided power into a single small network at voltage levels of maximum 10 kV. In the case of power failure, the supply was interrupted. There are still such small scale networks in some rural areas.

One of the great achievements of the power supply business has been the evolution to large High Voltage Alternating Current (HVAC) power grids, in which different interconnected networks maintain the same synchronous frequency.

A *power grid* is composed of several *interconnected networks* for the electrification of consumer centers. They are often referred to as *grid-connected* or *on-grid* networks. The grid commonly includes networks of several voltage levels for power transmission and distribution of power. Central power plants of different types supply power into the *grid.* In the case of failure of one power plant, the supply shortfall is balanced by increasing the output of the operating power plants and/or by putting in operation peaking or reserve power plants. Furthermore, through a meaningful dispatch of power plants upon their *merit order*[25], the overall electricity costs of the system can be optimized.

Small scale networks in remote rural areas which are not *grid-connected* are referred to as *isolated* networks or *off-grid* networks.

8.2.2 Voltage levels

Transport of electricity over long distances occurs at very high voltage. This is because the current flow through the lines causes a loss of electrical power due to the ohmic resistance of the wires; besides the cost implications, the lost power is transformed into heat, causing the wires to expand and sag; at higher temperatures the line may even be severely damaged.

This power loss is, according to the power law (8.15), proportional to the square of the current "I". By raising the voltage "U", the current is reduced proportionally according to the Ohm's law (8.14), and the power loss will drop to the square of the current.

Ohm's law $$U = R \cdot I \quad [V] \tag{8.14}$$

Power law $$P = U \cdot I = I^2 \cdot R \quad [W] \tag{8.15}$$

[25]) Merit order means dispatching of plants in a sequence of increasing marginal generation cost to meet increasing demand

The generators of large utility size power plants produce power at a voltage of 20 kV[26]. In order to reduce transmission losses, the generator's voltage is raised by the step-up transformer to the voltage level of transmission grid, e.g., to 380 kV (or higher). As shown in the following equations, derived from the Ohm's law and the power low, the current is hereby reduced to about 5% and the transmission power loss to about 0.28% of that of the 20 kV level.

$$I_{380} = \frac{U_{20}}{U_{380}} \times I_{20} = \frac{20}{380} \times I_{20} = 0.053 \times I_{20} \qquad (8.16)$$

$$P_{TL\_380} = \left(\frac{I_{380}}{I_{20}}\right)^2 \times P_{TL\_20} = (0,053)^2 \times P_{TL\_20} = 0.0028 \times P_{TL\_20} \qquad (8.17)$$

For this reason (three-phase) alternating current is applied for electricity transport in the grid because it can be easily transformed to higher or lower voltages. The grid consists of multiple networks with different operating voltages. The standard classification of voltages, according to ANSI [53] is:

Table 8-3: Typical voltage levels

Voltage Level	Voltage*) kV	USA kV	Europe kV
Ultra High Voltage - UHV	765 to 1100	-	-
Extra High Voltage - EHV	230 to < 765	345, 500, 765	230, 380
High Voltage - HV	69 to < 230	115, 138, 230	110
Medium Voltage - MV	0.6 to < 69	11, 15, 34.5	30, 20, 10
Low Voltage - LW	< 0.6	240 V , 120 V	230/400 V

*) ANSI Classification

The links between the voltage levels of the grid are the *substations*; they include the transformers which change higher voltages to lower voltages.

Figure 8-9 depicts the typical structure of the power system of Central European Countries, integrated into the European Network of Transmission System Operators for Electricity – Europe (ENTSO-E), formally UCTE[27].

[26]) The generators' voltage varies between 6 kV and 30 kV, depending on the power output

[27]) UCTE -Union for the Co-ordination of the Transmission of Electricity

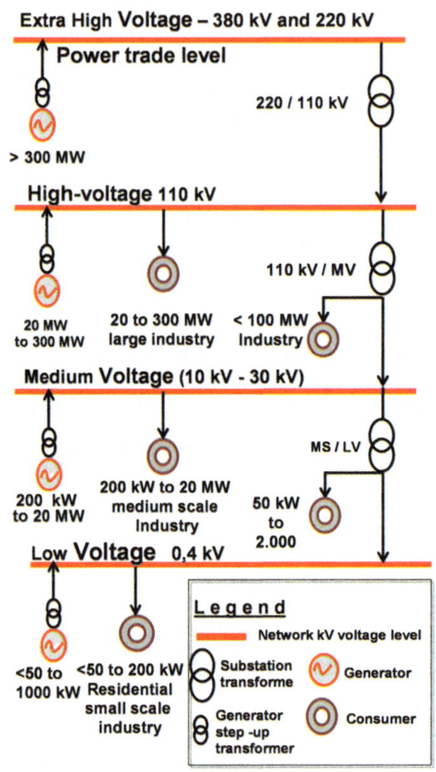

Generators of power plants produce electricity at voltage levels between 1 kV and 30 kV, depending on their size. Large, utility scale power plants feed their electricity into the power *transmission grid* which is operated at the extra high voltage levels of 380 kV and 220 kV. Cogeneration plants of district heating utilities, municipal utilities and industrial companies or power generators from renewable sources are connected to the HV, MV or even the LV voltage networks, depending on their size. Step-up transformers raise the voltage level to that of the corresponding grid.

The *power distribution* systems are operated at voltage levels of 110 kV to 0.4 kV. The links between the voltage levels are the *substations* which include the transformers that convert electricity to lower voltages.

Figure 8-9: Typical structure of the ENTSO-E power grid

8.3 Main Components of Electric Power Systems

Electric power systems are composed of many complex components, designed to operate together to safely provide quality and reliable electricity to the consumers. The main components are: Power generation plants, overhead lines, underground cables and substations including transformers, switch gears and several auxiliary, safety and protection systems.

8.3.1 Overhead lines

An overview of key technical parameters of overhead lines by voltage level is depicted in the table of Figure 8-10.

Voltage	kV	20	110	220	380
Number of power circuits	-	1 to 2	1 to 2	1 to 3	1 to 4
Approximate tower width	m	15	24	28	35
Approximate tower height	m	10	30	35	50
Capacity, per circuit, max.	MVA	16	140	490	1,700
Power loss per km *)	MW /km	0.09	0.22	0.34	0.68
Power loss in percent	-	0.54%	0.16%	0.07%	0.04%

*) Thermal limit at 80°C

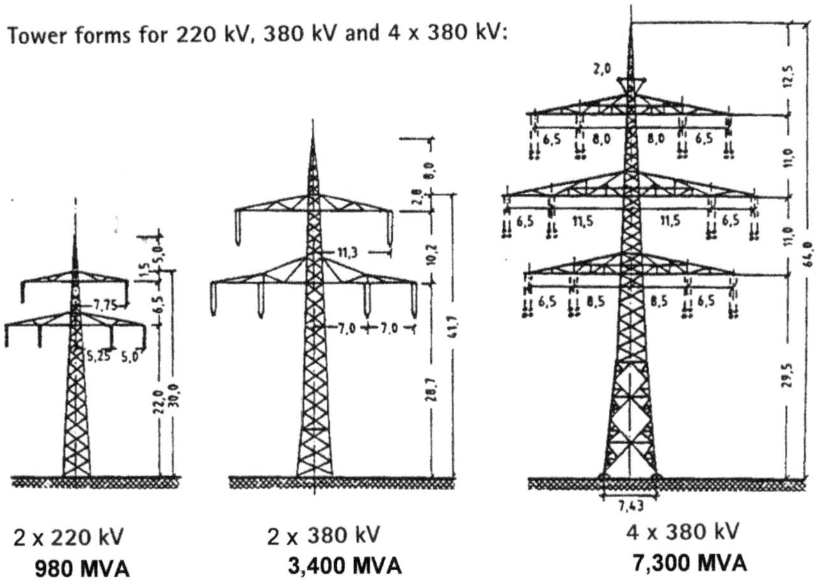

Figure 8-10: Overhead power lines, key technical parameters

Overhead lines are the most common system for power transport outside of residential areas. They are available for all voltage levels from 0.4 kV up to 1000 kV. They can suit to any terrain and are a relatively cost effective option. Overhead lines can include one or more systems (power circuits) with three wires per system and an earthling wire on the top of the tower. The wires are suspended from traverses of the high steel towers by strings of porcelain isolators. For distribution lines at lower voltages usually wooden poles are used instead of steel towers.

The transport capacity is usually increased by adding more circuits. A 380 kV line with 2 circuits (6 wires) will transmit 2×1700=3400 MVA with four circuits (12 wires) 4 ×1700 = 6,800 MVA. The power loss of a 100 km 380 kV line would be 100×0.68=68 MW or 4%.

The conductor material is commonly an aluminum alloy consisting of several strands reinforced with steel strands in the core. The conductors of high voltage lines are usually bundled up to 4 conductors per wire.

Please note, that overhead lines transmit AC current consisting of active and reactive power. Although the reactive power is a back-and-forth flow, it causes power and thermal losses in the lines. This is because losses of HVAC systems are higher compared to HVDC transmission systems which carry active power only.

Overhead lines are sensitive against adverse environmental conditions such as lightning strokes and ice loading. Due to their size and large space requirements (see Table 8-4 below), they have a negative impact on the landscape of natural scenery. Hence, the erection of new overhead lines finds very little acceptance by the population.

Table 8-4: Typical dimensions of 380 kV Line in m **[54]**

Tower height	50 - 65
Tower width	35
Minimal distance of wires from soil	8.5
Distance between wires	6.5
Length of isolator string	5
Distance between towers	400
Total route corridor *)	50 - 80

*) incl. protective strips, free from high growing trees, agricultural use allowable

The wires of overhead lines are not electrically insulated as the adequate isolation would be much too thick. Instead, air serves as the isolation medium. The distance between the wires must be sufficient in order to prevent short circuits or electric arc formation.

8.3.2 Underground power cables

Underground cables have several advantages versus overhead lines, such as: no negative impact on landscape, insensitive towards adverse weather conditions such as lightning strokes, snow load, storms, significantly lower hazard potential, and significantly reduced exposure to electro smog, better acceptance by the population. On the other hand the costs of high voltage cables are significantly higher, cost ranges of cables versus overhead lines are stated below:

- Low voltage 0.4 kV almost the same
- Medium voltage 10 to 20 kV 1.5 to 2 times higher
- High voltage 110 kV 3 to 8 times higher
- Extra high voltage 380 kV 20 to 30 times higher

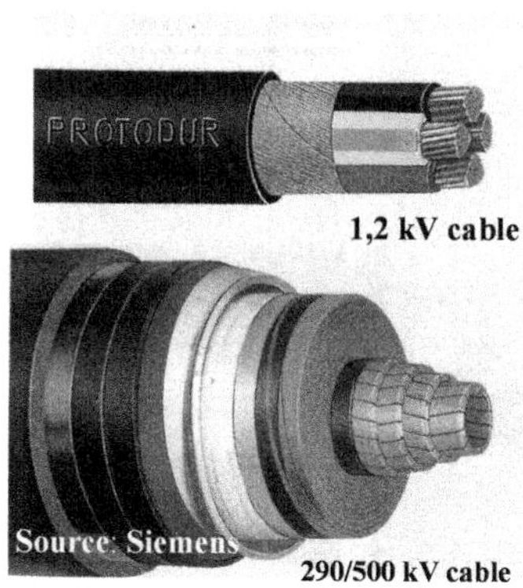

Figure 8-11: Power cables

Electrically isolated underground cables are used for power distribution and transmission inside of densely inhabited urban areas, in order to prevent safety hazards and increase acceptance by the population. They are deliverable for all voltages from 0.4 kV to 500 kV. Power cables consist of the copper conductor, electrical isolation, isolation shield and PE sheath. The XLPE-isolation of a 380 kV cable is about 3 cm thick.

8.3.3 Substations

Substations are the power link between the voltage levels of the grid. Their task is to convert and distribute the power according to the needs of the different consumer centers. Bulk supply substations receive electricity from the transmission system (220 kV and higher) and distribute it to the distribution system networks (e.g., in Europe 110 kV, 20 kV and 10 kV). Substations include transformers and switch gears.

8.3.3.1 *Transformers*

Transformers convert power into different voltages according to the requirements of the power system. With regard to the application they are classified as step-up and step-down transformers. *Step-up* transformers raise the input voltage of power plant generators to the voltage level of the power lines of the grid to which they are connected. *Step-down* transformers are installed in substations and reduce the voltage of the higher voltage power lines to lower voltages; they are referred to as *power transformers*.

Large transformers obtain efficiencies up to 99% in full load operation. In spite of low electrical loss rates, transformer operation causes significant heat losses due to the high amount of power that is being transformed; the generated heat must be dissipated by appropriate cooling systems. Most transformers are submerged into an oil bath, the oil is cooled down by natural or forced circulation of ambient air. For the type of cooling abbreviations are used such as: ONAN (Oil Natural Air Natural) and ONAF (Oil Natural Air Forced).

110 kV/20 kV/10 kV
40 MVA; ONAN Cooling

220 kV/115 kV/10kV
50 MVA; ODAF Cooling

400 kV/115 kV/30kV
75 MVA; ODAF Cooling

Figure 8-12: Power transformers

Distribution transformers, 50 to 2,500 kVA;
20 kV/10 kV/0.4 kV, Oil cooling, Source: ABB

Figure 8-13: Distribution transformers

Distribution transformers receive electricity from the medium voltage network (e.g.20 kV or 10 kV) and transform it to 380/230 V for distribution to the residential and small scale consumers. They are generally installed in prefabricated concrete containers in the streets.

Small transformers are equipped with radiator fins and are cooled down by ambient air.

8.3.3.2 *Switch gears*

Switch gears are typically installed in substations at the high and at the low voltage side of transformers. They include the several components which are necessary to control, protect and isolate electrical equipment and to ensure the reliability of power supply. Their key components are: busbars, circuit breakers, switchgears, break isolators, earthing switches, over voltages conductors.

Indoor switch gears are mainly used where limited space is available or for sites with adverse climatic conditions as near the sea, desert climates, industrial areas. Very effective systems are the gas insulated switch gears (GIS). They are composed from metal enclosed components where conductors and contacts are insulated by pressurized sulfur hexafluoride (SF_6) gas.

Siemens – 420 kV GIS Switch gear

Figure 8-14: Gas insulated indoor switch gear

A *busbar* is a conductor that serves as a common connection for two or more circuits. It is represented schematically by a straight line with a number of connections made to it. Standard bus bars are made of aluminum or copper.

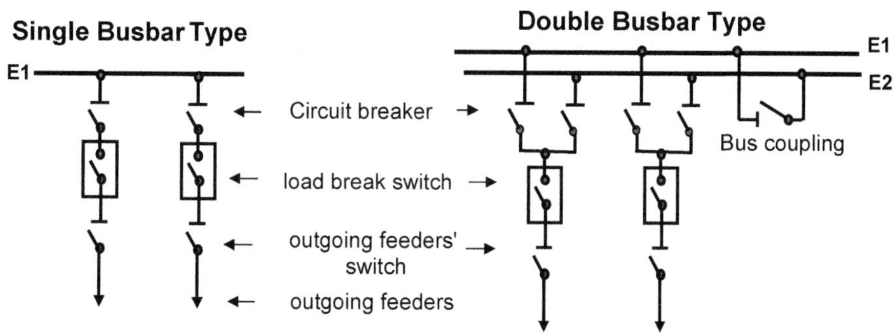

Figure 8-15: Busbar example

A double busbar provides a higher safety of supply. In case of failure of one busbar the supply is ensured by the other. Furthermore, repair work or maintenance can be carried out without operation interruption. Large or important plants may include up to four busbars.

A *circuit breaker* is a primary component of switch gears for their operation and safety. It can interrupt fault currents within 30 to 150 milliseconds even under short circuit conditions. For this purpose an activator serves for sufficient power. An electric arc is generating when the circuit breaker pulls apart the contacts and disconnects the circuit. This must be extinguished by appropriate agents. There are four major types: oil, SF_6 gas and vacuum.

A *load-break-switch* can switch the electrical circuit under nominal load but not under short circuit conditions.

An *isolator* is a non-load breaking switch. It is used to isolate equipment from high voltages e.g. for maintenance.

Earthing switches are isolators that put the systems potential into ground after the system is switched off to carry out tests or maintenance work.

Surge diverters protect electrical facilities from overvoltage, e.g., caused by lightning. They are installed at ingoing and outgoing overhead lines at substations.

Wave traps prevent high frequency to be transferred to the busbars.

8.4 High Voltage Direct Current Power Transmission

Besides the synchronous interconnection of power systems via high voltage alternating current (HVAC) linkages, the coupling of different power systems can be done by High Voltage Direct Current (HVDC) links – Figure 8-16.

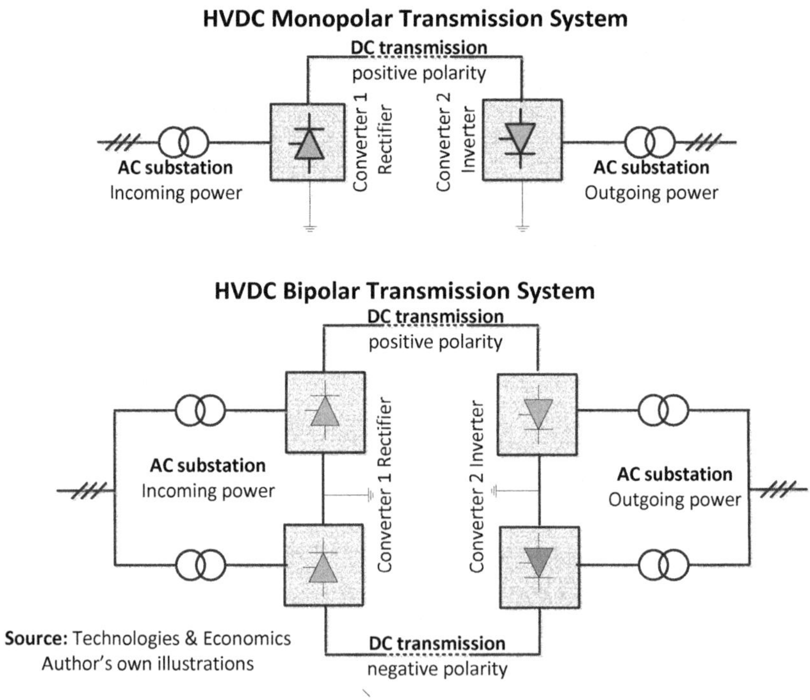

Figure 8-16: Schematics of HVDC transmission systems

The voltage is selected dependent on the transmission capacity. Large systems are operated at voltage levels up to 800 kV; higher voltage levels are technically feasible.
HVDC links consist of an AC substation and an AC/DC converter on each end of the HVDC transmission system. At the sending terminal the voltage of the AC power is raised to DC voltage level. The rectifier then converts the AC to DC. The DC-transmission network transfers the power to the receiving station where the inverter reconverts the DC to AC and the transformers step-down its voltage to that of AC receiving network.

In *monopolar* HVDC links one end of the converters is connected to the DC-transmission line and the other is grounded. The ground serves as the negative path. Monopolar links are often used for submarine power transfer over the sea.

In *bipolar* HVDC links there are two-pole converters, with one connected to the positive and the other to the negative polarity of the DC transmission lines. In case of failure of one of the poles the system can be kept in operation as monopolar link with the ground as return path.

With regard to the application there are three systems to be distinguished:

- *Back-to-back* links permit coupling of power systems which are either asynchronous or/and are operated in different frequencies (60 Hz and 50 Hz). Usually both terminals are installed at the same site or in short distance to each other.
- *Point-to-point* HVDC links are developed to transfer large power amounts over long distances
- Multi-terminal HVDC links, connecting more than two systems [55]

A further distinction with regard to the type of technology is shown in Table 8-5.

- LCC-technology (based on Line Commutated Converters)
- VSC-technology (based on Voltage Source Converters).

Line-commutated converters (*LCC*) are made with electronic switches that can only be turned on. *Voltage-sourced converters* are made with switching devices that can be turned both on and off. The LCC are used for back-to-back interconnections up to 500 kV and to point-to-point interconnections up to 800 kV. This is the classic technology and has been used for about 50 years so far.

Voltage-source converters (*VSC*) are used for connecting renewable energies, for bi-directional grids and multi-terminal networks. Some typical characteristics of the two systems are shown in the table below

Table 8-5: Characteristic features of technologies, source ABB **[56]**

Item	**LCC technology classic**	**VSC technology light® (ABB)**
Applications	Bulk power long distance transmission	Multiple areas of application
Relative size of system	4	1
Size example	1. 600 MW 200×120×22 m	550 MW, 120×50×11 m
Losses %	1.5 – 4,5	4 – 6
Multi-terminal operation	Limited, max. 3	Simple, unlimited
Typical layout	Outdoor	Indoor, except trafos
Scheduled maintenance	< 1% /a Invest	0,5 %/a Invest
Independent control of active and reactive power	No	Yes
Sea cable installation	Special ship, 3 available	Barge, <200 available

The advantages of HVDC versus HVAC transmission systems are: lower losses in the transmission lines as direct current has only ohmic resistance, no limitations in length as there is no reactive power consumption, considerably smaller footprint. The initial costs of converter stations are high.
The break-even distance for economic application of HVDC versus HVAC transmission are reported to be [56] [57]:

- Overhead lines longer than 700 km
- Submarine cables longer than 40 km

Overview of selected projects:
Point-to-point HCDC systems

- Itaipu 1&2 Brazil: 3 GW, 600 kV, 1984 and 3 GW, 1987, both about 800km each
- Rio Madeira, Brazil: 7 GW, 600 kV, 2400 km, 2013
- Cabora Basa Mozambique – South Africa, 2GW, 533 kV,1400 km
- The Baltic Cable, Sweden/ Germany, submarine HCDC 600MW, 450 kV, 250 km, 1994
- Coupling of asynchronous grids: ENTSO-E – Nordel

Back-to-back:

- Al Fadhili, Interconnector Saudi Arabia (60Hz) with 6 Gulf states (50 Hz), 3×600 MW, 222 kV, 2008

8.5 Load Structures and Performance Parameters

The load of the power grid (MW) varies in terms of time resulting from the power requirements of the different consumer groups and is to be covered by power plants, designed and operated for the respective load ranges. A precondition for an optimal dispatching of power plants is a reliable forecast of the expected load for the next operation period and of knowledge about the operational capabilities and electricity generation cost of the power plants.

Therefore, the load of the grid is continually metered and recorded by the transmission system operators (TSO), usually for every quarter of an hour. Based on these historical recordings, the dispatching of the power plants for the next period is planned.

The recordings are commonly described in load profiles with the y-axis indicating the load MW and the x-axis the time in hours. The area integral under the curves is the energy in MWh.

The load profiles, shown in Figure 8-17, are typical for countries with hot climates as, e.g., Saudi Arabia. In those countries the peak load occurs in summer caused by the high demand for air conditioning while winter is the off-peak period.

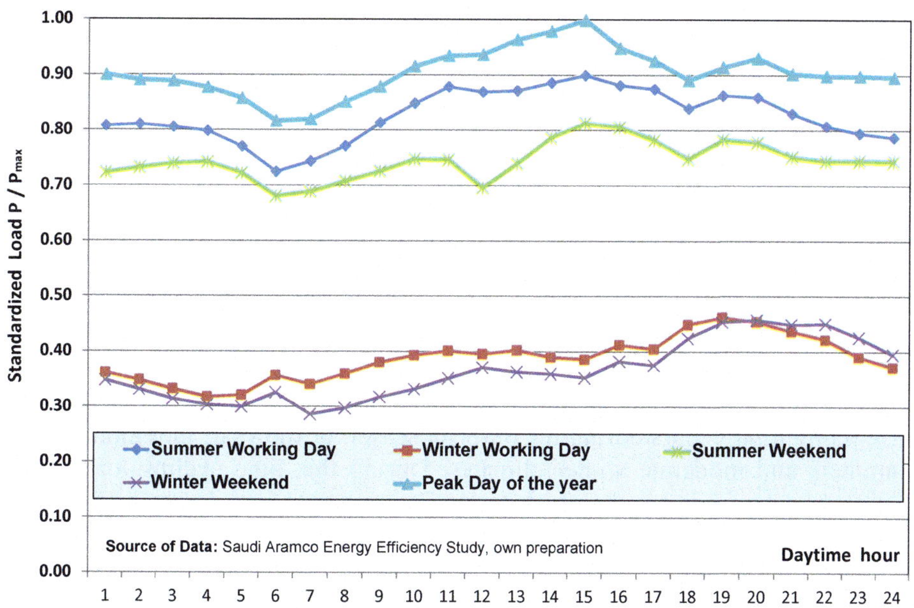

Figure 8-17: Daily load profiles, Saudi Arabia Central Operating Area

In contraries and in areas with cold winters the peak load occurs in the winter season as, e.g., in Central Europe – Figure 8-18.

In general the different loads can be classified as base load, cycling load, intermediate load and peak load.

A first assessment of the different loads can be derived with the help of *typical daily load profiles* as shown in Figure 8-17 and Figure 8-18 below:

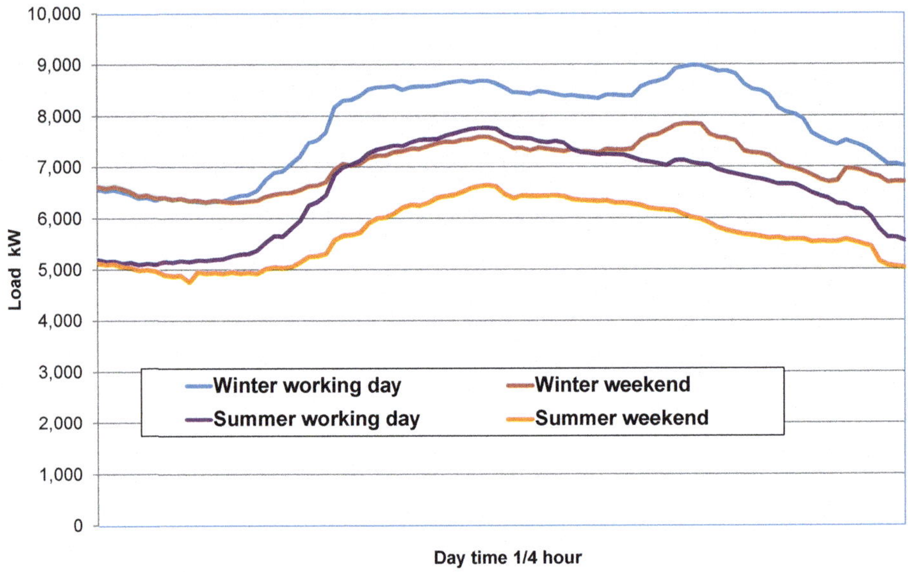

Figure 8-18: Daily load profiles, Austrian grid

Base load is the load that occurs throughout the year; in Figure 8-17 this is the winter load, in Figure 8-18 the summer weekend load. It is covered by base load power plants. These are run-of-river hydro power plants, nuclear power plants, lignite or new hard coal steam power plants. Base load power plants have the lowest marginal costs (= fuel costs) but they are inflexible with regard to load changes and are operated at almost constant load.

Cycling load occurs during the on-peak season of the year, in regions with hot summers and moderate winter climates. During this time cycling load remains almost constant and can be characterized as seasonal base load. In Figure 8-17 this is the load between winter load (off-peak load or base) and summer load (on-peak load).

Intermediate load occurs during the working time. In countries with industrial power demand the load of the grid is higher during the working time. So power plants have to start-up in the early morning at working days; they are out of operation but remain in hot state during the night and are shut-down during weekends.

Please note, that it is not always easy to distinguish between cycling and intermediate load. Most suitable and economic power plants for cycling and intermediate load are CCGT and oil fired steam power plants.

Peak load, from the point of view of power plant dispatching, is the intraday load change on top of cycling and intermediate load. *Intraday peak load* occurs during the whole year. It becomes better apparent in the unsorted annual load duration curve – Figure 8-19. In this curve the loads appear as they occur in real time.

Peak load power plants need quick start-up and load change capabilities. In countries with hydropower resources storage and pump storage, hydro power plants cover intraday peak loads, otherwise gas or diesel oil fired simple cycle gas turbines are to be run.

> Note, however, that the term peak load is also used for the maximum load of the year – $P_{max.}$– [kW or MW]. In order to distinguish the two terms from each other, we use the term *annual peak load* for the maximum load of the year and *intraday peak load* for the daily load deviations.

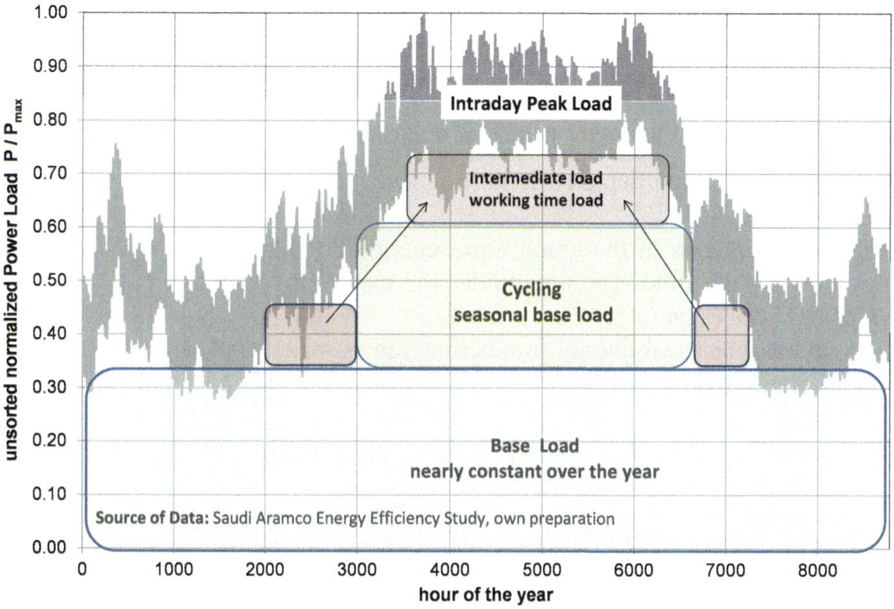

Figure 8-19: Annual load duration curve, normalized, unsorted

A better picture of the grid's load characteristics is provided by the *sorted annual load duration curve*, depicted in Figure 8-20. In this curve the load is sorted in descending order instead of real time order. So the different load ranges can be easier recognized in terms of load in MW as well as in terms of energy in MWh, which are the respective areas under the curve.

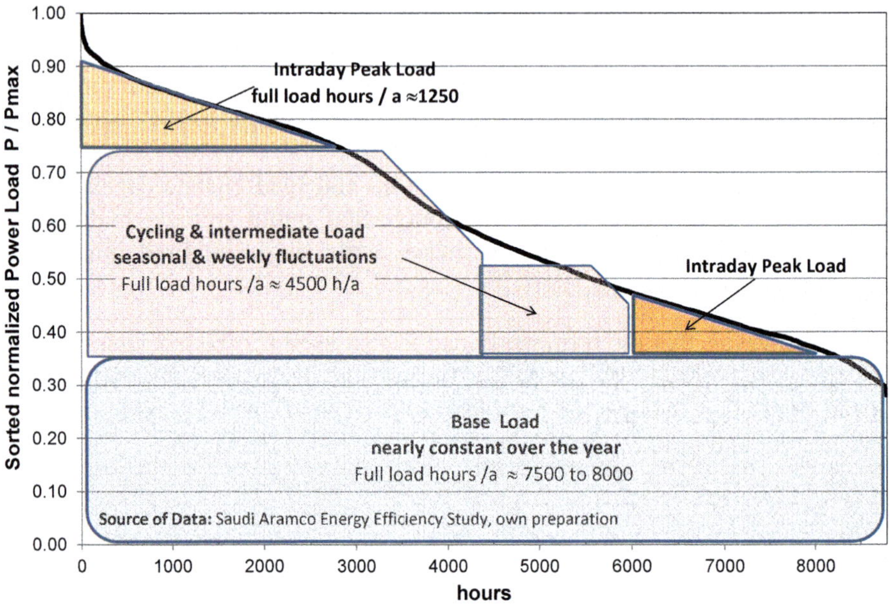

Figure 8-20: Sorted, normalized annual load duration curve

Load profiles are often depicted in normalized form (P/P_{max}) as shown in Figure 8-19 and Figure 8-20.

Usually, the y-axis of the graph represents the load in kW or MW and the x-axis stands for the time. The area under the curve corresponds to the energy in kWh or in MWh (area integral).

The annual electrical energy production can be calculated with the following formulas alternatively:

$$W_e = P_{\max}\left[\mathrm{MW}\right] \cdot t_{FLH}\left[\mathrm{h/a}\right] \quad \left[\mathrm{MWh/a}\right] \tag{8.18}$$

or

$$W_e = LF \cdot P_{\max}\left[\mathrm{MW}\right] \cdot t_{OPH}\left[\mathrm{h}\right] \quad \left[\mathrm{MWh}\right] \tag{8.19}$$

Where t_{FLH} denotes the equivalent full load hours and LF is the load factor during a reference operating time period. The former is used in Central Europe while LF is preferred in the English speaking countries. The relation between the two parameters is:

$$t_{FLH} = LF \times t_{OPH} \tag{8.20}$$

There is a distinction between operation hours t_{OPH} and full load operating hours t_{FLH}. The former denotes the reference time period (a year, month, week or day) for which the load is depicted, usually at varying load (P). Where t_{FLH} denotes the *equivalent* hours referred to the peak load (P_{max}) of the reference period. The latter is shorter. The amount of produced electricity (W_e) is in either cases the same.

Example 8-3: Load factor vs. full load hours
As shown in the table below, the load factor and the equivalent full load hours may change if the reference operating period is different. The load factor of an intraday peak may be high (e.g. 0.5) while during a year it is only about 0,143. In contrary the base load of one single day may become even 1.

Reference period	Hours	Base		Peak	
	t_{OPH}	LF	t_{FLH}	LF	t_{FLH}
Year	8,760	0.800	7,008	0.143	1,250
Month	720	0.850	612	0.150	108
Week	168	0.900	151	0.400	67
Day	24	1.000	24	0.500	12

In general the load factor provides better information about the utilization of the load than the full load hours if the reference operating time is shorter than a year.

Public power grids in developed countries have usually equivalent full load hours 5000 to 5500 h/a corresponding to load factors of 0.57 to 0.63 referred to the annual peak load (P_{max}). Representative equivalent hours for the different load ranges are: Base ≈7500 to 8000 h/a, cycling/intermediate ≈ 4500 to 5000 h/a, and peak ≈ 1250 to 1500 h/a.

8.6 Balancing Power Supply and Consumption in the Grid

8.6.1 Types of control power

The amount of electricity to be supplied to the consumer is not exactly known in advance; furthermore electricity, unlike other commodities, cannot easily be stored. However, a balance between generation and consumption is an indispensable precondition to maintain frequency and voltage stability in the grid. Hence, the amount of electricity fed into the grid must be at any time exactly the same as the amount that is extracted by the consumers. Deviations from the balance may

be caused by *incidents* such as changes of the load, e.g., by sudden connection or disconnection of large consumers or forced outage of power plants. Imbalances can also be caused by fluctuations of wind or solar power supply into the grid.
Transmission System Operators (TSO) are obliged to maintain the balance between supply and consumption in their control area by providing *control energy* into the system. A *control area* is part of an interconnected network for which a TSO is responsible. For operational and costs reasons a distinction is made between three types of controls for balancing supply and consumption namely: primary, secondary and tertiary control – Figure 8-21.

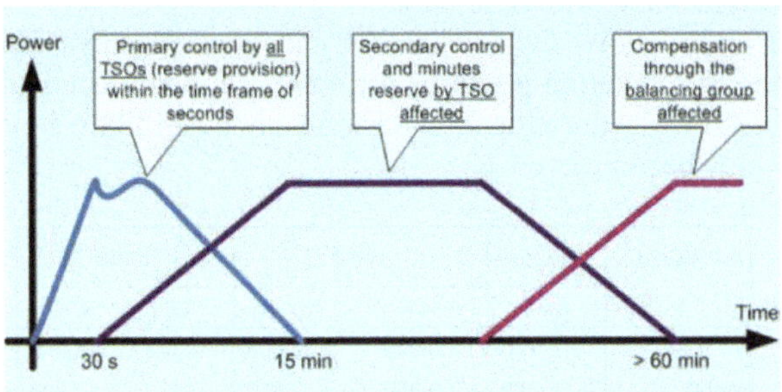

Figure 8-21: Time frame of control energy usage, Amprion GmbH [58]

The purpose of *primary control* is to maintain the frequency in the grid within an allowable range of ±0.2 Hz, referred to the target value of 50 Hz (valid only for ENTSO_E). It is mainly provided by base load power plants which are operated (spinning) at slightly reduced output (e.g., by throttling the steam flow to the turbine). They are capable to raise their output at full capacity within seconds. This reduced capacity rate is also-called *spinning reserve*. TSOs are obliged to acquire and maintain permanently the required spinning reserve for their control area. The spinning reserve is automatically activated by the TSO and must be made fully available within 30 seconds and is maintained up to maximum 15 minutes after occurrence of an incident causing a deviation of the frequency from the target value.

A *secondary control reserve* is needed if the system is or may be affected longer than 30 seconds after occurrence of an incident. It is used to replace the primary control reserve so it can be restored and maintained for frequency control. The amount of secondary control reserve must be capable to compensate a forced outage of the largest power plant unit in the control area of the TSO. It is automatically activated by the TSO. The time period per incident lasts 30 seconds up to 60 minutes. Secondary control reserve can be provided by pump storage power plants or other power plants within the control area which are operated in part load and can provide temporarily reserve capacity.

Tertiary control called also minutes' reserve is activated when the system is or may be affected longer than 15 minutes after occurrence of an incident. It is used to support and relieve the secondary control in the case of failure of the largest unit in the system. It can be provided by, e.g., gas turbine power plants which are maintained in hot stage and are capable for quick start-up. It is usually activated manually by the TSO after telephone notification to the suppliers of tertiary reserve for dispatch. It can be activated automatically or manually to replace the secondary control.

According to the ENTSO rules, the TSOs are responsible for the provision of control power for the first four quarters of an hour (60 minutes) after occurrence of an incident. For imbalances caused by deviations between forecast and actual consumption the *balancing group managers* are responsible (see also section 8.6.2).

Secondary and tertiary control power can be positive or negative. If the frequency decreases, there is a power deficit in the grid that must be compensated by input of *positive power reserve*. This means, reserve power capacities must be activated. If the frequency increases there is a surplus of power (e.g., caused by wind power) that must be compensated by a *negative power reserve*. Thus, operating power capacities must be shut down. Imbalances must be permanently compensated by activation and deactivation of appropriate power plant capacities.

Primary control reserve is always positive. The size of the primary control reserve corresponds to the default risk of the capacity loss within a grid. In general, the minimum primary reserve must be sufficient to compensate a forced outage of the largest unit within a system. Therefore grid size, number and capacity of the power plant units have to match each other. For small grids the primary control requirements may be large and costly if power plant units are too large compared to the size of the grid.

In contrary, the default risk of capacity loss in large interconnected networks is low. This is because of the ratio: size of the largest power unit and the size of the grid is small. In the synchronous interconnected grid of continental Europe (ENTSO-E RG CE) 29 TSOs are involved with a total installed capacity of about 700 GW. A total volume of required primary control reserve is only 3 GW (about 0.4%). All TSOs are obliged to make a contribution to the primary control reserve corresponding to their net annual electricity production in relation to the total production of the interconnected grid. The volume of primary control reserve that must be provided by the four German TSOs is approximately:

$$\text{Primary control reserve: } \Delta P_{PR} = 572/2{,}325 \times 3{,}000 = 738 \text{ MW}$$

The total net generating capacity in 2013 was about 186.6 GW [59] the primary reserve ratio is about 0.4% only.

8.6.2 Imbalance energy

In liberalized markets generators or electricity traders provide electricity to consumers within the control area of a TSO which are bundled in *balancing groups.* The managers of each balancing group have to provide load forecasts to the TSO for scheduling every quarter of an hour of the day ahead.
However, deviations from the expected consumption level always occur during the actual supply day. The *control area manager* of the TSO has to offset the deviations through the connection or disconnection of generation units (e.g., special backup power plants "minute reserve"). This energy is called *imbalance energy.*

There is a distinction between control power and imbalance energy. *Control power* (primary and secondary reserve) serves exclusively for frequency and voltage control in the control area. *Imbalance energy* serves to offset deviations between the forecast electricity consumption and actual consumption within a balancing group.

The costs of the control power are included in the use of system tariff. The costs of the imbalance energy are charged to the balance group that was responsible for the respective deviation of forecast and actual consumption.

8.6.3 Procurement of control and imbalance energy

All transmission system operators (TSOs)[28] of the synchronous interconnected grid of continental Europe (ENTSO-E RG CE) are obliged to maintain a permanent balance between power generation and load in their control area by keeping the required control power, and to provide balancing energy to the balancing groups (electricity producers and consumers). The close cooperation among the TSOs enables the overall amount of control power required to be minimized.

The TSOs procure their primary control, secondary control and minutes' reserve in an open, transparent and non-discriminatory control power market. Primary control and secondary control power are usually procured in a monthly cycle; minutes' reserve is daily called for tender. For tenders the shared IT-platforms have been installed by the TSOs. They have developed market-based solutions which also meet the requirements of a secure and stable system operation.

---

[28]) This section actually describes the control power market of Germany; however, the proceedings are, in general, the same for all TSOs involved in the (ENTSO-E RG CE) grid. The German grid includes four TSOs. The text is taken from the Website of the TSO Amprion and is slightly adjusted [58] and generalized.

Procurement is ensured through competitive bidding on a tender basis in the national control power markets where a large number of suppliers (generators as well as consumers) participate, especially for minutes' reserve. Via pooling also small suppliers can participate in the call for tenders.

Costs for the provision of primary and secondary control power and minutes' reserve are included in the use of system tariffs.

8.7 Economics of Power Transmission and Distribution

8.7.1 Tariff models for wheeling power in the grid

Several models are applied for formulation of power wheeling tariffs, known also as use of system charges in liberalized electricity markets. Three of the most important are briefly presented below. These are:

- The postage stamp model
- The contract path model
- The MW-miles model

The **postage stamp model** is the simplest, most transparent and most common method. It is best applicable in fully intermeshed networks. The central idea is, that in a meshed transport system with many generators and customers, it is not possible to define where the power comes from. Consequently, the consumers have to share the cost of the voltage levels of the grid, which are involved in their supply. The main characteristic of the model is that there is no distance component in the tariff; only the power demand and the voltage level of the consumers are relevant for the tariff and the wheeling cost.

In the postage stamp model the wheeling tariff is determined by dividing the annual cost "*AC*" of the relevant kV levels of the grid by the maximum power "P_{max}" during the billing period (year or month). The annual wheeling costs of individual consumers are determined by multiplying the wheeling tariff with the coincidental maximum power demand (peak demand) of the individual consumer *Pi*.

$$C_i = \frac{\sum AC}{P_{\max}} P_i \quad [\$/a] \tag{8.21}$$

Where:

C_i: The annual wheeling cost of the individual consumer ($/a)
ΣAC: The sum of the annual costs of the transmission grid level ($/a)
P_{max}: The annual maximum power of the grid level (kW)
P_i: The maximum coincident power demand of the individual consumer

The **contract path method** is based on the assumption that the power transfer is confined to flow along a specified, fictitious, electrically continuous path through the wheeling company's transmission system, agreed upon by transaction participants. Changes in power flows in facilities that are not within the specified path are ignored.

Correspondingly, the considered wheeling costs are limited to those facilities that are within the specified path. Contract path pricing may be selected to minimize transmission charges. However, it does not reflect actual power flows through the transmission grid, including loop and parallel path flows. The recovery of embedded capital costs is thus limited to those facilities, which lie along this assumed path.

The annual wheeling cost of an individual consumer "*i*" are defined as follows:

$$C_i = \left[\frac{\sum AC_{lines}}{P_{\max}} + \frac{\sum AC_{trafos}}{P_{\max}} \right] \times P_i \quad [\$/a] \tag{8.22}$$

Where:

C_i: The annual wheeling cost of the individual customer ($/a)

ΣAC_{lines} The sum of the annual cost of the lines of the path ($/a)

ΣAC_{trafos}: The annual cost for transformation ($/ya)

P_{max}: The annual maximum power of the grid in this path (kW)

P_i: The maximum coincident power of the individual consumer

The wheeling tariff includes also a distance component. The method is mainly applied in grids with large operational areas and less intermeshed networks.

The **MW-miles method** calculates annual wheeling costs by using changes in MW power flows, due to the power wheeling in all facilities of the wheeling company and the lengths of the transmission lines. Two power flow studies are carried out successively for every year with and without the wheeling power flow to the consumers in all transmission lines.

The changes in kW power flow of a transmission line ΔP_i, due to each transaction, is then multiplied by the line length in miles (or also km) giving the kW-miles of the line. Then they are added up over all lines in the grid to obtain a measure of how much each transaction uses the grid. Different transactions are then charged in proportion to their utilisation of the grid.

The load level can be either at peak load level or any appropriate load levels. Although change in load flow could be calculated across all hours of the year, most transmission charges are calculated at peak system load conditions. The cost of transmission per megawatt-mile is the total cost averaged over megawatt miles of usage.

The annual wheeling cost of an individual consumer "i" are calculated with the formula:

$$C_i = \frac{\sum AC}{\sum(\sum \Delta P_i) \times l_i} \times \sum \Delta P_i \times l_i \quad [\$/a] \tag{8.23}$$

Where:

C_i:	The annual wheeling cost of the individual consumer ($/a)
ΣAC:	The sum of the annual cost of the transmission grid ($/a)
$\Sigma(\Sigma \Delta P_I) \times l_i$:	The sum of the MW-miles of all consumers
$\Sigma \Delta P_I \times l_i$:	The MW-miles of the individual consumer
P_i:	The maximum coincident power of the individual consumer
l_i:	The line lengths in miles

The MW-miles-method is applied in some cases for very large individual consumers. The method is probably the most accurate one but it needs extremely time consuming and expensive load of flow calculations with special software tools.

8.7.2 Costing definitions and concepts

The most fundamental costing methods, applied in the context of use of network system charges, can be classified into three major groups:

- Embedded cost method
- Short-run marginal cost (SRMC) method
- Long-run marginal most (LRMC) method

The *embedded cost* method considers the cost of all existing facilities only. These costs are quite practical, readily available and consistent. Any costs for reinforcement of existing facilities or for new facilities are considered only if they are due during the validity period of the tariff agreement.

The *Short Run Marginal Cost Method* is defined as the marginal cost of supplying an additional unit of energy when the installed capacity of the system remains the same. Consequently, only the variable operation cost are considered, which can be directly attributed to customers for whom production increase is required.

The *Long-Run Marginal Cost (LRMC)* method is defined as the marginal cost of supplying an additional unit of energy when the installed capacity of the system is allowed to increase optimally in response to the incremental increase in demand. It incorporates both capital and operating costs which reflect the cost of system expansion. The LRMC provides a tariff today based on the present value of future investments required to support an incremental increase in demand at different locations in the system, based on peak scenarios of future demand and supply growth.

For the application of this method an extension plan for the whole system including demand forecast, investment for new facilities and projections of the O&M cost for a period of 10 to 20 years is required.

8.7.3 Main cost items

In liberalized markets, purchase of power is practised on a short term basis (e.g., for the day or week ahead) and also the validity period of power purchase agreements is short (one to three years). Tariffs therefore reflect the costs structures of the current system and not those of the longterm structures. For use of system charges the annual costs of the system are required. They are taken from the Profit and Loss statements of the companies of the most recent year.
If actual data are not available, first an estimation of the assets, based on the specific cost for lines and transformers for actual investments and from budget offers, may be used.

In general, the following items are considered in the calculation of the annual costs of the transmission and distribution networks:

- Depreciation of assets
- Interest for borrowed capital
- O&M costs
- Personnel cost
- Cost for losses
- Cost for ancillary services
- Other costs

There are different approaches from country to country how to calculate the annual costs. As transmission and distribution are natural monopolies, the use of system tariffs are regulated. The regulatory authorities of each country define the scheme of computation.

With regard to the grid operation costs, the following points are of particular importance:

Costs for losses: The electricity, wheeled through the grids and their voltage levels, is subject to respective losses in the lines and transformer stations. For the required energy and load to be delivered at each extraction point in the network, the respective downstream losses have to be taken into account. In order to balance these losses, the Transmission Company has to buy additional energy from generators. The costs of this additional energy are paid by the Transmission Company and included afterwards in the transmission system tariff in the form of energy charge ($/kWh).

Costs for ancillary services: Ancillary services for the grid are the following:

- Frequency control
- Voltage control
- Restoration of supply

For frequency control the transmission grid operator has to ensure that sufficient spinning and backup reserves are available to balance the power and stabilize the frequency in the case of forced outage of generation units or sudden changes of the load. The costs for ancillary services are included in the transmission system tariff.

8.8 Tariff Formulation for Wheeling Power in the Grid

8.8.1 Basic requirements

Use of system tariffs for wheeling power in the grid must meet the following requirements:

- cost-covering revenues for the supplier including a reasonable profit margin
- allocation of the costs to the consumers according to their cost responsibility
- non-discrimination of consumers or consumer groups
- provision of incentives to the consumers to save energy and reduce capacity requirements
- simple and understandable structures
- affordability for the consumers

The formulation of the use of system tariffs is needed first:

- to determine the costs of each voltage level of the grid, broken down in fixed and variable costs
- to assess the coincident peak load of all connected consumers to each voltage level of the grid

8.8.2 Roll-over costs to subordinate voltage levels

The basis of the tariff formulation, independent of the model used, is the allocation of the costs of the entire grid to the different voltage levels according to the cost responsibility of the connected consumer groups. The costs of higher voltage levels are thereby *rolled over* to the subordinate levels, as far as they are not attributable to users of the same voltage level of the system. The consumers at each voltage level share the costs of the grid at their voltage level with consumers of subordinate voltage level. The total costs are composed from the rolled over costs from the higher level plus the transformation costs from the higher voltage level and the costs of the respective voltage level.

The general approach is visualized in Figure 8-22: As no consumers are connected in EHV, all the costs are rolled over to the HV-level. To the voltage levels HV and MV the own consumers and the consumer of the lower voltage level are connected. In the LV-level only the own consumers are connected.

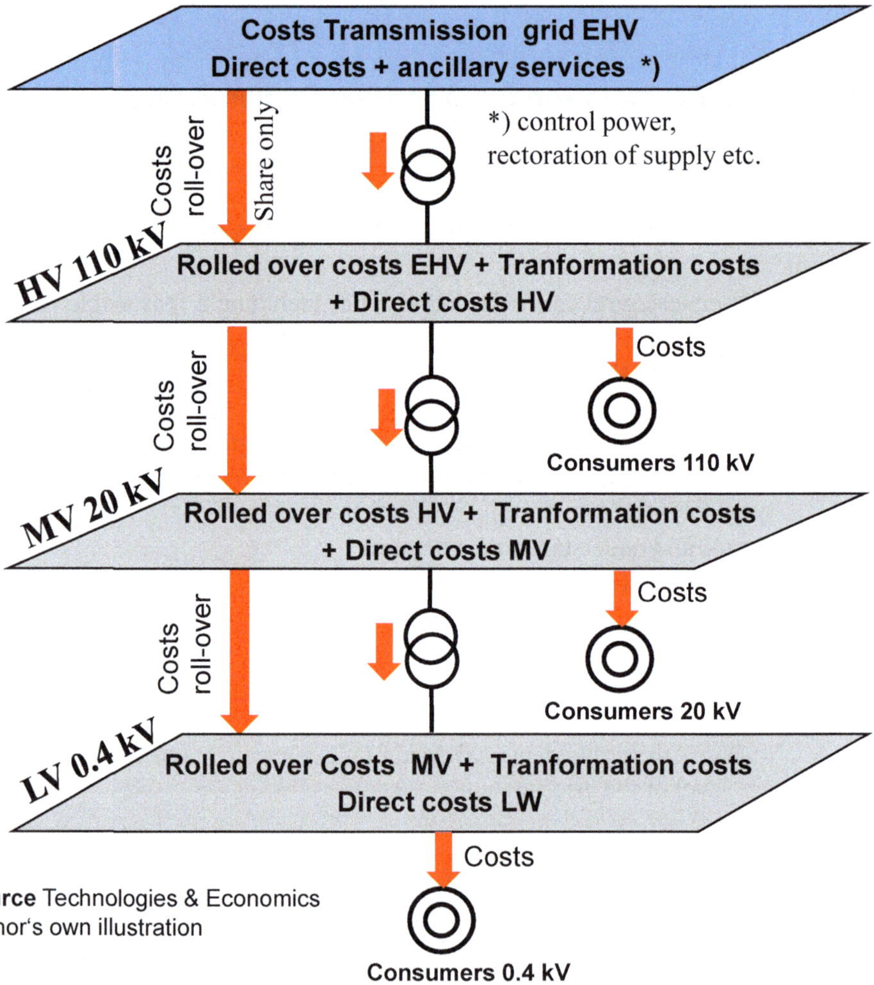

Figure 8-22: Costs' roll-over from higher to lower kV-levels

8.8.3 Coincidence functions of the consumers

For tariff formulation, each distribution system operator has to develop a coincidence graph of his consumer, as shown in Figure 8-23. The graph shall represent the coincidence factors versus full load hours for each individual consumer. The coincidence factor reflects the actual coincident load of the consumer during the peak load of the voltage level at which the consumer is connected. The full load hours are the quotient of the annual electricity consumption of the consumer divided by his actual load.

The graph consists of two straight line sections, one for the lower "g_1" and one for the higher "g_2" full load hours. Their straight line equations are also embedded in the graph.

The coincidence functions must meet the following requirements

- The sum of the coincident load of the individual consumers must be equal to the peak of the voltage level and
- The sum of the charges of the individual consumers must be equal to the sum of the annual costs of the voltage level

The compliance with these requirements has to be validated iterative with repeated trial balances.

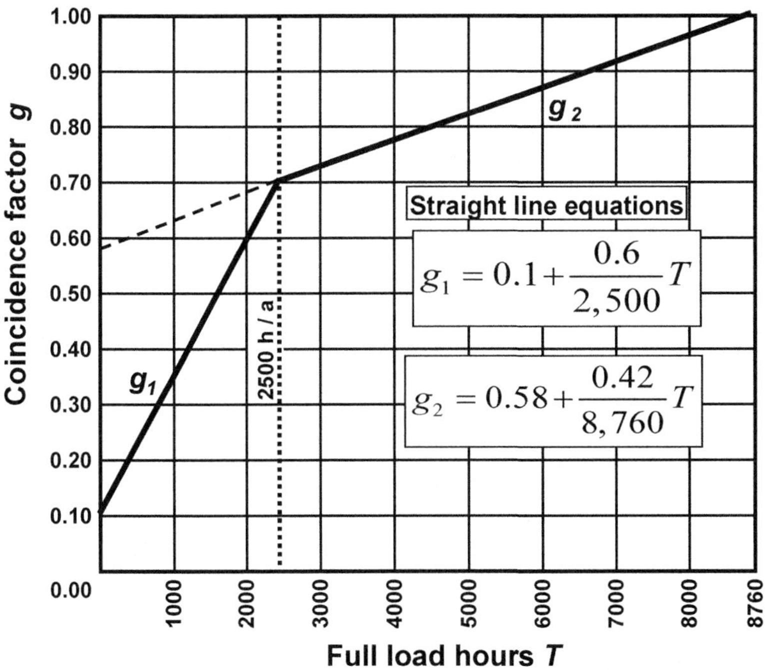

Figure 8-23: Typical coincidence functions of the consumers

8.8.4 Use of system tariff formulation

The formulation of the use of system tariffs is conducted stepwise as follows (see for a better understanding also actual calculation in Example 8-4):

- Estimation of the costs for each voltage level of the system in \$/a; they include the rolled-over costs from the higher voltage level, the transformation costs and the direct costs of the own voltage level (see graph in the example)

- Calculation of the annual peak load of each voltage level. This is composed of the peak load of the direct connected consumers plus the peak load of the consumers of the subordinate voltage level (rolled-over peak)

$$P_{\max\_VL} = \left\{\sum Cf_i \cdot P_{ic}\right\}_{connected} + \left\{\sum Cf_i \cdot P_{ic}\right\}_{subordinate} \quad [\text{kW}] \quad (8.24)$$

Where:

P_{max_VL}: the annual peak load of the voltage level
P_{ic}: the connected load of the individual consumers
Cf_i: the coincidence factor of the individual consumers

- Cost allocation of the total cost of the voltage level between the consumers connected to the same voltage level and the consumers of the subordinate voltage level (roll-over) based on their coincident peak loads

- Calculation of the *annual average use of system charge* USC_{VL} of the voltage level based on the annual costs allocated to the directly connected consumers and their coincident peak load

$$USC_{VL} = \frac{\Sigma C_{VL\_connected}\,[\$/\text{a}]}{P_{\max\_VL\_connected}\,[\text{kW}]} \quad [\$/\text{kWa}] \qquad (8.25)$$

- Finally, the *use of system tariff* UST_{ci} is computed by multiplying the annual use of system charge of the voltage level USC_{VL} with the coincident g_i equation and the load P_{ci} lf of the individual consumer

$$UST_{ci} = USC_{VL} \cdot g_i \cdot P_{ci} \qquad (8.26)$$

The *use of system tariff* of the individual consumers of the grid shall reflect two cost items: the contribution of the individual consumer to the peak load (kW) and to the energy consumption (kWh) of the voltage level to keep the system in safe operation. The tariff shall therefore consist of a capacity charge (\$/kWa) and an energy charge (\$/kWh)

- Conduct a validation test and verify if the use of system tariffs charged to the consumers of voltage levels cover the cost. If there is a deviation, the coincidence functions are to be readjusted.

Example 8-4: Use of system tariff formulation

The example shall demonstrate the tariff formulation with the method described above. The numbers are not representative for any actual system.

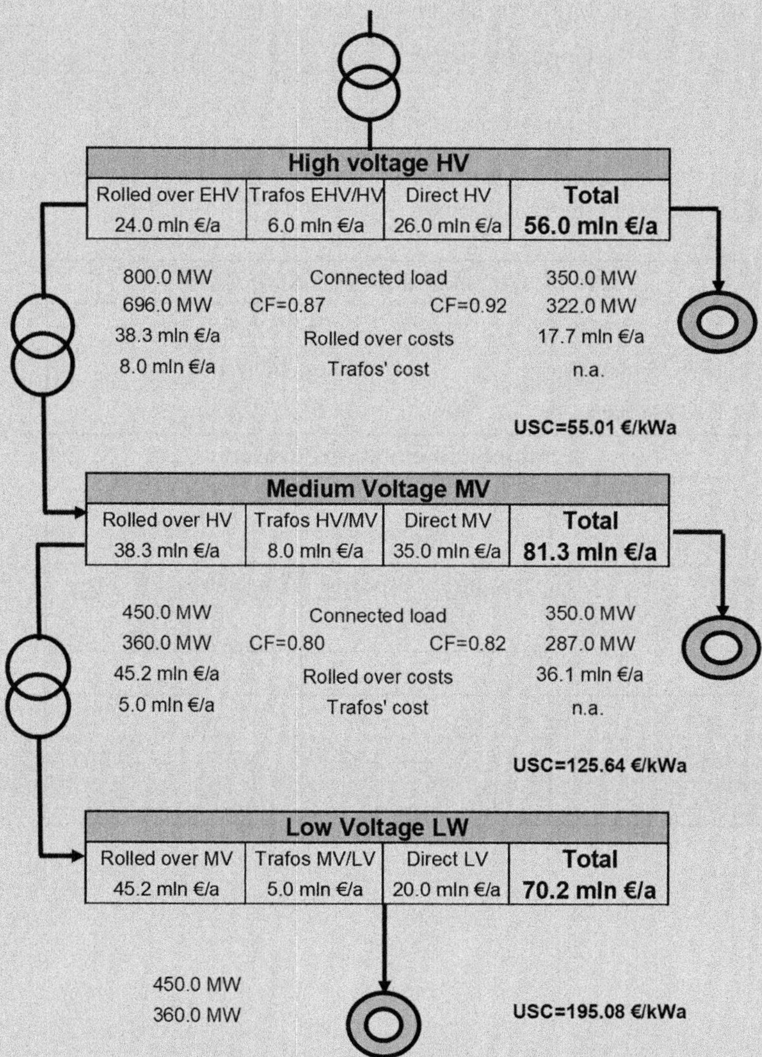

The tariff for a consumer of the HV level with 2,500 h/a is calculated as follows:

$$UST_{HV} = 55.01 \cdot P_{ci} \cdot (0.1 + \frac{0.6}{2500} \cdot T) \text{ or}$$

$$UST_{HV} = 5.5 \cdot P_{ci} + 0{,}0132 \cdot P_{ci} \cdot T \text{ and with } P_{ci} \cdot T = W_e \left[\frac{\text{kWh}}{\text{a}}\right]$$

$$UST_{HV} = 5.5\left[\frac{\$}{\text{kWa}}\right] \times P_{ci}\left[\text{kW}\right] + 0.0132\left[\frac{\$}{\text{kWh}}\right] \cdot W_e\left[\frac{\text{kWh}}{\text{a}}\right]$$

Based on the last equation the tariff can be expressed as follows:

$$\text{Capacity price} = 5.5\left[\frac{\$}{\text{kWa}}\right]$$

$$\text{Energy price} = 1.32\left[\frac{\text{ct}}{\text{kWh}}\right]$$

If we assume that the coincident graph is valid for all the voltage levels, the tariffs will be as shown in the table below.

Use of system charge by voltage level				
HH US charge	55.01 €/kWa			
MV US charge	125.64 €/kWa			
LV US charge	195.08 €/kWa			
Straight line equations' factors				
Constant factor	0.1		0.58	
Slope factor	0.6		0.42	
Tariffs	**Capacity**	**Energy**	**Capacity**	**Energy**
	2,500 h/a		8,760 h/a	
	€/kWa	Ct/kWh	€/kWa	Ct/kWh
HH	5.50	1.32	31.91	0.26
MV	12.56	3.02	72.87	0.60
LV	19.51	4.68	113.15	0.94

9 Electricity Trading

9.1 From Monopoly to Market Economy

The energy supply sector in general and the power industry in particular were traditionally dominated by vertically integrated monopolies. A single company took care of all the business fields such as generation, transmission, distribution and selling of electricity. Consumers had no choice of suppliers; they had to buy electricity from their regional utility. Since the beginning of the 90s of the last century, however, the entire energy sector is worldwide undergoing a major transition from monopoly to a market based business. The overall objective of this development has been to lower costs and to foster competition.

The international approach for the design of the legal, regulatory, and institutional framework includes the following aspects [60]:

- The *privatization* and restructuring of state-owned energy utilities
- The *unbundling* of vertically integrated utilities into separate business units for generation, transmission, distribution, and trade
- The *free customer* choice of supplier
- The introduction of *competition* in power generation
- The establishment of an independent transmission system operator (TSO)
- The fair, transparent and non-discriminatory access to the grid for all market players
- The *regulation* of the naturally monopolistic business fields, namely transmission and distribution, along with the establishment of an independent regulatory authority (regulator)

The liberalized markets involve a variety of players as:

- Generators, producing electricity. They are mainly utilities owning several power plants, Independent Power Producer (IPP) and small scale power producers
- Transmission System operators (TSO). The TSOs operate the very high voltage network (the grid), usually ≥ 380kV, that transports electricity to the distribution networks and to very large consumers. They are among other responsible for voltage and frequency stability, settlement of imbalances in the grid, black start and reserve services

P. Konstantin and M. Konstantin, *The Power Supply Industry*,
https://doi.org/10.1007/978-3-319-72305-1_9

- Distribution system companies (DisComs) operate the lower voltage networks (i.e. $\leq$110 kV down to 0.4 kV), transport electricity from the transmission system to households, businesses and factories
- Suppliers who buy electricity and sell it to groups of consumers
- Traders and marketers and other players without physical demand of electricity providing risk management, hedging and brokerage.

9.2 Power Market Models

In the course of the liberalization process of energy markets, a large number of different market models have been established around the world. However, from all these models it is possible to distinguish three basic types of market structures:

- The Single Buyer model
- The electricity pool or centralized market model
- The fully competitive market model

9.2.1 The Single Buyer model

In this model – Figure 9-1, the Single Buyer entity buys electricity from power generators (GenCos and IPPs) and sells it to distribution utilities and large consumers in its service area. In most cases the Single Buyer and the transmission system company (TransCo) are the same entity. All power generators have to sell the produced electricity to the Single Buyer. Distribution companies (DisCos) and retail services are disaggregated but are only able to purchase electricity from the Single Buyer. They do not have a free choice of their power supplier.

The Single Buyer makes a longterm contract with power generator companies and IPPs [61]. Long-term contracts are necessary to attract investors to invest the required large amounts of capital in power generation plants. The contracts are generally of life-of-plant type, indicating sale of all capacity of generating units for its lifetime.

The Single Buyer model is a way of attracting private participation in the generation sector, especially in developing countries. It also provides some limited competition on the generation level and is a first step towards liberalization.

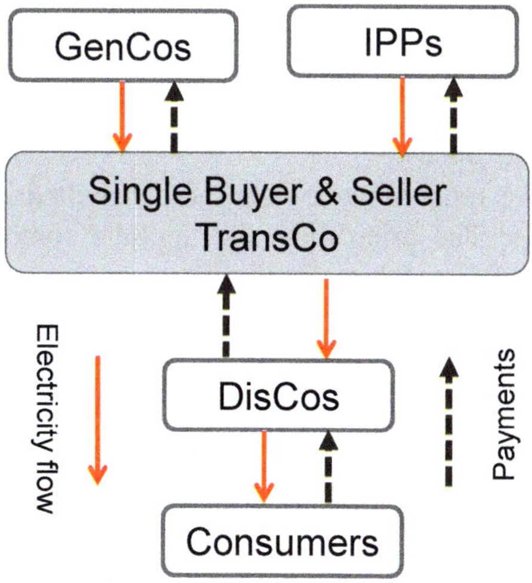

Author's own illustration

Figure 9-1: Single Buyer & Seller Model

The tariffs of the Single Buyer must be regulated because it has monopoly over the DisCos and *monopsony* (buyer's monopoly) over the power generators. Disadvantages are the longterm power purchase agreements with IPPs and regulated tariffs for final consumers in all levels.

9.2.2 The Power Pool model

Power pools [62] require generators to submit bids indicating the amount and price of electricity they are prepared to deliver for each hour[29] of the day ahead. In a *mandatory pool* (Figure 9-2) all generators can deliver electricity only through the pool, as the pool is the Single Buyer. In a *voluntary pool* generators have also the opportunity to enter into bilateral contracts with suppliers and have to request dispatch from the pool administrator.

The key characteristics of the pool[30] are:

- The pool itself is an administrative entity and takes no risks; it collects the bids of generators and sends dispatch notifications to the system operator
- Dispatch is done by the Transmission System Operator and optimized on a day-ahead-basis. Ancillary services are provided either based on the grid code or by commercial transactions or both

[29]) or shorter periods, quarter of an hour, half an hour depending on the model
[30]) See also [62] Electricity markets for more details

- The System Operator predicts demand and dispatches generators against this demand assumption in general with no input from buyers
- Buyers draw their demand from the grid without contracts and any commitments in advance
- The generators receive money for the dispatched energy based on one or several prices. The price may also include some disincentive for non-performance of dispatch instructions
- The transmission system operator receives money from the Single Buyer
- The pool pricing is based on marginal economics including also some form of capacity compensation

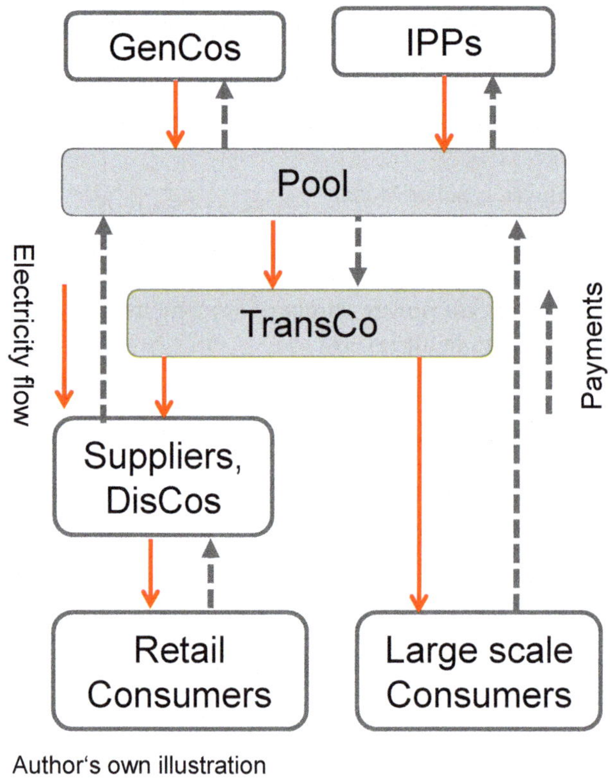

Figure 9-2: Pool Model with generator competition

The pool is the intermediate step from monopoly to competitive markets; the main difference to the Single Buyer model is that generators have to submit offers which are subject to competition and not based on longterm contracts between the Single Buyer and the generators.

Distribution companies and very large consumers buy electricity from the pool and pay the pool price reflecting mainly the costs of generation and the use of system costs. The final consumers still have no choice of supplier. They are served by the distribution company of their service area.
Generators submit bids for supplying a volume of electricity (MWh) at a specific price based on (audited) marginal cost for usually each hour[31] of the day ahead. The pool operator accepts bids from generators, and compiles an initial dispatching schedule for each hour of the day ahead, starting with the *must-run* plants[32], and then the bids in ascending order of marginal cost, until the predicted demand is met (Figure 9-3). The initial schedule ignores physical limitations of the system and is called the *unconstrained schedule or market schedule*. Generators are 'in-merit' when their bids are successful and 'out-of-merit' when unsuccessful. The in-merit plant with the highest price defines the *System Marginal Price (SMP)* and is valid for all bidders.

On the day of trading, the system operator usually readjusts this schedule to deal with system constraints as imbalances, congestion, and ancillary services. This is called the *constrained schedule*. A plant, e.g., that is in merit in the initial schedule but cannot be dispatched, is replaced by a plant that is out of merit but can be dispatched – Figure 9-3.

This is relevant for the remuneration of the supplied electricity. Plants which are in-merit receive the system marginal price. A plant that is out-of-merit in the unconstrained schedule but runs in the constrained schedule is referred as "constrained on". This plant usually receives its offered price referred as "pay-as bid". Constrained off plants usually do not receive any payment.

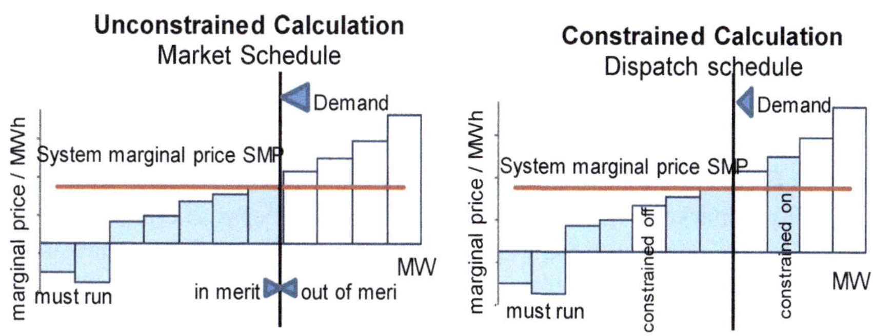

Figure 9-3: Dispatching schedules

[31] In some pools for half an hour or quarter of hour

[32] *Must-run plants* have priority in the schedule; these are, e.g., run of river hydropower, nuclear, renewable, plants providing steam for seawater desalination. Must-run plants do not have to submit bid commitments into the pool.

The *Pool Purchase Price (PPP)* comprises two components [62]. All the dispatch generators receive the System Marginal Price (*SMP*) plus a compensation for capacity costs. The latter is designed to provide an incentive for future investment in generation capacity; it is determined by the regulator by assessing the loss of load probability (*LOLP*) and the value of loss of load (*VOLL*). The *PPP* is the reference price that the pool pays to the generators. In the UK model the PPP is then calculated with the following formula:

$$PPP = SMP + \left[LOLP \times (VOLL - SMP) \right] \quad \left[\frac{\text{CU}}{\text{MWh}} \right] \tag{9.1}$$

Buyers additionally pay the "uplift". This stands for the transmission system costs including, e.g., ancillary services, losses, and other system related items. The *pool selling price (PSP)* then is:

$$PSP = PPP + uplift \quad \left[\frac{\text{CU}}{\text{MWh}} \right] \tag{9.2}$$

The PSP varies for each (half) hour during the trading day, due to the fluctuation of demand, and may be extremely volatile. Buyers and sellers try to hedge against this price volatility by entering into bilateral contracts [63] called *contracts for differences (CfD)*. The buyer and the seller agree a specific volume and price, called the "strike price" in the CfD. If the PSP is higher than the strike price, the seller pays the buyer the difference; in contrary if the PSP is lower, the buyer pays the seller the difference. Bilateral contracts are traded in forward markets that usually coexist with the pool.

Transmission and distribution services are natural monopolies and the use of system fees are regulated. There are different regulations regimes referred to as price-based regulation, cost-based regulation, performance-based regulation, benchmarking regulation etc. For details[33] see [64].

9.2.3 The fully competitive power market model

The market structure of the fully competitive model is depicted in a simplified form in Figure 9-4 and in more detail in Figure 9-5. The model is a multi-seller multi-buyer market. All market participants have access to competing generators either directly or through their choice of supplier. The model requires a complete separation of generation, transmission, distribution business fields as well as sales. Both, transmission and distribution networks must provide a fair and non-discriminatory access to all market participants.

[33]) Michael Kraus, Liberalized Energy Markets

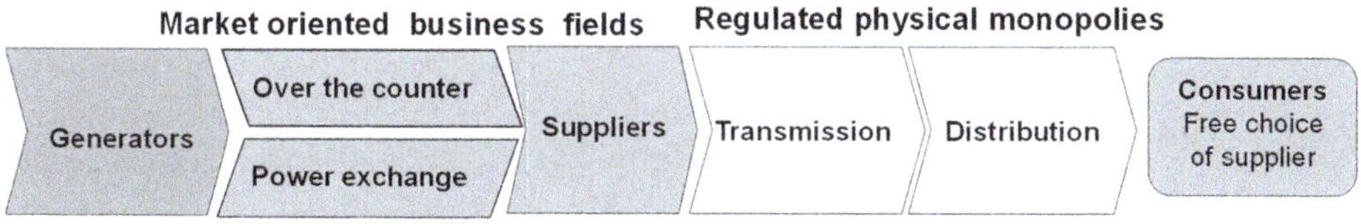

Figure 9-4: Simplified structure of the fully competitive market

The trading takes place over the counter (OTC) and in the power exchange. Suppliers and traders buy electricity directly from generators and/or from the power exchange and transfer it via the transmission and distribution networks to the consumers.

The main tasks of the system operator is to dispatch the negotiated contracts based on schedules provided by suppliers, traders and large consumers and generators, to provide ancillary services as voltage and frequency stability and to offset imbalances between predicted and actual demand. This requires the system operator to purchase additional power from power generators in the event of under-scheduling of demand or by compensating power generators who have to reduce generation in the case of over-scheduling. The additional power or reduction of power is usually procured in the balancing market of the power exchange (see also 9.3.3.2). The use of system fees for the transmission as well as for the distribution networks is regulated,

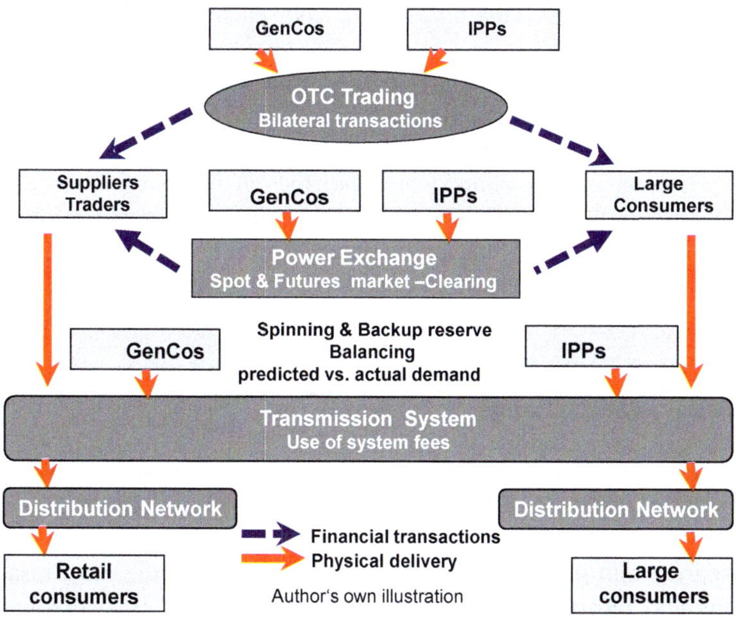

Figure 9-5: Structure of the fully competitive market

9.3 Market Places for Electricity Trade

9.3.1 The OTC market

In the *OTC market* the electricity trading takes place over bilateral contracts between generators and suppliers, which may be retailers providing electricity to final consumers, or large consumers (e.g., steel, chemical, aluminum industries) or municipal utilities. OTC trade preferably deals with physical delivery of electricity to cover demand that occurs in foreseeable time and magnitude, e.g., for base or intermediate load.

There is a wide range of bilateral contract types. However, contract duration and pricing are the issues of particular importance. The contract duration is typically two years but may take up to five years or even longer. The pricing mechanism depends on the duration of the contract. In short-term contracts (e.g., one year) the price is fixed while in contracts with longer duration the price will be indexed. The indexation may be based on the development of the fuel prices or the prices in the power exchange or both. This is, however, only the price for generation and does not include the use of system charge for transmission and distribution that are charged separately by the TSO.

Bilateral contracts can provide several benefits for both contract partners, notably among others:

- The price is less volatile and provides increased planning security for sellers and buyers
- Contracts with longer duration facilitate investment decisions for new power generation capacities

Worth mentioning in this context are also the *all-inclusive contracts* [40] between suppliers and medium and small scale business and residential consumers. For such type of consumers the proceedings for bilateral contracts with generators and TSO are too complex and costly. The price of this type of contract includes all items for generation, use of system, and all the other cost items.

9.3.2 The Power Exchange

More efficient electricity trading in liberalized markets is accomplished with the establishment of a Power Exchange in parallel to the bilateral trade. A Power Exchange is a neutral market place where market participants have equal access to a transparent and reliable electronic trading environment. All transactions are anonymous and information about competition, liquidity is available and price setting is clear and transparent. Furthermore, the Power Exchange or an associated entity provides clearing services for transactions. The retail trade is no longer regulated because small consumers can change their retailer for better price options. Further to electricity the trade includes other energy related products as natural gas, coal and carbon certificates.

Fully developed Power Exchanges provide two trading platforms: the spot market and the derivatives market [65] [66].

9.3.3 The spot market in Power Exchange

The spot market of the Power Exchange itself includes two trading platforms: the day-ahead and the continuous intraday trade.

9.3.3.1 *The day-ahead trade on the spot market*

In the *spot market* of Power Exchange standardized contracts for *physical delivery* of power are traded. Based on the delivery period of the power, the standardized products can be classified in hour contracts and block contracts. Hour contracts imply the delivery of power with a constant power over one given hour. Block contracts imply the delivery of power with a constant power for a given number of hours. The following standardized products are commonly traded. The quotation is in CU/MWh:

Table 9-1: Type of spot market contracts

Type of Contract	Contract volume
Day, base load block, Mon to Sun	1 MW × 24 h = 24 MWh
Day, peak load block, Mon to Fri 8:00 –20:00 h	1 MW × 12 h = 12 MWh
Weekend, base load,	1 MW × 48 h = 48 MWh
Day, hour contract	0.1 MW × 1 h = 0.1 MWh
Day, hour block contracts	0.1 MW × n h = n× 0.1 MWh

The *day-ahead* trading usually takes place in form of a daily auction. Sellers (generators) and buyers submit bids for every hour of the next day, indicating volume (MWh) and price (€/MWh) they are prepared to sell or buy. All bids are sorted and aggregated to a market demand and supply graph – Figure 9-6.

The intersection of the supply and the demand curves defines the *market clearing price (MCP)* and the *market clearing volume*. All volumes up to the market clearing volume (called *in merit* volumes) are priced at the same market clearing price. All sellers' bids at a price equal or lower to the market clearing price and all buyers' bids at a price higher or equal to the market clearing price are executed the day ahead. The deadline for bidding – called *gate closure* – is usually midday 12:00 h of the day before the trading day. All bidders can submit and/or change their offers up to the gate closure. All bids are registered in the order book of the Power Exchange and are made available to the bidders.

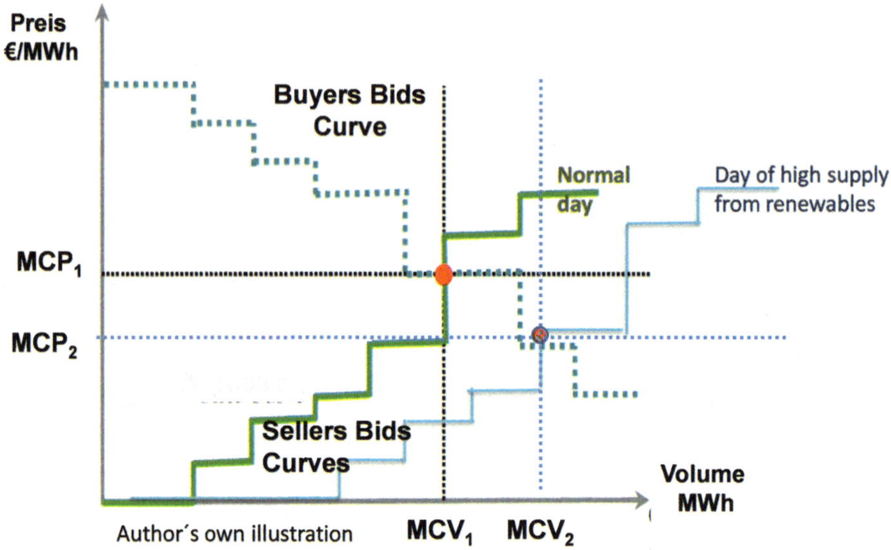

Figure 9-6: Market clearing at a power pool

Note: Electricity is offered to the market based on short run marginal cost. The marginal cost of electricity from renewable energies is practically zero. So if there is a surplus production from renewables during windy and sun-rich days the bids-curve is moved to the right and the market clearing price $MCVP_2$ becomes lower.

9.3.3.2 ***The continuous intraday trade on the spot market***

In the *continuous intraday trading* the Power Exchange provides the opportunity to the buyers (suppliers and large consumers) to buy additional volumes if they have underscheduled their demand, or to sell excess ordered volumes if they have overscheduled their demand. On the other hand, sellers can bid still available volumes to compensate imbalances in their service area. All these bids are executed at the same day to offset imbalances between the scheduled and actual demand in the respective service areas.

There are some distinct differences between auctioning for day-ahead and intraday trading. During intraday trading, participants have access to the order book and can check the energy balance in their services area. The incoming bids from both sides are immediately checked and matched in pairs based on their time/volume priority according to the matching rules of the Power Exchange. The price is not uniform for the transactions even for the same hour as the sellers price is taken "as bid".

9.3.4 The derivatives market in the Power Exchange

Trading in a competitive environment exposes participants to operational and financial risks. There are two main types of risks: the volume risk and the price risk. The *volume risk* arises due to the fact that electricity is a non-storable commodity but the demand must be predicted and the volumes (MW, MWh) must be scheduled in advance. The *price risk* arises as the price can change at any time and is extremely volatile while demand and supply must match which each other at any time. Deviations can often occur, e.g., due to oversupply from renewable resources. If the buyer overestimates demand and underestimates price, this may have serious financial consequences. Hence, risk management becomes a primary task in electricity trading. The power market provides a number of financial instruments called derivatives to hedge (reduce risks) against possible risks. Hedging refers to a strategy of reducing risks.

Derivatives are contracts between two parties who have opposite views on the market development; they are called derivatives, which means they derive their value from something else referred to as the underlying *asset*. This is again something valuable or a commodity that an entity owns to generate income. The *underlying asset* in the power trade is a quantity of electricity with a defined load profile.

An overview of the derivatives market is shown in Figure 9-7. Derivatives are traded in the Power Exchanges or OTC, the period between conclusion of agreement and fulfillment is longer than one week, while in the spot market it is the day ahead or intraday.

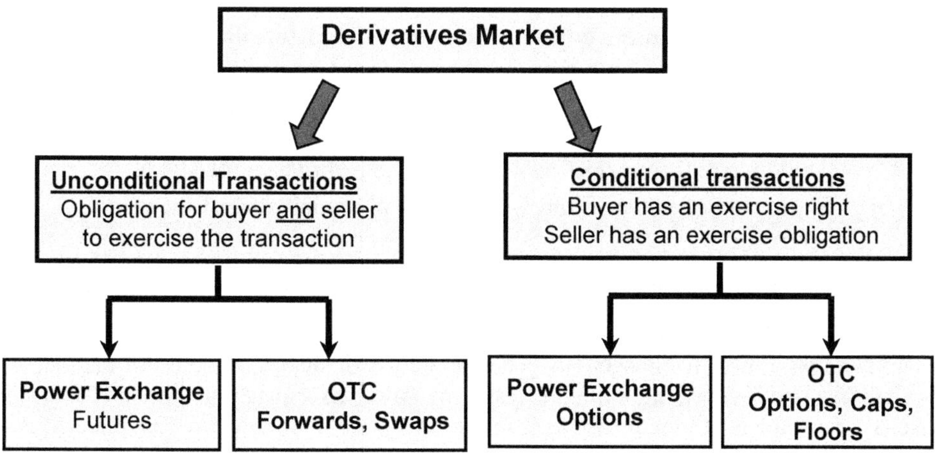

Figure 9-7: Trading contracts in derivatives market of Power Exchange

There is a main distinction between unconditional and conditional transactions. *Unconditional transactions* (futures, forwards) include the obligation for both parties to buy or sell a given underlying asset at a certain time in future at a price specified today. *Conditional transactions* include the right to buy or sell an underlying asset in future time, at a price specified today.

9.3.4.1 ***Futures and Forward contracts***

Power Futures contracts contain the obligation for the buyer and the seller to buy or sell a certain quantity (volume) of power, with a defined load profile, at a certain future time, at a certain price fixed in advance. The sale of futures contracts is used to hedge against falling electricity prices (short hedge). The purchase of futures contracts is used to hedge against increasing electricity prices (long hedge). Futures are standardized contracts traded at the Power Exchange. The clearing house of the exchange guarantees that the contract will be duly fulfilled.

In Power Exchanges commonly futures contracts with cash settlement instead of physical delivery are traded. This means, buyer and seller agree to settle the price difference between the price agreed in the contract and the futures market price at the conclusion of the transaction (see Example 9-1).

The main contract specification of power futures contracts are [65].

- The *delivery period:* calendar week, month, quarter, or year
- The *delivery rate* is 1 MW for each hour of the delivery period
- The *contract volume*: Number of contracts × delivery rate × number of days × hours per day [MWh]
- The *load profiles* are:
  - ⇨ Base load futures: constant delivery 24 h/d for all days of the period. Example, base load month contract (Sept.): 1×1 MW × 30 days × 24 h/day = 720 MWh
  - ⇨ Peak load futures: delivery 12 h/d, Monday to Friday. Example, peak load week future: 1×1 MW × 5 days × 12 h/day = 60 MWh
- Expiration of power futures is reached at the last day of the trading.

Example 9-1: Futures transaction

A power generator is planning to sell a 35 MW base load slice next September in the spot market of the Power Exchange. His generation costs, including some profit margin, are 45 €/MWh. He is expecting falling prices in the Power Exchange, therefore he concludes (sells) 35 contracts of month base load futures at a price of 45 €/MWh. The planned revenue then will be:

Contract Volume: 35 MW×30 days ×24 h/day =25,200 MWh
Expected revenue: 25,200 MWh ×45 € /MWh = 1,134,000 €

Case 1: As planned, he sells the contracted volume but the average price on the spot market in September is 39 € / MWh and his actual revenue falls short:

Actual revenue: 25,200 MWh × 39 €/ MWh = 982,800 €
Shortfall of revenue = 1,134,000- 982,800 = 151,200 €

In this case the *buyer* of the futures contract has to compensate the shortfall by paying the 151,200 € to the power generator (seller). So the actual revenue plus the received compensation is equal with the expected revenue.

Case 2: In contrary, if the average spot market price were 51 €/ MWh, the power generator would have to reimburse the buyer of the future with the amount of 151,200 €.
Note: In practice the settlement of futures occurs for each day of the delivery period with the actual price of the day and a credit or debit is charged to the holder's account and summed up to the final settlement. This is called the *variation margin.*

Futures positions can be neutralized by *closing-out* that means exercising a reversing transaction with exactly the same conditions as the original at some time prior to expiration date.

Forward contracts – Figure 9-7 – are similar to futures contracts; however they are traded over the counter. Unlike futures, forward contracts are not standardized and can be customized according to the needs of the involved parties. Due to the OTC nature, a clearing house is not necessarily involved in the transaction; this may increase the counter party default risk. The settlement may occur on a cash or delivery basis.

9.3.4.2 *Power option contracts*

Option Contracts: An Option is a contract between two parties under which the buyer of the option gets the right (not the obligation) to buy or sell a given quantity of an underlying asset at a certain future time for a price agreed in advance (*exercise* price or *strike* price). In order to enter the option contract, the buyer has to pay a *premium* to the seller. In return, the seller assumes the "obligation" to sell or buy the specified asset at the strike price providing that the buyer *exercises* his right.

The main differences between options and futures contracts are:

- The buyer of the option has the right not the obligation to exercise the transaction while in futures contracts both parties are obliged to exercise the transaction
- The buyer of an option has to pay the premium to the seller while futures contracts do not include a premium payment.

There are two types of options, call options and put options (both refer to the buyer of the option):

- A *call option* gives the buyer the right to *buy* an underlying asset. Buyers of calls expect that the price of the asset will increase.

- A *put option* gives the buyer the right to *sell* an underlying asset. Buyers of puts expect that the price of the asset will fall.

Example 9-2: Basic idea of an option contract in every day situation

A family is looking for an old house and they discover one that they would like to purchase. They negotiate a price of 250,000 € (*strike price*). However, the owner currently lives in the house and plans to move out after one year. One year is a long time and the buyer and seller may find better opportunities during this time and do not want to conclude a formal purchase contract yet. Instead, seller and buyer make a deal. The owner of the house grants to the buyer the "right" to buy the house (the underlying asset) in one year (delivery time) and assumes the "obligation" to sell the house (the owner sells a call option, he is the writer). The buyer (the holder of the call option) is not obliged, he has the "right" to buy the house. The seller requires in return an advance payment of 10,000 € (the premium). For the case that the buyer for some reason does not want to buy the house (does not exercise his right) in one year (delivery time), he would lose his down payment.

Case 1: In the course of the year it is found out that the house earlier belonged to a famous actor and its market value is rising to over 500,000 €. The buyer decides to exercise his right to buy the house and the seller is obliged to sell the house.

Case II: In the course of the year it comes out that the building structure is obsolete and the house urgently needs a costly refurbishment. So the buyer decides to waive his right to buy the house and prefers to lose the advance payment that remains with the seller.

Example 9-3: Sale of a buy option for power

A power generator is planning the operation of his plant for the month of June in January and wants to sell a still available base load slice of 30 MW in the wholesale market. His generation costs are 40 US$/MWh. The base load future price for June is currently far below his generation cost. He decides to sell a *call option* for 42 US$/MWh; he gets a premium of 1.20 US$/MWh and receives:

$$\text{Premium: } 30\ \text{MW} \times 1.20 \times 31\ \text{days} \times 24\ \text{h/day} = 26{,}740\ \text{US\$}$$

Case I: The spot price at expiration is under 42 US$/MWh. The option is *out-of- the-money*. So the buyer of the call does not exercise his right and prefers to lose his premium that remains with the seller (writer). The power generator can even generate additional revenues by selling the electricity on the spot market in June whenever prices are higher than his generation cost of 40 US$/ MWh.

Case II: The spot price at expiration is 46 US$/MWh. The option is *in-the-money*. The *holder* of the call option exercises his right and buys the option. The power generator receives a sell position for June at the strike price of 42 US$/MWh. If the generator would close out the futures position by buying a call option he would lose the received premium. So he decides to provide physical fulfillment via his plant. So he can secure the received premium and even earn additional revenue in the spot market by stopping generation whenever the spot price falls under 40 US$/MWh.

The average spot price in June is e.g. 46 US$/ MWh, the holder of the call options would earn the following revenue:
Revenues: 30 MW×30 days×24 h/day× (46-42-1.2 US$/MWh) = 60,480 US$

Example 9-4: Large industrial consumer buys a call option

An industrial consumer is planning his power procurement for June and needs a base load slice of 30 MW. The production department gives the order to the portfolio manager to proceed with the procurement of 30 MW base load slice for the month of June with a price of 41 US$/MWh or lower. The portfolio manager buys a call option (expiring in May) on the June future with a strike price of 39.50 US$/MWh and a premium of 1.50 US$/MWh. He pays:

Premium: 30 MW×30 days×24 h/d×1.50 US$/MWh = 32,400 US$

Case I: The June future price at expiration of the option is 45 US$/MWh. The option is in-the-money, so the portfolio manager exercises the option and receives the June future at the exercise price of the option (39.50 US$/MWh). By buying the call option, the portfolio manager ensured a maximum price of 41 US$/MWh (premium + exercise price), even though the future price did not move in favor of the company.

Total costs: 30 MW×30 days×24 h/d × (39.50 +1.50) US$/MWh = 885,600 US$

Case II: The June future price at expiration of the option is 37 US$/MWh. The option is out-of-the-money, so the portfolio manager does not exercise the option as this would not be profitable. Instead he secures the power requirement via a futures position at the current market price. By buying the call option, he paid for an insurance against unfavorable price movements which eventually he didn't need.

Total costs: 30 MW×30 days×24 h/d × (37+1.50) US$/MWh = 831,600 US$

9.3.4.3 *Terminology of derivatives*

To understand the derivatives market, it is necessary to know the associated terminology. Therefore, it is worth explaining the terms associated with the derivatives market transaction along with the working principles of the market instruments. The transactions are also demonstrated in Example 9-1 to Example 9-4 for a better understanding of the market instruments.
A position is a contract to buy or sell a financial instrument such as an option. It is also used to define the ownership status of a financial instrument (long or short position).

Trading Participants: Buyers of options are called *holders,* seller of options are called *writers.* Call holders and put holders (buyers) are not obliged to buy or sell; they have the choice to exercise their right. In contrary, call writers are obliged to buy or sell. A *holder* of an option is said to have *long position* by buying an option, the term long implies position ownership. A writer of an option is said to have *short position* by selling an option. Sellers can, e.g., sell a borrowed position without owning.

The *strike price* is the price at which an option can be purchased or sold. A *call option* for power can be exercised with profit if the price in the spot market is higher than the strike price, then the option is said to be *in-the-money*. Vice-versa, a *put option* will be *in-the-money* when the price in the spot market is lower than the strike price.

The value of an option: In order to enter the option contract, the buyer has to pay a *premium* to the seller.

Expiration date is the time at which an option can be traded for the last time. It depends on the type of option traded at a Power Exchange. If the holder chooses not to exercise his right, the option expires and becomes worthless.

Exercising and assignment: Exercising means that the option holder buys or sells the underlying asset at the strike price; the option seller is then *assigned* (required) to take the other side of the trade and sell or buy the options to the strike price. *American options* can be exercised at any time between the date of purchase and the expiration date. *European options* can only be exercised at the end of their lifetime. The name has nothing to do with the geographic location of the exchange.

Opening and Closing out: Opening a position means buying or selling option contracts. Option positions can be neutralized by closing-out that means exercising a reversing transaction with exactly the same conditions as the original. A long position (buying) of call options for 30 contracts base load (each 1 MW) for June with a strike price of 45 US$/MWh can be neutralized by selling call options 30 contracts base load for June with a strike price of 45 US$/MWh.

9.3.5 **Clearing**

Each Power Exchange has its own clearing house that is a separate entity responsible to enable a smooth processing of all financial transactions [65] [68]. In derivatives' markets the clearing houses guarantee that the buyers and sellers fulfill their obligations related to their futures' and options' contracts. Their main tasks are administrating trading accounts, regulating issues necessary for the settlement, such as payment method and overseeing a proper delivery of the underlying asset.

The clearing house acts as the central contractual partner for the trading participants and assumes the opposite position of each side of the trade and takes the counterparties' risk. On the other hand, the trading participants have to deposit margin payments (securities) for liabilities entered under their transactions which are sufficient to cover their debit balance. The clearing house collects and maintains margin payments accounts and is responsible to all participants for the fulfillment of the contracts.

All participants of the Power Exchange are required to clear their transactions through the clearing house. A contract is formally concluded after liabilities and obligations of the contract parties of financial transactions are registered and confirmed.

When a market participant enters a power futures' contract, he has to deposit a minimum amount of money into his account in the clearing house, called *the additional margin* (called also initial margin). This is used to cover any losses that may incur for closing out an open position of the market participant under the assumption of the most unfavorable price development. When the futures' contract is closed out, the participant will be refunded the initial margin plus or minus any gains or losses that occur over the span of the futures' contract.

The buyer of an option position has to pay a premium to the seller and does not pose any further risk. For this reason he does not have to pay a margin in the clearing house. In contrary, the seller has the obligation to open a corresponding futures position at the strike price if the buyer exercises the option. Therefore he has to deposit the *premium margin* and *the additional margin*. The amount of the margins depends on the positions held in options and on the futures on which they are based.

9.3.6 **Portfolio management**

Electricity procurement managers of large companies or suppliers of a pool of clients optimize the purchase of the expected electricity consumption profile in a portfolio consisting of several procurement transactions as shown qualitatively in Figure 9-8. Base load can be predicted quite well and is usually purchased in longterm bilateral contracts with power generators. However, uncertainties surrounding production patterns, consumers' behavior and weather effects make it difficult to issue an accurate forecast for peak load during the day and partly for intermediate load during working days during the week.

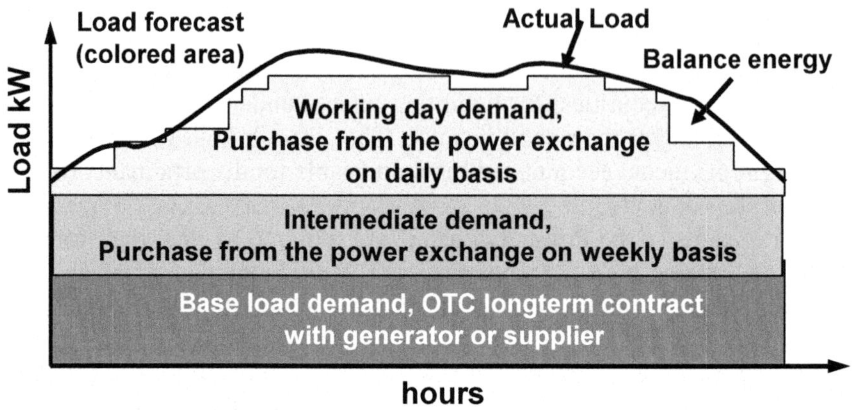

Figure 9-8: Portfolio management, daily load profile

On the other hand, the transmission system operator (TSO) requires from purchasers and generators to submit demand schedules for each ¼ hour of the day ahead to ensure that the supply and demand are sufficiently balanced.

Electricity demand fluctuates constantly, and actual load will never be completely congruent with the schedule demand. Changes in load, that are not offset by the scheduled demand, require the TSO to balance the difference by purchasing additional electricity from generators in case of underscheduling or to request generators to reduce output in case of overscheduling of demand. This additional or reduced power is purchased intraday in the balancing market. The TSO has to determine market clearing price for the balancing electricity for every quarter of an hour and will charge the resulting higher cost ex-post to the parties responsible.

10 Development and Implementation of Projects

10.1 Project Definition

According PMBOK® of the US Project Management Institute [69][34], a project is defined as a temporary venture undertaken to create a unique product or service. In the energy supply business the project may, e.g., be a new power plant, a power transmission network or the provision of engineering services. The key project constraints are:

- Unique with a defined scope and objective
- Fixed time frame, with a defined "beginning" and "end"
- Requiring a high efficient form of organization
- Availability of human and capital resources
- Multidiscipline and complex, requiring excellent team work
- Aimed to provide deliverables (returns) during its commercial operation phase.

The project constraints are often visualized in the project triangle:

The *scope* comprises of the project objective, the project outcomes and the product definition. For projects in the energy supply business, the project objective outlines, e.g., the type of plant, the capacity and the mode of operation, usually described in detail in the project specifications. The product definition includes the quality of electricity that the project must provide after it is completed, such as availability, voltage level, frequency fluctuation margin etc. These parameters are specified in the project performance warranties.

---

[34]) PMBOK®: A Guide to the Project Management Body of Knowledge

P. Konstantin and M. Konstantin, *The Power Supply Industry*,
https://doi.org/10.1007/978-3-319-72305-1_10

Resources comprise project budget, human resources, land, cooling water, equipment, software tools etc. Project budget shall ensure that the project will be implemented within the given budget, preferably not even exhausting it. Adequate human resources are a precondition to achieve the required project and product quality. All other resources have to be ensured to remain available during the project lifetime.

The *time frame* with a defined beginning and end is the most stringent project constraint. It is possible, for instance, to increase the budget, if necessary, with some additional loan or to engage more construction staff, but the time frame is usually fixed and absolutely binding.

10.2 Project Phases

A project in the energy supply business undergoes several phases during its lifetime as are noted below:

Table 10-1: Project phases

Phase	Objective	Outcome
Project inception	Definition of overall objective and scope	TOR, Inception report
Project definition	Technical feasibility and economic viability	Feasibility study report
Project planning	Conceptual design	Design report Technical Specifications
Procurement	Tendering procedure	Contract award to a contractor
Project implementation	Construction and erection	Turn-key plant
Commissioning and trial operation	Functionality tests of system components and entire plant	Plant ready for commercial operation
Acceptance	Plant takeover by owner	Punch list *)
Warranty	Close all punch list items	Project close, return of performance guarantee
Operation	Production of goods	Commercial usage

*) Items which do not conform with the project specifications and require immediate clearance

Project inception: During this phase of the project, the following key issues have to be defined: technical and economic requirements as well as objectives (e.g., rising power demand requires the construction of a new power plant of a certain size), definition of power plant type, selection of project site, definition of basic technical data and time table for construction. Based on the established information, the *Terms of Reference (ToR)* are issued by the owner.

This phase is concluded with the elaboration of the *Inception Report* that also includes, besides all the established information and data, approaches to a solution resulting from initial impressions of site conditions and following discussions between owner and his consultants.

The Inception Report is fundamental to the further project work. It has to be discussed in detail and approved by the owner, so that subsequently there will be no room for misunderstandings or doubt regarding the objective of the project.

Project definition: This phase mainly contains the preparation of a techno-economic feasibility study based on the investor's ToR and the inception report. For large projects usually first a prefeasibility study, followed by a bankable feasibility study, is elaborated.

In the *prefeasibility study,* technical concepts for several alternatives and options are developed and evaluated. The findings of this study result in a conclusion as to whether or not a capital investment is technically feasible and economic viable. The technical concepts must allow an estimate of capital expenditures (CAPEX) with an accuracy margin of ± 20%, an OPEX estimate and the establishment of preliminary technical-financial evaluation of the options. The study report comprises a description of the technical concepts, the comparative analysis and finally the selection of the technically and economically most favorable option for follow-up.

The subsequent *bankable feasibility study* focuses especially on the economics of the preferred option. Besides a detailed technical conception and performance analysis over the lifetime of the project, it includes financial modelling with cash-flow analysis on year-by-year basis, sensitivity analysis and risk analysis in a *bankable* form. The result of cash-flow analysis must provide sufficient evidence that the projects' returns ensure a full repayment of loans in due time and manner and allow also the expected dividend payments for the equity investors. The results must permit a decision by the creditors to finance the project and to attract the interest of investors.

Project planning: Based on the findings of the feasibility study, the basic engineering for the recommended option is carried out to a sufficient degree of detail permitting elaboration of specifications for tendering and procurement. The degree of detail highly depends on the selected scheme of procurement contracting, namely *Lots' wise* contracting or *EPC* contracting. The former requires drafting of specifications for delivery and installation for main project work groups such as civil works, boiler plant, flue gas cleaning facilities, mechanical, electrical works etc. This may comprise between 10 and 40 lots for a power plant project. For EPC contracts (Engineering, Procurement and Construction) specifications are prepared for a *turn-key* plant.

For utility size projects EPC contracting is usually preferred. The advantage is that the construction risks are fully allocated to the EPC contractor while for lots' wise contracting the owner bears the construction risks.
During the basic engineering stage, the plant processes are established and optimized, taking into account the constraints and boundary conditions specified by the owner or resulting from legislation, such as: plant performance parameters, environmental protection standards, operational reliability and time scheduling.

An integral part of project management during project planning, is risk assessment and risk mitigation. This includes risk identification, analysis of the consequences in the case of occurrence, risk management and mitigation.

The outcome of this phase is the *Design Report* and *Technical Specifications* including all documents which are necessary for subsequent detail planning by an EPC contractor or lots' contractors. They include the basic description of the plant, its technical design data and documents, standard quality requirements as well as scope of supply and boundary limits.

Procurement: The design report and technical specifications provide the basis for starting the *tendering* procedure. Besides the technical quality and quantities to be supplied, the tendering documents also include legal and commercial conditions. They will be laid down in the so-called *Conditions of Contract.* After preparation of the required documents, the following proceedings are undertaken:

- Prequalification of suitable bidders
- Shortlisting of qualified bidders
- Invitation to tender for shortlisted bidders
- Issue of tendering documents
- Bid preparation
- Technical-economic evaluation of bids
- Selection of the most favorable bid(s)
- Contract negotiations with the most favorable bidders
- Award of contract to the best placed bidder

Note, however, that for evaluation of bids, a technical and financial analysis over the project lifetime is required. This is because bids with low upfront costs may have a poor performance (e.g., lower efficiency) resulting to higher lifetime costs.

Project implementation: The main engineering activities during this phase are carried out by the contracted suppliers. They perform the detailed engineering, calculation and dimensioning of plant components for all mechanical, electrical, instrumentation and control equipment as well as for the civil part.

The owner's consultant and his team perform essentially review of the documents of detailed engineering and construction and site management supervision involving among other:

- Supervision of detailed design and manufacturing to ensure compliance with the provisions of the contract
- Clarification of interfaces between the various works and existing plant installations
- Compliance with the requirements of the technical specifications, standards, guidelines etc., quality control
- Budgetary and costs control

During the *construction and erection phase*, equipment components, specified during the engineering phase and supplied by the manufacturers, are assembled to a functioning plant.

The site management supervises and monitors construction and installation works. The principal tasks of the site managing staff are:

- Preparation of a general erection schedule
- Monitoring of a functional health and safety system
- Coordination of installation sequences, times and deadlines
- Monitoring supplies and services of suppliers on site with regard to compliance with working drawings, completeness and conformity with the contractual conditions
- Checks and monitoring of construction progress
- Coordination of special functional tests
- Inspections and check measurements at plant components
- Progress check and release of invoices for payment.

Commissioning and trial operation: After completion of the erection, individual components and the plant as a whole are tested and put into operation. This starts with the *functionality tests* of main system components with regard to their safety standards, operational functionality and efficiency. After conclusion of functionality tests, the *commissioning* of the entire plant follows. This involves a faultless interaction of all system components so that the operational readiness of the entire plant is ensured.

After successful commissioning, the *trial operation* of the entire plant follows; the purpose is testing the functionality of the plant for commercial operation and compliance with guaranteed performance parameters. This may take several months for large power plants.

Acceptance- and punch-list: At the end of the trial operation numerous acceptance checks and inspections are carried out and compliance with the terms and conditions of the technical specifications is testified by reports and certificates. Finally the accepted plant is taken over and put into commercial operation by the owner.

The take-over certificate is in most cases conditional. This means that all deviations, redefinitions, refurbishments on other contract obligations, which are not to the full satisfaction of the owner, are summarized in a punch list. Punch lists contain only minor deficiencies of the plant which are not related to the safe and reliable operation of the plant. All punch list items (up to several thousand individual points) need to be cleared during the warranty period.

Warranty period and project closure: During the warranty period the owner already operates the plant. Especially during the closure of the punch list items, a small team of the EPC contractor remains on site. Afterwards, the contractor may leave a warranty engineer on site to deal with the potential warranty claims.

In case of malfunctions of any system or component, the owner's engineer will perform a root cause analysis assessing if the malfunction was related to the operating conditions or operating mistakes or whether it was related to a design failure or quality issue manufacturer.

Subject to the detailed contract conditions, but in most cases after one year, the warranty period ends out and the contractor can be released.

Upon **final acceptance** and certification by the owner of services rendered, a *project closure* document is produced and the project is formally closed out. The contractor's performance guarantee (bank guarantee) is returned and from that time onward the full responsibility for the plant and its performance is with the owner.

10.3 Project Management during Project Implementation

Figure 10-1 below depicts an overview of the main activities, which are carried out from project startup up to the beginning of commercial operation for PP projects. The descriptions refer mainly to EPC contracts, which are common for power plant projects.

The implementation phase of large, utility scale projects requires a high degree of expertise and experience. Common practice is that investors engage a consultant to coordinate and carry out the complex activities during project implementation in close cooperation with the owner. The consultant team comprises an engineer consultant as the leader of the team who is usually supported by lawyers, bankers and other experts for special issues.

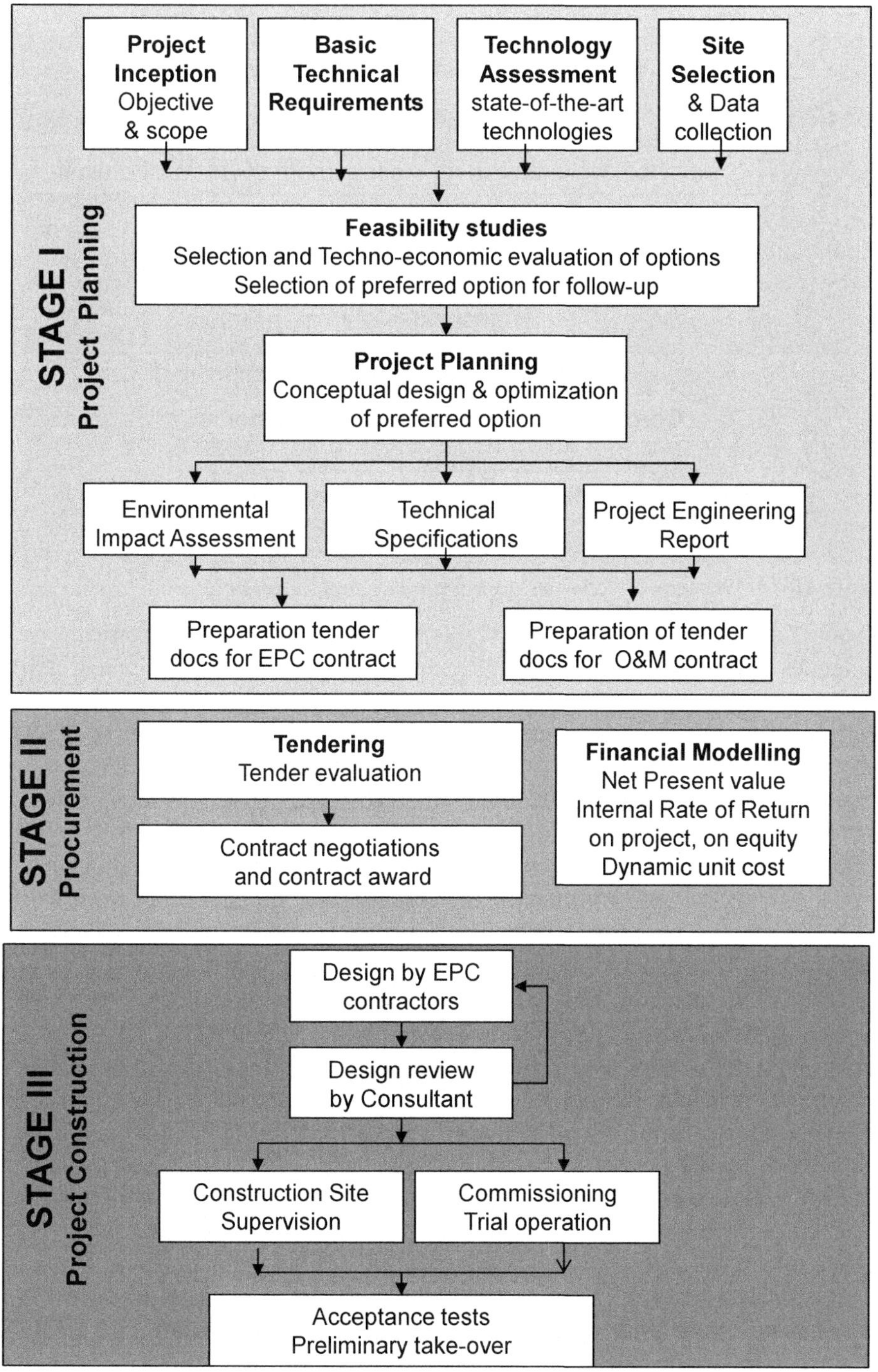
Project Inception
Objective & scope
Basic Technical Requirements
Technology Assessment
state-of-the-art technologies
Site Selection
& Data collection
Feasibility studies
Selection and Techno-economic evaluation of options
Selection of preferred option for follow-up
STAGE I
Project Planning
Project Planning
Conceptual design & optimization of preferred option
Environmental Impact Assessment
Technical Specifications
Project Engineering Report
Preparation tender docs for EPC contract
Preparation of tender docs for O&M contract
STAGE II
Procurement
Tendering
Tender evaluation
Contract negotiations and contract award
Financial Modelling
Net Present value
Internal Rate of Return
on project, on equity
Dynamic unit cost
STAGE III
Project Construction
Design by EPC contractors
Design review by Consultant
Construction Site Supervision
Commissioning
Trial operation
Acceptance tests
Preliminary take-over

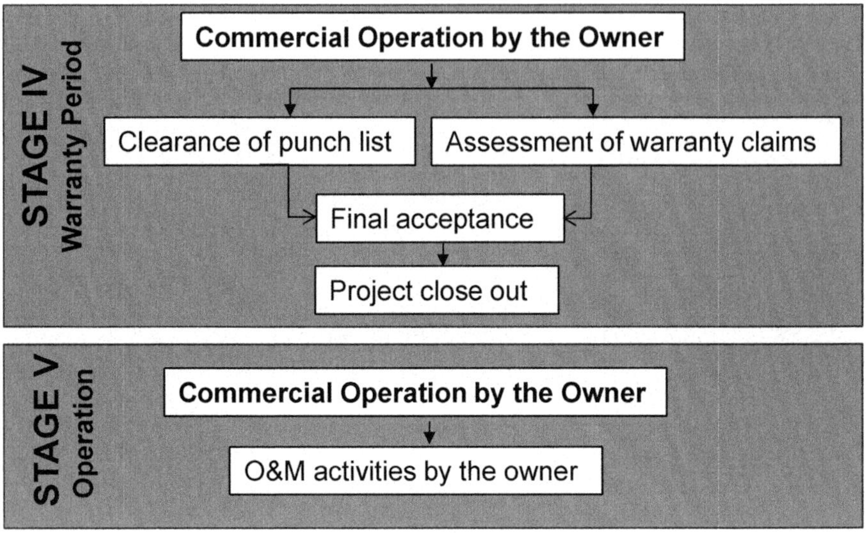

Figure 10-1: Overview of activities during project implementation

In stage I and II the risks are mainly potential gaps in technical specifications and the tender documents that may result to additional unforeseen equipment deliveries or services during the construction phase. Such kinds of omissions are charged with considerably higher costs by the contractors and will be borne by the owner. Possible are also delays in construction permits and operational licensing processes. The owner should therefore engage a reputable consultant with expertise and references in similar size and types of projects.

Some preparatory works, such as tie-in to the local electrical and water grid for provision of power and utilities for the construction site and other infrastructure works are to be organized and shall be accomplished before the actual construction work can start. Incomplete site preparation may result in delays of the entire project and allow claims depending on the provisions of the contract. These issues are a crucial part of EPC contracts.

During construction – stage III – all risks are formally borne by the EPC contractor. Nevertheless, they may have serious consequences for the whole project and must be prevented. These are, e.g., delays in delivery of main components, delays due to strikes and shortages of skilled construction staff etc. Therefore, preventive mitigation measures and contingencies for acceleration of works after occurrence of such events must be considered in the time schedule.

Difficult to manage are risks which may arise from *force majeure*. These may include strikes, local uprisings, acts of terror or sabotage, natural disasters but also changes in taxation by governments or customs processing and clearance. These require special provisions for mitigation.

After completion of construction the commercial operation starts. Stage IV is the guarantee period that usually takes one year. During this time the punch list items are cleared. Finally, in stage V the actual commercial operation begins and the full responsibility of the plant and its performance is with the owner.

10.4 Key Agreements

The contractual agreements of large projects in the energy supply business are numerous and complex. They are interrelated with each other and must allow an acceptable risk allocation for all project participants.

10.4.1 Agreements for procurement and construction

The International Federation of Consulting Engineers, known as FIDIC (acronym for the French name *Féderation Internationale Des Ingénieurs-Conseils*), provides a range of contract templates and business practice documents for the Construction Industry (website http://www.fidic.org).

The FIDIC-Contracts are tailored for large construction projects, including international partners. They are very much comprehensive and detailed. They are compulsory or recommended for projects financed by the Worldbank and other multinational development banks and organizations[35]. Currently, the following templates are known:

- Works of Civil Engineering Construction, *"The Red Book"*
- Construction Contract & Subcontract, *updates "The Red Book"*
- EPC/Turnkey Projects, *"The Silver Book"*
- Electrical & Mechanical Works, *"The Yellow Book"*
- Design-Build and Turnkey, *"The Orange Book"*
- Plant & Design-Build Contract, *updates Yellow and Orange Book"*
- The Short Form of Contract for small projects, *"The Green Book"*

The acronyms refer to the color of the cover page of the contract templates.
FIDIC also publishes a Client/Consultant Model Services Agreement, *"The White Book"*, often used when appointing a consultant as the employer's representative for the above contracts.

FIDIC also runs a bookshop for publishing the contract templates [70].

[35] Partly retrieved form https://de.wikipedia.org/wike/FIDIC

10.4.2 Agreements for operation phase

The key agreements for the operation phase are shown in Figure 10-2 below. They mainly apply to power generation projects; however, reference is also taken to different types of power sector projects.

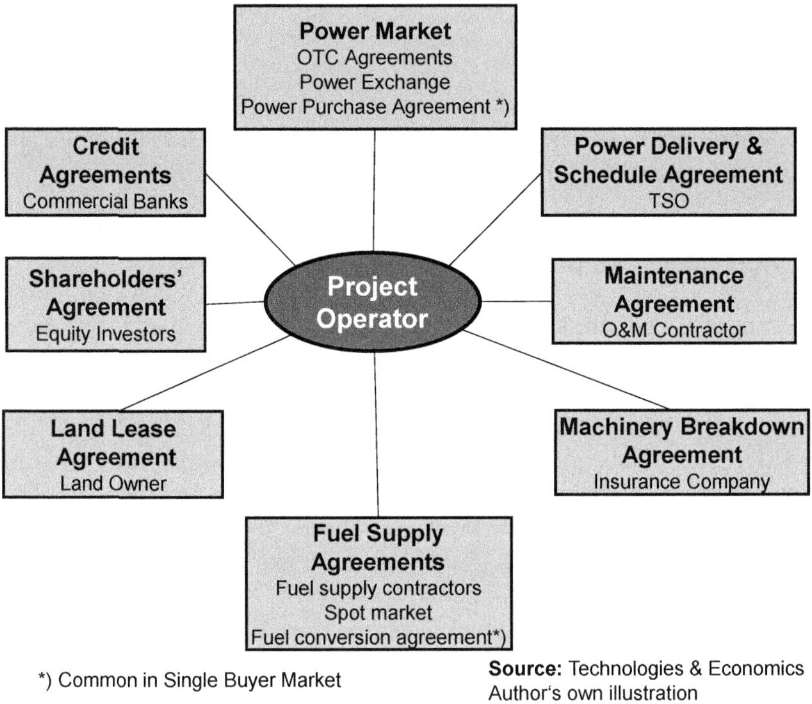

Figure 10-2: Key agreements of large power plant projects

Power market agreements: A most important issue for the financial performance of a project is to ensure product sales on the market place with adequate prices to cover cost and to ensure some reasonable profit margin for the equity investors. In this respect the power market model is of crucial importance.

In a *competitive market* power generators conclude power purchase agreements over the counter (OTC) for the highest possible share of their production, and also sell electricity in the Power Exchange. However, OTC agreements have a relatively short lifetime of one up to three years. Longer lifetimes may be obtained with attractive price concessions for the buyers. Furthermore, the contracts include price adjustment formulas often linked to spot market prices of the Power Exchange.

Selling electricity in the Power Exchange is a risky business. Power generators have to place bids for the day ahead with a fixed volume and price. Only bids up to the market clearing price MCP (see Figure 9-6) are *in merit* and will be executed the next day. Furthermore, prices in the Power Exchange are extremely volatile.

The space of volumes in merit is restricted by *must run* plants and power generation from renewable sources. Especially in systems with a high share of renewables, the volumes for fossil generated power are often limited due to excess production from renewables, and fossil fueled plants must be often shut-off.

In order to reduce risks, Power Exchanges offer instruments for hedging risks in the derivatives market such as futures' and options' contracts (see section 9.3.4). Nonetheless, power generation from fossil resources remains a risky business in the competitive market.

A privileged position has power generation from renewable energy. Transmission system operators are obligated to take-off power from renewable energy and remunerate with a feed-in tariff that in most cases is cost-covering and remains constant over the lifetime of the project. Furthermore, power generators from renewable resources are exempt from the obligation to provide generation schedules for each hour of the day-ahead.

In a *Single Buyer market* power generator companies or IPPs (Independent Power Producers) conclude a *power purchase agreement* with the Single Buyer for the operational lifetime of the project from the very beginning. This also includes the *power delivery scheduling agreement.* The Single Buyer is obliged to *take-or-pay* the produced electricity to the contractual agreed conditions regarding volumes and prices. On the other hand the power generator has the *delivery obligation*; this includes also a penalty clause for the case of non-performance. However, this risk is mainly allocated to the O&M contractor in the maintenance agreement and in the machinery damage insurance agreement. The risk is insofar limited for the power generator.

Investments in *transmission and distribution* projects are a relatively low risk business. Because of the natural monopoly nature, this business field is regulated and the investors know from the beginning market and price conditions.

Fuel supply agreements: Fuel prices are also very volatile in *liberalized markets.* Power generators usually try to secure a large part of their fuel demand with medium term supply contracts with fuel producers or traders, and buy the other part on the spot market in order to have some flexibility. Note, however, that for coal besides the FOB (free on board) price at the port of origin, also the oversea transportation costs are sometimes extremely high so that the CIF cost may become double as high as the FOB price [70]. Hedging of price risks with futures' and options' contracts are also common practice.

The risks are again different in *Single Buyer* markets and in competitive markets. In the former, power generators and IPPs sign a so-called *fuel conversion agreement* with the fuel supplier, who in most cases is a state owned monopoly company (e.g., Saudi Aramco). The power generator assumes the obligation to convert the delivered fuel into electricity according the terms of the agreement. A clause in the contract obliges the power generator to obtain the guaranteed fuel conversion efficiency during commercial operation. In the case of lower efficiency, the generator has to pay a penalty; on the other hand he will obtain a credit in the case of higher fuel conversion efficiency. This is, however, a relatively marginal risk and can be mitigated with a proper maintenance of the plant.

Land lease agreement: Land lease cost is an important cost item mainly for renewable energy power plants, because the land requirements are extremely high. For fossil fueled power plants the lease cost is not a significant cost factor. Problematic may be sometimes to obtain acceptance from population for construction sites. This is especially the case for waste-to-energy plants and for transmission line projects and in some cases also for wind farms.

If the site is property of the owner property, cost for leasing is allowed to be included in the tariffs in regulated markets.

Credit Agreements: About 70% to 75% of the capital expenditures for projects in the power sector are financed by loans from commercial banks. The duration of the loans is usually 15 years with a fixed interest rate. Banks thoroughly scrutinize the financial performance of projects before approving loans. First obligation of the power plant operators is to pay interest and annual redemption in due time and manner.

Banks on the other hand try to minimize risks by forming consortiums consisting of a number of banks to distribute risks. They also mitigate the risk exposure with appropriate risk premiums in the interest rates [71].

Shareholders agreement: Finally, equity investors expect that the free cashflow allows the expected dividend yield every year or the commercial operation of the project. They bear the major risk throughout the lifetime of the project.

Maintenance agreements: Maintenance of power plants is usually done by experienced contractors. Also suppliers of main components offer maintenance contracts. Details are described in sections 3.2.7 for gas turbines and 3.4.3 for internal combustion engines. Important is to know that they are remunerate based on equivalent operating hours (EOH), which include running hours plus equivalent hours for start-ups, fuel change etc.

Machinery breakdown agreement: This is an insurance contract covering losses resulting from accidental breakdown of any type of equipment according the terms of the contract. It also may cover reduction of revenues, if it is ensured. A longterm maintenance contract for the plant is required.

11 Case Studies

Introduction and Notes:

In this chapter, all Case Studies included in the book are presented as hardcopy along with some brief description of scope and purpose. Softcopies of the Case Studies can be downloaded from the author's websites:

www.PK-Energy-Practical-Knowhow.com
or
www.PK-Energie-Praxiswissen.com.

Note, however, softcopies usually show the calculations in the spreadsheets only. Explanations and descriptions of the Case Studies are available in the respective chapters of the book.

- For certain Case Studies **Add-Ins** (macros) for financial calculations or/and the software tool **FluidEXL** for calculation of water/steam properties are required. Both are available for download on the author's website. This is also noted on the cover page of each Case Study
- Readers are advised to study the brief instructions for Add-Ins, software tools and frequently used Excel® functions in the **Toolbox Section** of the book
- Detailed descriptions of Case Studies, including explanations for calculations in the spreadsheets, are presented in the respective chapters of the book. Reference is given in the cover page of each Case Study
- Comments and recommendations for improvement from readers or pointing out any errors are highly appreciated and will be considered in forthcoming updates.

Note: The Case Studies below are in black & white, in the website they are presented colored!

P. Konstantin and M. Konstantin, *The Power Supply Industry*,
https://doi.org/10.1007/978-3-319-72305-1_11

11.1 Rankine Cycle in T-s diagram calculated with FluidEXL

PK Panos Konstantin

The Power Supply Industry

Case Study

Rankine Cycle in T-s Diagram Calculated with FluidEXL

Notes:

1. Cells with black characters include inputs
2. Cells with red characters include formulas
3. Download of the FluidEXL software tool is required !
4. Read FluidEXL brief instruction in the **Toolbox** of the book !
5. Read introduction in the Case Study chapter of the book !

Last update October 2017

Introduction and Objective of this Case Study

The purpose of the Case Study is to help readers of the book to become familiar with the software tool FluidEXL^{Graphics} for calculation of water/steam properties.

A brief operating instruction of FluidEXL is available **Tool Guide 1** in the **Toolbox Section** of the book. Users are advised to read this instruction before starting with calculations.

The application of the tool is practiced on the example of a Rankine cycle with super critical steam parameters and steam reheat depicted in T-s diagram below. (**Note:** internal turbine efficiency = 100% for simplification).

The calculations are conducted in an Excel® spreadsheet, where most of the functions of FluidEXL used in examples of this book are used.

A second spreadsheet with the same structure is provided for practicing. Input values are included while the cells with calculations are empty. The user shall calculate the required water/steam properties using corresponding functions from FluidEXL and can compare the results of his calculations with those of the first spreadsheet.

T-s Diangram: Rankine Cycle, live steam 285bar/600°C, Reheat 60bar/620°C, Condensation 0.05 bar/33°C

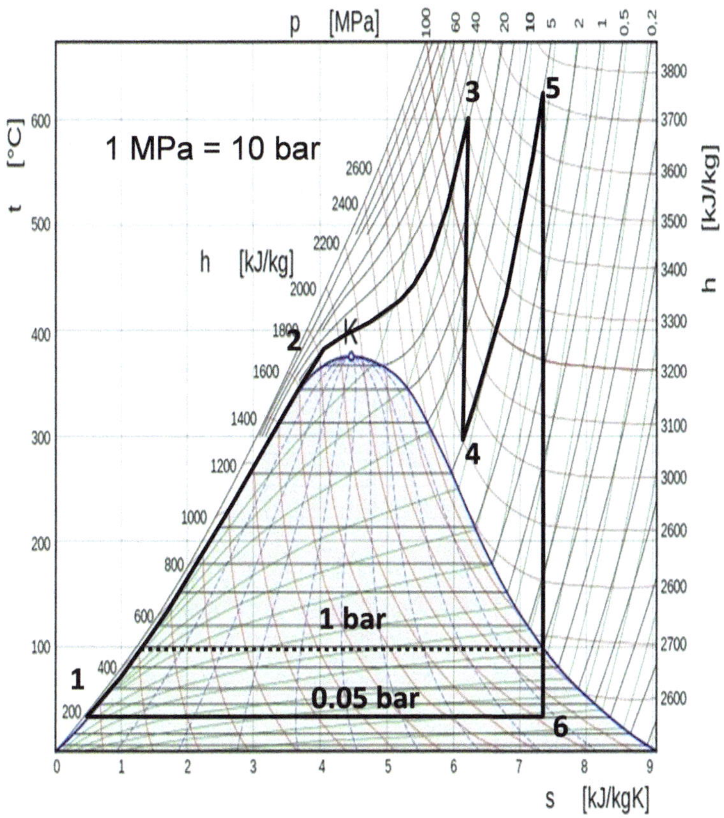

Spreadsheet 1: Spreadsheet with calculated water/steam properties

Point	Pressure p	Temperature t	Enthalpy h	Entropy s	Vapor fraction x
	bar	°C	kJ / kg	KJ / kg K	kg/kg
1	0.05	32.9	137.8	0.48	0.00
3	285	600.0	3461.0	6.27	1.00
4	60	336.9	3005.8	6.27	1.00
5	60	620.0	3705.8	7.22	1.00
6	0.05	32.9	2202.3	7.22	0.85
1	0.05	32.9	137.8	0.48	0.00

Item	Symbols	Unit	Value
Mass flow	m	kg	1
Heat input 1 - 3	Q_{in}	kJ/kg	3,323
Heat input 4 - 5			700
Condenser 6 - 1	Q_{out}	kJ/kg	2,065
Mechanical work	$W = Q_{out} - Q_{in}$	kJ / kg	1,959
		kWh / t	544
Cycle efficiency	$\eta = W / Q_{in}$	%	48.7%
Steam content of exhaust steam		%	85.2%

Spreadsheet 2: User's spreadsheet for practicing FluidEXL

Point	Pressure p	Temperature t	Enthalpy h	Entropy s	Steam content x
	bar	°C	kJ / kg	KJ / kg K	kg /kg
1	0.05				0.00
3	285	600.0			1.00
4	60				1.00
5	60	620.0			1.00
6	0.05				
1	0.05				

Item	Symbols	Unit	Value
Mass flow	m	kg	
Heat input 1 - 3	Q_{in}	kJ / kg	
Heat input 4 - 5			
Condenser 6 - 1	Q_{out}	kJ / kg	
Mechanical work	$W = Q_{out} - Q_{in}$	kJ / kg	
		kWh/t	
Cycle efficiency	$h = W / Q_{in}$	%	
Steam content of exhaust steam		%	

Note: All empty cells shall be calculated

11.2 Modeling and Calculation of a Simple Rankine Cycle

PK Panos Konstantin

The Power Supply Industry

Case Study
Modelling & Calculation
of a Simple Rankine Cycle
with FluidEXL

Notes:
1. Cells with black characters include inputs
2. Cells with red characters include formulas
3. Dowload of FluidEXL is required
4. Read FluidEXL brief instruction in the Toolbox of the book
5. Red Inttroduction in the Case Study chapter of the book

Last update October 2017

Objective of the Case Study

Thermodynamic cycles are commonly calculated with professional software tools available on the market such as Fichtner's KPRO®. However, some background knowledge about the thermodynamics of cycle calculations is indispensable for users of such kind of tools, especially for those with little experience.

A stepwise guidance for calculation of the Rankine cycle depicted below is provided in the toolbox section of the book – **Tool Guide 3**. Users are advised to read Guide and follow the instructions during practicing the example in this case study. For calculations of water/steam properties FluidEXL is used – brief instruction see **Tool Guide 1**.

Note: The following mass and energy flow diagram and the tables of all calculations have been conducted in a first Excel spreadsheet. A second spreadsheet of the same structure with empty calculation cells is provided for practicing, where the calculation shall be conducted by the user.

Mass and energy flow diagram of the cycle

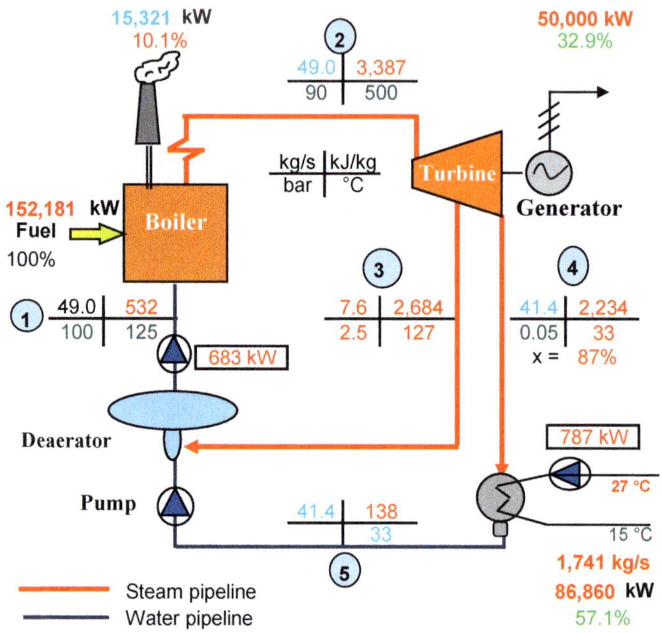

Make Inputs only in this box		
Live steam	t/h	176.5
Pressure	bar	90
Temperature	°C	500
Boiler feed water temperature	°C	125
Extraction to deaerator:		
Pressure	bar	2.5
Temperature	°C	computed
Condeser pressure	bar	0.05
Terminal termperature difference	K	6
Cooling water inlet temperature	°C	15
Turbine thermodynamic efficiency	%	85%
Boiler efficiency	%	92%
mechanical turbine efficiency	%	98%
Generator efficiency	%	96%

Calculated Cycle				
Live steam			kg /s	49.0
Pressure			bar	90
Enthalpy			kJ/kg	3,387
Entropy			kJ/kg*K	6.660
Saturation temperature			°C	303
Isentropic expansion				
Condenser pressure			bar	0.05
Enthalpy			kJ/kg	2,030
Enthalpy difference, isentropic			kJ/kg*K	1,357
Real expansion				
Turbine thermodynamic efficiency			%	85%
Enthalpy difference			kJ/kg	1,154
Condensate enthalpy			kJ/kg	2,234
Temperature exhaust steam			°C	33
Steam content of the exhaust steam			%	87%
Extraction				
Enthalpy, isentropic Expansion			kJ/kg	2,559
Enthalpy difference, isentropic			kJ/kg	828
Enthalpy difference, real			kJ/kg	704
Enthalpy real			kJ/kg	2,684
Mass flow			kg /s	7.6
Temperature			°C	127
Saturation temperature			°C	127
Power generation			**kW_e**	**50,000**
Condensing steam	98%	96%	kW_e	44,975
Extraction steam	98%	96%	kW_e	5,025
Fuel energy input to boiler		92%	**kW_t**	**152,181**
Condenser				
Cooling water outlet			°C	27
Mass flow exhaust steam			kg /s	41.4
Heat reject			kW_t	86,860
Cooling water mass flow		4.2 kj/kgK	kg /s	1,741
Cooling water pump	76%	35 m	kW_e	787
Feed water pump				
Mass flow			kg /s	49.0
Pressure head			bar	100
Water density			kg /m3	944
Pump motor Power			kWe	683

For Calculation by the User					
Live steam			kg /s	49.0	
Pressure			bar	90	
Enthalpy			kJ/kg		FluidEXL
Entropy			kJ/kg*K		FluidEXL
Saturation temperature			°C		FluidEXL
Isentropic expansion					
Condenser pressure			bar	0.050	
Enthalpy			kJ/kg		FluidEXL
Enthalpy difference, isentropic			kJ/kg*K		FluidEXL
Real expansion					
Turbine thermodynamic efficiency			%	85%	
Enthalpy difference			kJ/kg		FluidEXL
Condensate enthalpy			kJ/kg		FluidEXL
Temperature exhaust steam			°C		FluidEXL
Steam content of the exhaust steam			%		FluidEXL
Extraction					
Enthalpy, isentropic Expansion			kJ/kg		FluidEXL
Enthalpy difference, isentropic			kJ/kg		FluidEXL
Enthalpy difference, real			kJ/kg		FluidEXL
Enthalpy real			kJ/kg		FluidEXL
Mass flow			kg /s		FluidEXL
Temperature			°C		FluidEXL
Saturation temperature			°C		FluidEXL
Power generation			**kW_e**		
Condensing steam	98%	96%	kW_e		calculate
Extraction steam	98%	96%	kW_e		calculate
Fuel energy input to boiler		92%	**kW_t**		calculate
Condenser					
Cooling water outlet			°C		calculate
Mass flow exhaust steam			kg /s		calculate
Heat reject			kW_t		calculate
Cooling water mass flow		4.2 kj/kgK	kg /s		calculate
Pump motor Power	76%	35 m	kW_e		calculate
Feed water pump					
Mass flow			kg /s	49.0	
Pressire head			bar	100	
Water density ρ			kg /m3		FluidEXL
Pump motor Power			kWe	#DIV/0!	calculate ρ

11.3 Demo – Development History of Steam Rankine Cycle

PK Panos Konstantin

Power and Energy Systems

Technologies & Economics

Case Study
Development History
of Steam Rankine Cycle
Presented in T-s Diagram

Notes:
Internal efficiency = 1 for simplification

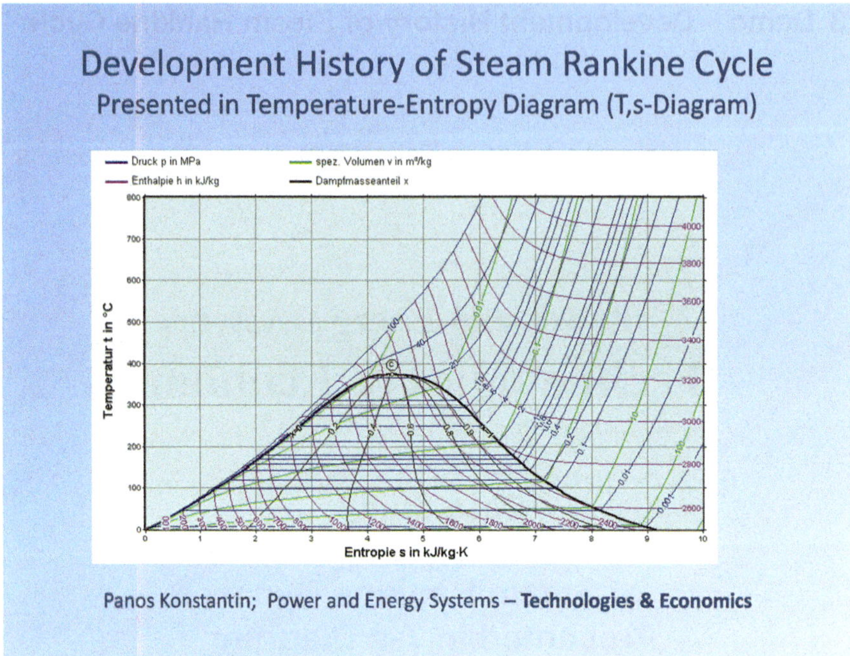
Development History of Steam Rankine Cycle
Presented in Temperature-Entropy Diagram (T,s-Diagram)
Druck p in MPa
Enthalpie h in kJ/kg
spez. Volumen v in m³/kg
Dampfmasseanteil x
Temperatur t in °C
Entropie s in kJ/kg·K
Panos Konstantin; Power and Energy Systems – Technologies & Economics

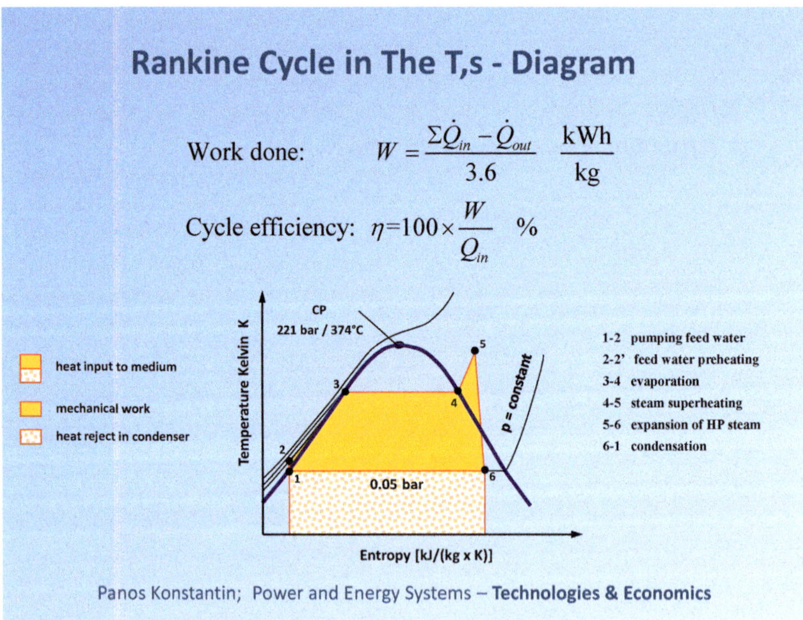
Rankine Cycle in The T,s - Diagram
Work done: $W = \frac{\Sigma \dot{Q}_{in} - \dot{Q}_{out}}{3.6}$ $\frac{kWh}{kg}$
Cycle efficiency: $\eta = 100 \times \frac{W}{Q_{in}}$ %
heat input to medium
mechanical work
heat reject in condenser
CP
221 bar / 374°C
Temperature Kelvin K
p = constant
0.05 bar
Entropy [kJ/(kg x K)]
1-2 pumping feed water
2-2' feed water preheating
3-4 evaporation
4-5 steam superheating
5-6 expansion of HP steam
6-1 condensation
Panos Konstantin; Power and Energy Systems – Technologies & Economics

Rankine Cycle Open Type without Condenser

Steam locomotive 20 bar/212 °C, Exhaust steam 1 bar

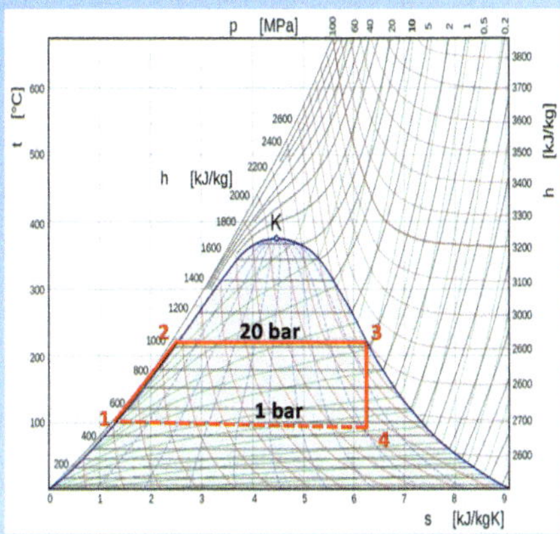

1 → 2: Heating to boiling temperature 212 ° C
2 → 3: Evaporation at 20 bar/212.6 °C
3 → 4: Expansion at 1 bar and 99.6 °C

Energy flow	Symbols	Unit	Value
Heat input 1 - 3	Q_{in}	kJ / kg	2,381
Heat reject 3 - 4	Q_{out}	kJ / kg	1,877
Mechanical work	$W = Q_{out} - Q_{in}$	kJ / kg	504
		kWh/t	140
Cycle efficiency	$\eta = W / Q_{in}$	%	21.1%

Panos Konstantin; Power and Energy Systems – **Technologies & Economics**

Rankine Cycle

Steam Turbine Superheat 60 bar/400 °C, Exhaust Steam 1 bar
Small scale PP without condenser

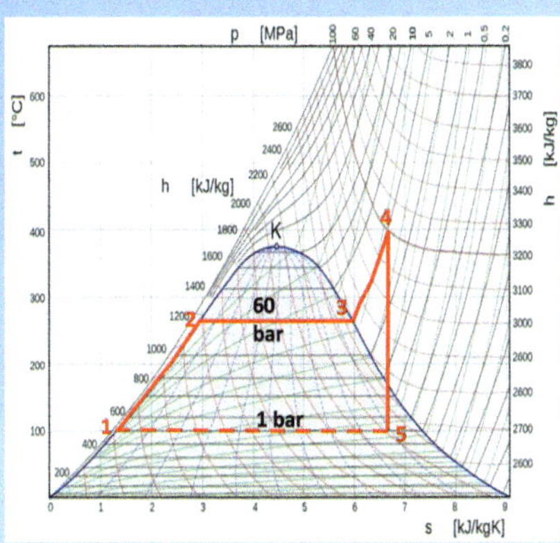

1 → 2: Heating to boiling temperature 275.6 ° C
2 → 3: Evaporation at 60 bar/275.6 °C
3 → 4: Superheating at 400° C
4 → 5: Expansion at 1 bar and 99.6 °C

Point 5: Steam content of exhaust steam 87%

Energy flow	Symbols	Unit	Value
Heat input 1 - 4	Q_{in}	kJ / kg	2,761
Heat reject 1 - 4	Q_{out}	kJ / kg	1,953
Mechanical work	$W = Q_{out} - Q_{in}$	kJ / kg	807
		kWh/t	224
Cycle efficiency	$\eta = W / Q_{in}$	%	29.2%

Panos Konstantin; Power and Energy Systems – **Technologies & Economics**

Rankine Cycle

Steam Turbine 180 bar/400°C, condensation at 1 bar

Measure: Increasing pressure but keeping temperature at the same level
Result: higher efficiency but too low steam content in exhaust steam

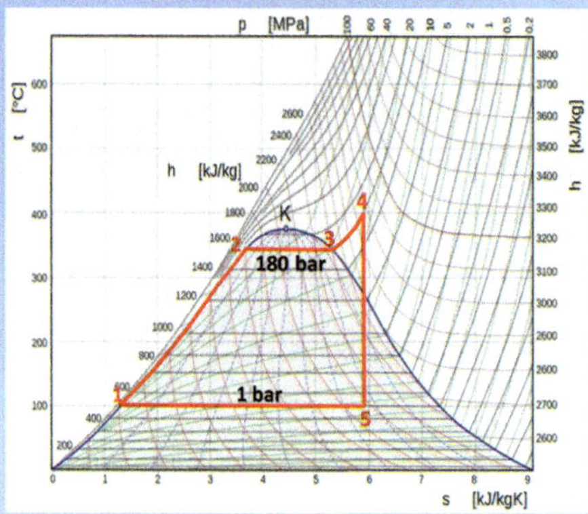

1 → 2: Heating to boiling temperature 357° C
2 → 3: Evaporation at 180 bar /357 °C
3 → 4: Superheating at 400° C
4 → 5: Expansion at 1 bar and 99.6 °C

Energy flow	Energy flow	Unit	Value
Heat input 1 - 4	Q_{in}	kJ / kg	2,469
Condensation heat 5 - 1	Q_{out}	kJ / kg	1,635
Mechanical work	$W = Q_{out} - Q_{in}$	kJ / kg	834
		kWh/t	232
Cycle efficiency	$\eta = W / Q_{in}$	%	33.8%
Steam content of exhaust steam		%	72%

Panos Konstantin; Power and Energy Systems – **Technologies & Economics**

Rankine Cycle Power Plant

Steam turbine 180 bar/520°C, condensation at 1 bar

Measure: Keeping pressure constant and Increasing temperature to obtain higher steam content
Result: higher efficiency but steam content still low

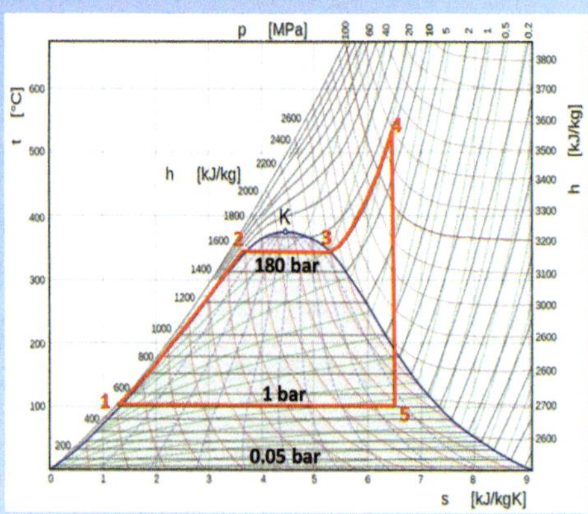

1 → 2: Heating to boiling temperature 357° C
2 → 3: Evaporation at 180 bar
3 → 4: Superheating at 520° C
4 → 5: Expansion at 1 bar and 99.6 °C

Energy flow	Energy flow	Unit	Value
Heat input 1 - 4	Q_{in}	kJ / kg	2,913
Condensation heat 5 - 1	Q_{out}	kJ / kg	1,863
Mechanical work	$W = Q_{out} - Q_{in}$	kJ / kg	1,050
		kWh/t	292
Cycle efficiency	$\eta = W / Q_{in}$	%	36.1%
Steam content of exhaust steam		%	82.5%

Panos Konstantin; Power and Energy Systems – **Technologies & Economics**

Rankine Cycle Closed Type with Condenser

Power Plant 180 bar/520 to 0.05 bar/33°C

Measure: Lowering condensing pressure to vacuum level

Result: Substantial efficiency improvement, however steam content far too low

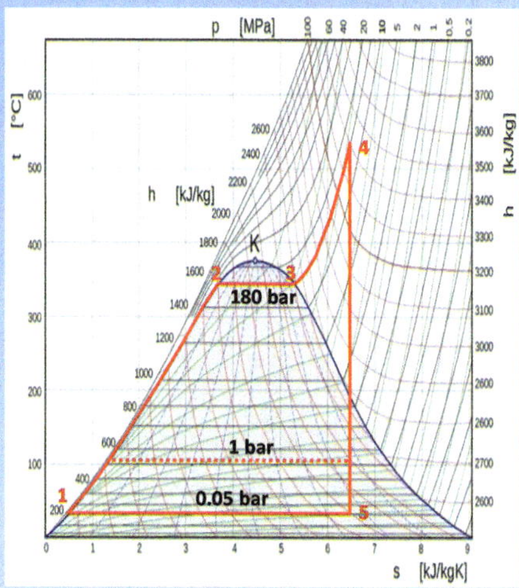

1 → 2: Heating to boiling temperature 357° C
2 → 3: Evaporation at 180 bar
3 → 4: Superheating at 520° C
4 → 5: Expansion 180 bar to 0.05 bar
5 → 1 Condensation at 0.05 bar, 33 °C

Energy flow	Energy flow	Unit	Value
Heat input 1 - 4	Q_{in}	kJ / kg	3,193
Condensation heat 5 - 1	Q_{out}	kJ / kg	1,782
Mechanical work	$W = Q_{out} - Q_{in}$	kJ / kg	1,411
		kWh/t	392
Cycle efficiency	$\eta = W / Q_{in}$	%	44.2%
Steam content of exhaust steam (too low !)		%	73.6%

Increased cavitation and destruction of blades !!!

Panos Konstantin; Power and Energy Systems – **Technologies & Economics**

Rankine Cycle with Reheat, Subcritical

Power Plant 180 bar/520; Reheating to 520 °C; Expansion 0.05 bar/33°C

Measure: Reheating of steam

Result: Substantial efficiency improvement, and acceptable steam content

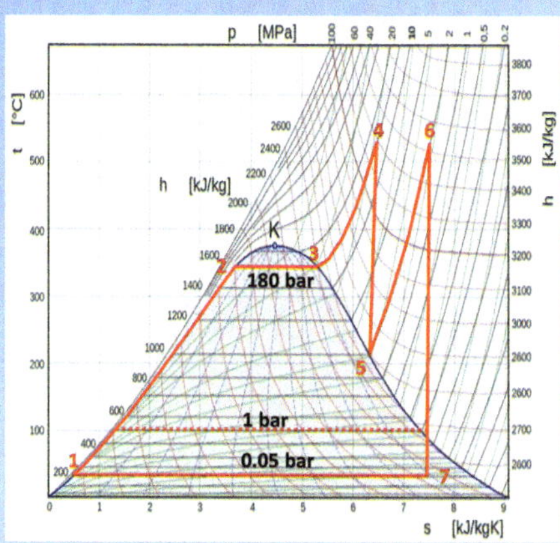

1 → 2: Heating to boiling temperature 357° C
2 → 3: Evaporation at 180 bar
3 → 4: Superheating at 520° C
4 → 5: Expansion in high pressure turbine at 30 bar
5 → 6: Reheating at 30 bar / 520° C
6 → 7: Expansion to Condensing pressure 0.05 bar, 32.9 °C

Energy flow	Symbols	Unit	Value
Heat input 1 - 4	Q_{in}	kJ / kg	3,193
Heat input 5 - 6			640
Condenser 7 - 1	Q_{out}	kJ / kg	2,086
Mechanical work	$W = Q_{out} - Q_{in}$	kJ / kg	1,747
		kWh/t	485
Cycle efficiency	$\eta = W / Q_{in}$	%	45.6%
Steam content of exhaust steam		%	86.1%

Panos Konstantin; Power and Energy Systems – **Technologies & Economics**

Rankine Cycle with Reheat, Super critical

Power Plant 285 bar/600 °C; Reheating to 620 °C; Expansion to 0.05 bar/33°C

Measure: Super critical steam parameters, no condensation

Result: Substantial efficiency improvement , however higher material requirements

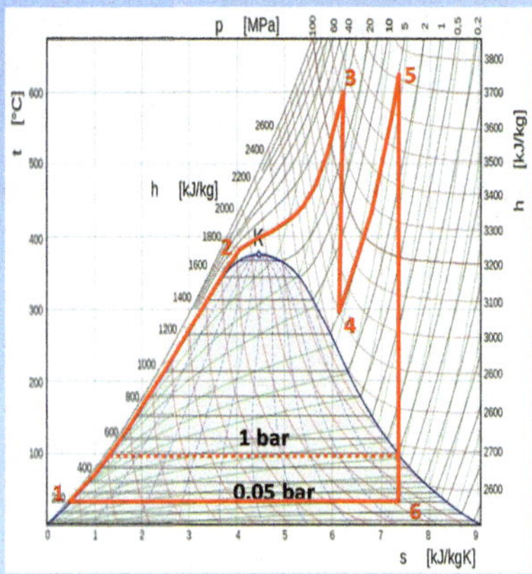

1 → 3: Heat input up to 285 bar and 600° C
3 → 4: Expansion 285 bar to 60 bar
4 → 5: Reheating at 60 bar to 620° C
5 → 6: Expansion in Condensing turbine to 0.05 bar, 32.9 °C
6 → 1: Condensation

Energy flow	Symbols	Unit	Value
Heat input 1 - 3	Q_{in}	kJ / kg	3,323
Heat input 4 - 5			700
Condenser 6 - 1	Q_{out}	kJ / kg	2,065
Mechanical work	$W = Q_{out} - Q_{in}$	kJ / kg	1,959
		kWh/t	544
Cycle efficiency	$\eta = W / Q_{in}$	%	48.7%
Steam content of exhaust steam		%	85.2%

Panos Konstantin; Power and Energy Systems – **Technologies & Economics**

Sub-Critical vs. Ultra-Super-Critical Rankine Cycle in T,s - Diagram

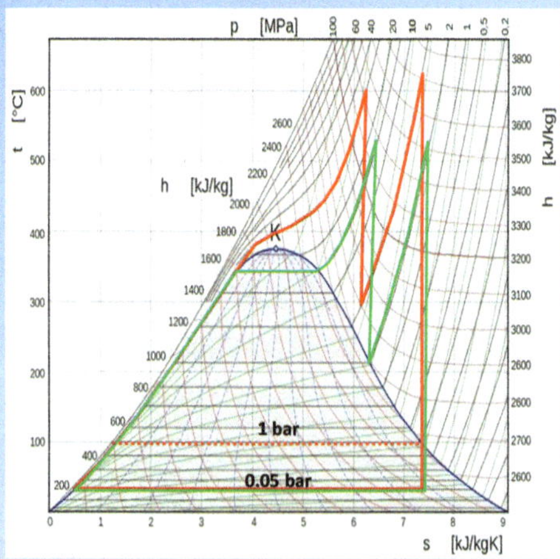

Subcritical (SC) - Ultra Super Critical (USC)

Panos Konstantin; Power and Energy Systems – **Technologies & Economics**

11.4 Integrated Techno-Economic Model for Fossil PPs

PK **Panos Konstantin**

The Power Supply Industry

Case Study
Integrated Model
Generation Cost Calculation
for Fossil Fueled Power Plants

Notes:

1. Cells with black characters include inputs
2. Cells with red characters include formulas
3. Download of Add-Ins (macro) from website required
4. Read description of Case Study in section 3.5 of the book !

Last update October 2017

Spreadsheets

- Summary of results
- Input technical parameters
- Input technical parameters
- Input financial parameters
- Input WACC, discount rate
- CAPEX estimate
- Equivalent operating hours EOH estimate
- Generation Costs Calculations
- Graph Levelized electricity generation cost vs. full load hours
- Graph Structure of electricity generation costs

Summary of results

Item	Unit	Steam USC coal	Steam SubC coal	CCGT nat. gas	IC Engine HFO	IC Engine LFO	GT LFO
Energy balance							
Number of units	-	1	1	1	24	20	2
Power output net	MW	744	555	404	402	335	329
Net electricity production	GWh_e /a	5,566	4,152	1,959	2,009	502	478
Fuel consumption	GWh_t / a	11,843	10,127	3,320	4,481	1,120	1,462
Financial constraints							
Life time	a	35	35	25	25	25	20
Construction time	a	5.0	5.0	2.5	2.0	2.0	1.5
Discount rate (WACC), real terms	% / a	6.5%	6.5%	6.5%	6.5%	6.5%	6.5%
Fuel price	US$ / MWh_{LHV}	19.3	19.3	39.5	38.0	61.3	61.3
CAPEX, 2014 US$, ±25%	**mln US$**	**2,440.5**	**1,564.2**	**404.3**	**415.0**	**282.3**	**120.6**
Annual power generation costs	**mln US$ / a**	**462.6**	**348.5**	**179.7**	**227.5**	**101.7**	**109.9**
Annualized CAPEX	mln US$ / a	177.7	113.9	33.1	33.9	23.1	10.9
OPEX fixed	mln US$ / a	51.2	35.4	8.6	8.0	6.1	3.2
OPEX variable, incl. fuel costs	mln US$ / a	233.6	199.2	138.0	185.6	72.5	95.8
Power generation cost, levelized							
Capacity (fixed OPEX + Annualized capex)	US$ / (kWa)	307.7	269.0	103.2	104.3	87.1	42.8
Energy (variable cost)	US$ / MWh	42.0	48.0	70.5	92.3	144.3	200.4
Composite cost, excl. CO_2-cost *)	**US$ / MWh_e**	**83.10**	**83.93**	**91.75**	**113.21**	**202.35**	**229.85**
Composite cost, incl. CO_2-cost *))**	**US$ / MWh_e**	**86.74**	**88.10**	**93.46**	**116.34**	**205.32**	**233.92**
*) referred to full load hours	h /a	7,481	7,481	4,850	5,000	1,500	1,455
**) Spec. Emission cost	US$ / t_{CO2}	5.0	(for example, spreadsheet Input financial)				

Input technical parameters

Item	Unit	Steam USC coal	Steam SubC coal	CCGT nat. gas	IC Engine HFO	IC Engine LFO	GT LFO
Key technical parameters							
Selecte type	-	USC	SubC	Siemens SGT6 8000H 1GTx1ST	Wärtsilä W18V50DF	Wärtsilä W18V50DF	Siemens SGT5-2000E
Load segment	-	base	base	intermediate	intermediate	peak	peak
Number of units	-	1	1	1	24	20	2
Unit power output RSC, gross		800.0	600.0	410.0	17.0	17.0	166.0
Plant power output, RSC, gross	MW_e	800.0	600.0	410.0	408.0	340.0	332.0
Auxiliary power demand	-	7.0%	7.5%	1.5%	1.5%	1.5%	1.0%
Power output, net	MW_e	744	555	404	402	335	329
Electrical efficiency, gross	-	48.0%	42.0%	60.0%	45.3%	45.3%	34.7%
Rated efficiency, gross	-	48.0%	42.0%	60.0%	45.3%	45.3%	34.7%
Degradation *)	-	1.0%	1.0%	1.0%	0.5%	0.5%	2.0%
Average annual efficiency	-	47.0%	41.0%	59.0%	44.8%	44.8%	32.7%
Average heat rate	MJ / kWh_e	7.66	8.78	6.10	8.03	8.03	11.01
CO_2 emission factors of fuels	kg/MWh_{LHV}	342	342	202	281	266	266

*) due to wear, deterioration and part load

Input financial parameters

Item		Unit	Steam USC coal	Steam SubC coal	CCGT nat. gas	IC Engine HFO	IC Engine LFO	GT LFO
Financial constraints	Escalation *)	no						
Life time	n.a.	a	35	35	25	25	25	20
Construction time	n.a.	a	5.0	5.0	2.5	2.0	2.0	1.5
Number of operating staff	n.a.	-	90	90	35	25	25	25
Discount rate (WACC), real terms	n.a.	% / a	6.47%	6.47%	6.47%	6.47%	6.47%	6.47%
Fixed O&M cost	0.5 %/a	%EPC / a	1.5%	1.5%	1.0%	1.0%	1.0%	0.5%
Costs of personnel	1.0 %/a	US$ / (cap*a)	80,000	80,000	80,000	80,000	80,000	80,000
Insurance	0.5 %/a	%EPC / a	0.5%	0.5%	0.5%	0.5%	0.5%	0.5%
Fuel price ref. to LHV	0.5 %/a	US$ / MWh_{LHV}	19.26	19.26	39.51	38.02	61.32	61.32
Non-fuel variable cost	0.5 %/a	US$ / MWh_e	1.00	1.00	0.50	1.50	1.50	0.50
Maintenance contract, fixed **)	0.0 %/a	US$ / MWa	n.a.	n.a.	5,500	7,875	7,875	5,500
Maintenance contract, variable **)	0.0 %/a	US$ / MWh_e	n.a.	n.a.	2.67	5.95	5.95	4.00
CO_2 emission allowance price	0.0 %/a	US$ / t_{CO2}			5.0			

*) Escalation rates in real terms, exlcuding inflation key in "yes" or "no"

**) for Gas turbines and ICE, for CCGT 2/3 of variable part of the price referred to the total output

Input crude oil barrel price		➡	80 US$/bbl	1.62 MWht/bbl				
Fuel		**Crude oil**	**Steam Coal**	**Steam Coal**	**Natural gas**	**HFO**	**LFO**	**LFO**
Ratio referred to crude oil *)	-	1.00	0.39	0.39	0.80	0.77	1.24	1.24
Specific cost free PP site	US$ / MWh_{LHV}	49.38	19.26	19.26	39.51	38.02	61.32	61.32

Input WACC, discount rate

Item	Equity	Loan
Asset shares	**30%**	**70%**
Expected returns after tax		
Risk free rate of return / interest	5.0 %/a	5.0 %/a
Venture risks premium	6.0 %/a	1.0 %/a
Country risk premium (depends on country) *)	0.0 %/a	0.0 %/a
Cost of capital in nominal terms, after tax	**11.0 %/a**	**6.0 %/a**
Corporate tax 25%	3.7 %/a	0.0 %/a
Cost of capital in nominal terms, before tax	**14.7 %/a**	**6.0 %/a**
***$WACC_n$* in nominal terms, before tax**	**8.60 %/a**	
./. Expected Inflation rate	2.00 %/a	
***$WACC_r$* inflation adjusted**	**6.47 %/a**	

CAPEX estimate

Item		Unit	Steam USC coal	Steam SubC coal	CCGT nat. gas	IC Engine HFO	IC Engine LFO	GT LFO
Constrains								
Specific capex, EPC		US$ /kW	2,750	2,350	950	980	800	350
Construction time		a	5.0	5.0	2.5	2.0	2.0	1.5
Bank Interest rate, nominal (see WACC)		%!a	6.00%	6.00%	6.00%	6.00%	6.00%	6.00%
Installments per year		-	4	4	4	4	4	4
CAPEX, 2014 US$, ±25%		mln US$	**2,440.5**	**1,564.2**	**404.3**	**415.0**	**282.3**	**120.6**
Overnight cost, EPC		mln US$	2,200.0	1,410.0	389.5	399.8	272.0	116.2
Owner cost	15%	mln US$	330.0	211.5	58.4	60.0	40.8	17.4
Interest during construction IDC	70% debt	mln US$	240.5	154.2	14.8	15.1	10.3	4.4

Equivalent operating hours EOH estimate

Item	Unit	Steam USC coal	Steam SubC coal	CCGT nat. gas	IC Engine HFO	IC Engine LFO	GT LFO
Full load hours							
Full load hours, for load segment	h / a	7,500	7,500	5,000	5,000	1,500	1,500
forced outages *)	%	0.25%	0.25%	3.00%	0.00%	0.00%	3.00%
Actual full load hours	h / a	7,481	7,481	4,850	5,000	1,500	1,455
Equivalent operating hours							
Startups	-	1 / month	1 / month	1 / week	multiunit plant		daily
Number of startups	-	12	12	52	unlimited	unlimited	300
Addition for startups * 10 h/start	h / a	120	120	520	-	-	3,000
Equivalent operating hours *)	h / a	7,620	7,620	5,520	5,000	1,500	4,500
O&M contract for GT/ICE							
Fixed price O&M contract	US$ / a	n.a.	n.a.	5,500	7,875	7,875	5,500
Variable price O&M contract **)	US$ / MWh_e	n.a.	n.a.	2.67	5.95	5.95	4.00

*) for multiunit IC engine PP no forced outages and full load hours = EOH

**) Variable price is referred to the electricity production of the gas turbines which is assumed to be 2/3 of the total (2/3x4=2.67)

Generation Costs Calculations

Item	Unit	Steam USC coal	Steam SubC coal	CCGT nat. gas	IC Engine HFO	IC Engine LFO	GT LFO
Energy and emission balance							
Power output net	MW	744	555	404	402	335	329
Full load hours, actual *)	h / a	7,481	7,481	4,850	5,000	1,500	1,455
Net electricity production	GWh_e / a	5,566	4,152	1,959	2,009	502	478
Annual average, gross efficiency	-	47.0%	41.0%	59.0%	44.8%	44.8%	32.7%
Equivalent operating hours	h / a	7,601	7,601	5,370	5,000	1,500	4,455
Fuel consumption	GWh_t / a	11,843	10,127	3,320	4,481	1,120	1,462
CO_2 emission factors	kg /MWh_t	342	342	202	281	266	266
CO_2 emissions	kt / a	4,050	3,463	671	1,259	298	389
Financial constraints							
Life time	a	35	35	25	25	25	20
Construction time	a	5.0	5.0	2.5	2.0	2.0	1.5
Number of operating staff	-	90	90	35	25	25	25
Discount rate (WACC), real term	% / a	6.5%	6.5%	6.5%	6.5%	6.5%	6.5%
Fixed O&M cost	%EPC / a	1.5%	1.5%	1.0%	1.0%	1.0%	0.5%
Insurance	%EPC / a	0.5%	0.5%	0.5%	0.5%	0.5%	0.5%
Costs of personnel	US$ / (cap*a)	80,000	80,000	80,000	80,000	80,000	80,000
Fuel price ref. to LHV, based on	US$ / MWh_t	19.26	19.26	39.51	38.02	61.32	61.32
Non-fuel variable cost	US$ / MWh_e	1.00	1.00	0.50	1.50	1.50	0.50
Maintenance contract, fixed	US$ / MW a	n.a.	n.a.	5,500	7,875	7,875	5,500
Maintenance contract, variable	US$ / MWh_e	n.a.	n.a.	2.67	5.95	5.95	4.00
CAPEX, 2014 US$, ±25%	**mln US$**	**2,440.5**	**1,564.2**	**404.3**	**415.0**	**282.3**	**120.6**
Annualized CAPEX	**mln US$ / a**	**177.7**	**113.9**	**33.1**	**33.9**	**23.1**	**10.9**
OPEX, fixed, levelized	**mln US$ / a**	**51.2**	**35.4**	**8.6**	**8.0**	**6.1**	**3.2**
Maintenance	mln US$ / a	33.0	21.2	3.9	4.0	2.7	0.6
Personnel	mln US$ / a	7.2	7.2	2.8	2.0	2.0	2.0
Insurance	mln US$ / a	11.0	7.1	1.9	2.0	1.4	0.6
OPEX, variable, levelized	mln US$ / a	**233.6**	**199.2**	**138.0**	**185.6**	**72.5**	**95.8**
Fuel costs	mln US$ / a	228.1	195.0	131.2	170.4	68.7	89.7
Non-fuel variable cost	mln US$ / a	5.6	4.2	1.0	3.0	0.8	0.2
GT - maintenance	mln US$ / a	0.0	0.0	5.9	12.1	3.0	5.9
Levelized total annual costs	**mln US$ / a**	**462.6**	**348.5**	**179.7**	**227.5**	**101.7**	**109.9**
Power generation cost, Levelized							
Capacity (fixed OPEX + annualized C	US$ / (kW*a)	307.7	269.0	103.2	104.3	87.1	42.8
Energy (variable costs)	US$ / MWh	41.98	47.97	70.46	92.34	144.30	200.41
of which fuel cost, only	US$ / MWh	40.98	46.97	66.96	84.80	136.76	187.54
*Full load hours, effective *)*	*h / a*	*7,481*	*7,481*	*4,850*	*5,000*	*1,500*	*1,455*
Composite cost, excl. CO_2 cost	**US$ / MWhe**	**83.10**	**83.93**	**91.75**	**113.21**	**202.35**	**229.85**

*) forced outages during operation considered **) incl. allowance for wear, deterioration and part load

Power generation costs, inlcl. CO_2 costs

Annual costs for CO_2 emissions	mln US$ /a	20.25	17.32	3.35	6.30	1.49	1.95
Specific cost for CO_2 emissions	US$ / MWh	3.64	4.17	1.71	3.13	2.97	4.07
Composite cost, incl. CO_2 costs	**US$ / MWhe**	**86.74**	**88.10**	**93.46**	**116.34**	**205.32**	**233.92**

Graph Levelized electricity generation cost vs. full load hours

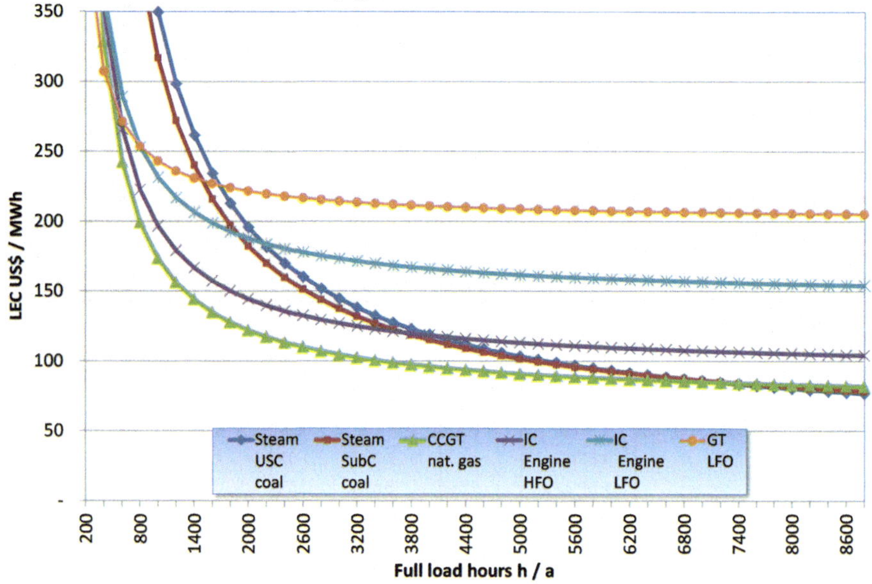

Graph Structure of electricity generation costs

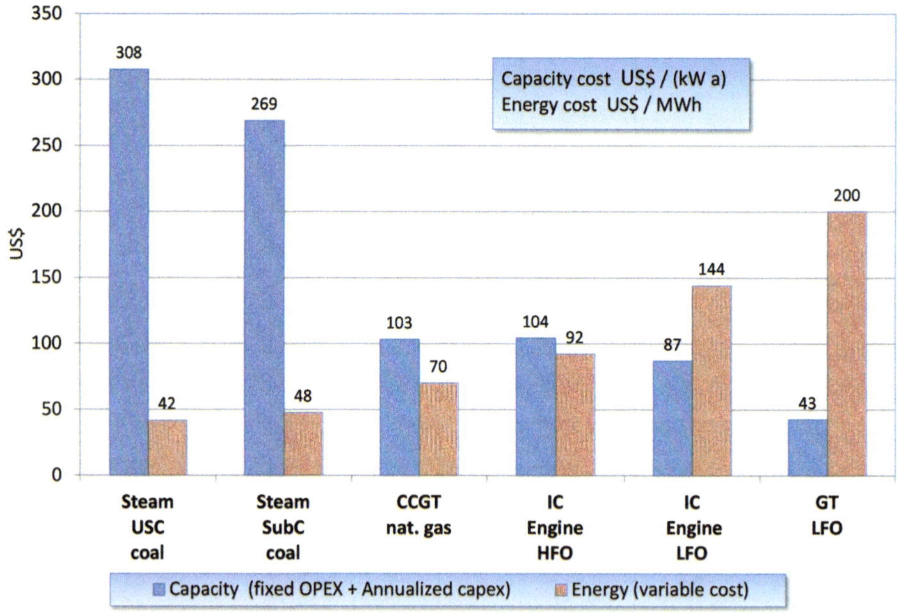

11.5 Integrated Model – Techno-Economics of Nuclear PPs

PK **Panos Konstantin**

The Power Supply Industry

Case Study
Integrated Model for
Nuclear Generation Cost Calculation

Notes:
1. Cells with black characters include inputs
2. Cells with red characters include formulas
3. Download of Add-Ins from website is requred
4. Description of Case Study in section 4.2 in the book

Last update October 2017

Summary of results

Item		Unit	Value
Power and Energy balance		-	
Rated power each, total, gross		MW_e	2,400
Thermal reactor power, total		MW_t	6,800
Electricity generation, net	7,500 h/a	GWh_e / a	16,740
Fuel consumption, in thermal units		GWh_t / a	47,430
metric tons nuclear fuel		t / a	35.4
Technical-financial constraints			
Service life for calculation		a	50
Discount rate, on real terms (WACC)		% / a	7.1%
Cost of nuclear fuel *)		US$ / kg UO_2	1,651
Reserve funds for decommissioning, waste disposal		US$ / MWh_e	3.65
Capital expenditures (CAPEX), US$ 2013 **)		**Mln US$**	**13,720**
Annual costs, US$ 2013		**Mln US$ /a**	**1,483**
Annualized CAPEX		Mln US$ /a	1,002
Fixed Operating expenses (fixed OPEX)		Mln US$ /a	362
Variable operating expenses (variable OPEX)		Mln US$ /a	120
Capacity cost ref. to net power		**US$ / (kW a)**	**611**
Energy cost, ref. to net electricity production		**US$ / MWh_e**	**7.14**
Composite cost		**US$/ MWh_e**	**88.60**

*) Average 2013, book, Engineering Economics

) **Source: eia - U.S. Energy Information Administration, updated capital cost estimates 2013,

**) including EPC price, owner's expenses and interest during construction

Weighted Cost of capital – Discount rate

Item		Equity	Debt
Asset shares		**30%**	**70%**
Expected returns after tax			
Risk free rate of return / interest		5.0 %/a	5.0 %/a
Venture risks premium *)		7.5 %/a	1.0 %/a
Country risk premium (depends on country)		0.0 %/a	0.0 %/a
Cost of capital in nominal terms, after tax		**12.5 %/a**	**6.0 %/a**
Corporate tax	25%	4.2 %/a	0.0 %/a
Cost of capital in nominal terms, before tax		**16.7 %/a**	**6.0 %/a**
***$WACC_n$* in nominal terms, before tax**		**9.20 %/a**	
./. Expected Inflation rate **)		2.00 %/a	
***$WACC_r$* inflation adjusted**		**7.1 %/a**	

*) 1.5 percentage points higher compared to conventional PPs due to the longevity risk of investment

**) Benchmark for inflation for longterm investment, worldwide

Nuclear fuel cost calculation

Component	Explanation	Units needed for 1 kg UO_2*)	US$/Unit **)	Total
Uranium oxide U_3O_8	This is the form Uranium is offered in the market place. It includes 0.7% of the fissile isotope U-235.	8.90 kg U3O8	72.64	646
Conversion in UF_6	The Uranium oxide is converted in gaseous form in Uranium hexafluoride (UF_6).	7.50 kg U	7.92	59
Enrichment	Uranium hexafluoride (UF_6) is in enriched in centrifuges to enriched UF_6 with a concertation of 3 to 4 percent U-235.	7.30 kg SWU	91.83	670
Fuel fabrication	The enriched UF_6 is converted in Uranium dioxide (UO_2), the actual nuclear fuel, in form of powder. It is compressed in pellets and filled in thin pipes bundle up in fuel assemblies.	-	-	275
Nuclear fuel	Assemblies of nuclear fuel	**1 kg UO_2**	-	**1,651**

*) **Source:** World Nuclear Association, information library, July 2015
http://www.world-nuclear.org/info/Economic-Aspects/Economics-of-Nuclear-Power/
**) Power & Energy Systems Engineering Economics, average 2013

CAPEX estimate

Item	Unit	Value
Technical parameters		
Number of units	-	2
Type	Westinghouse AP 1000	
Nominal capacity, total	MW	**2,234**
Cost Components, US$ October 2013		
EPC	**mln US$**	**10,127**
Civil structure material and installations	mln US$	1,792
Mechanical equipment, supply and installation	mln US$	3,519
Electrical Equipment, supply and installation	mln US$	652
Project indirects *)	mln US$	2,818
Fee and contingency	mln US$	1,346
Owner Cost (excluding project finance)	mln US$	2,228
Total project cost (excl. project finance, and IDC))	**mln US$**	**12,355**
Specific cost		
Total project EPC	US$ / kW	4,533
Owner Cost (excluding project finance)	US$ / kW	997
Total project cost (excluding project finance)	US$ / kW	5,530

*) includes, engineering, distributable costs, scaffolding, construction management, and start-up
Source: eia - U.S. Energy Information Administration, updated capital cost estimates 2013

Cost estimate for decommissioning and waste disposal

Item		Unit	Value
Basic constraints			
Installed capacity		MW	2,234
Electricity production		GWh / a	16,740
Capital expenditures, US$ 2013		mln US$	12,355
Total annual costs		mln US$	1,483
Cost for nuclear waste disposal			
Per MWh_e		US$ / MWhe	1.50
Annual amount		mln US$ / a	25
in percent of the total annual costs		-	1.7%
Reserves for decommissioning			
Required fund in US$ 2013 *)	650 US$/kWe	mln US$	1,452
in percent of CAPEX		-	11.8%
Future value after 30 yrs, in US$ 2013			
Per MWhe **)		US$ / MWhe	2.15
Annual amount		mln US$ / a	36
Bank interest rate for deposits		% / a	3.50%
Value in 30 years in US$ 2013	30 years	mln US$	1,858
in percent of initial CAPEX **)		-	15%
per kW installed capacity		US$ / kW	832
Value in 50 years in US$ 2013	50 years	mln US$	4,715
in percent of initial CAPEX **)		-	38%
per kW installed capacity		US$ / kW	2,110

*) OECD survey 2013, decommissioning nuclear power plants

**) Considering uncertainty due longevity risk of investment it is assumed that an amount

This corresponds to a fee of 2.15 US$/MWhe

Cost breakdown (see next spreadsheet)

Item	Unit	Value
CAPEX, incl. IDC *)	**US$ /kW**	**5,716**
Levelized electricity cost	US$ / MWh	**88.60**
Annualized CAPEX		59.83
Fixed OPEX		21.63
Fuel costs		3.49
Variable non-fuel cost **)		3.65

*) IDC: Interest during construction 6.0 years

**) Mainly costs for decommissioning and nuclear waste disposal

Integrated electricity generation cost calculation model

Item		Unit	Value
Technical Parameters			
Number of units ,Westinghouse AP1000		-	2
Rated power each, gross		MW_e	1,200
Net power output, each	7%	MW_e	1,116
Thermal reactor power, each		MW_t	3,400
Electrical efficiency, gross		%	35.3%
Fuel burnup rate		MWd / kg	60
Energy balance			
Full load equivalent operation time	FLH	h / a	7,500
	days	d / a	312.5
Electricity generation, net		GWh_e / a	16,740
Fuel consumption, in thermal units		GWh_t / a	47,430
metric tons nuclear fuel		t / a	35.4
Technical-financial constraints			
Service life for calculation		a	50
Construction time		a	6.0
Discount rate, on real terms (WACC)		% / a	7.1%
Operating staff		Persons	400
Costs of personnel		US$ / (Pers. a)	100,000
Fixed O&M cost referred to EPC		% / a	1.5%
Insurance referred to EPC		% / a	0.5%
Cost of nuclear fuel *)		US$ / kg UO_2	1,651
Cost for nuclear waste disposal		US$ / MWh_e	1.50
Reserve funds for decommissioning		US$ / MWh_e	2.15
Capital expenditures (CAPEX), US$ 2		**Mln US$**	**13,720**
EPC price **)		Mln US$	10,127
Owners expenditures **)		Mln US$	2,228
Interest during construction		Mln US$	1,365
Annual costs		**Mln US$ /a**	**1,483**
Annualized CAPEX		Mln US$ /a	**1,002**
Fixed Operating expenses (fixed OPEX)			**362**
Cost of personnel		Mln US$ /a	40
Fixed O&M costs		Mln US$ /a	152
Insurance		Mln US$ /a	51
Variable operating expenses (variable OPEX)		**Mln US$ /a**	**120**
Fuel direct costs		Mln US$ /a	58
Costs for waste disposal		Mln US$ /a	25
Reserve funds for decommissioning		Mln US$ /a	36
Capacity cost ref. to net power		**US$ / (kW a)**	**610.9**
Energy cost, ref. to net electricity production		**US$ / MWh_e**	**7.14**
Composite cost		**US$/ MWh_e**	**88.60**

*) Average 2013, Engineering Economics

**) source U.S. Energy Information agency – eia, 2013

11.6 Cashflow and IRR Analysis of a Wind Farm Project

PK Panos Konstantin

The Power Supply Industry

Case Study
Integrated Model
Cashflow and IRR Analysis
of a Windfarm Project

Notes:

1. Cells with black characters include inputs
2. Cells with red characters include formulas
3. Description of Case Study in section 5.2.6 of the book "economics of Windpower"

Last update October 2017

Item	Unit	Value
Key technical parameters		
Number of Wind turbines	-	40
Power output of each WT	kW	3,000
Installed capacity wind farm	MW	120
CAPEX	mln €	188.7
of which loan, 15 year maturity	mln €	151.0
Results for Exceedance Probability	**P50**	
Sales of electricityn, net	GWh/a	364
Electricity price, 1st year *)	€/MWh	69.30
Internal Rate of Return		
on investment	-	8.1%
on equity before tax	-	12.2%
on equity after tax	-	10.0%
Cashflow analysis, 1st year results		
Revenues	mln €	25.8
-Operating expenses	mln €	9.1
Operating income (EBIDA)	mln €	16.6
- Depreciation, interest on loans, corporate tax	mln €	16.9
Net income	mln €	- 0.3
+ Depreciation	mln €	9.4
Cashflow bevor principal repayment	mln €	9.2
- repayment of loan	mln €	10.06
Free cashflow	**mln €**	**- 0.89**
Debt coverage ratio (DSR), over load life	-	1.42

Electricity yield calculation

Base Case yiel for a single Wind Turbine			
Annual average wind speed w:		5.5 m/s	30 m above ground
Hub height H_N:		149 m	
Average wind speed at hub height w_N:	Z0=0.10	7.05 m/s	Roughness class 2
power output of one singel wind turbine		3000 kW	
Availability		0.97	

w_N	m / s	0.0	1.0	2.0	3.0	4.0	5.0	6.0	7.0	8.0	9.0	10.0	11.0	12.0	**Sub - total**
P_{el}	kW	0	0	3	49	115	339	628	1,036	1,549	2,090	2,580	2,900	3,000	
t	h / a	0	273	520	721	861	933	941	894	806	693	570	450	341	
W_{el}	MWh	0	0	2	34	96	307	573	898	1,210	1,404	1,426	1,265	992	**8,206.6**
Continuation															
w_N	m / s	13.0	14.0	15.0	16.0	17.0	18.0	19.0	20.0	21.0	22.0	23.0	24.0	25.0	**Sub - total**
P_{el}	kW	3,000	3,000	3,000	3,000	3,000	3,000	3,000	3,000	3,000	3,000	3,000	3,000	3,000	
t	h / a	249	175	118	77	49	30	17	10	5	3	1	1	0	
W_{el}	MWh	724	509	345	225	142	86	51	29	16	8	4	2	1	**2,142.1**

P_{el} kW: Manufacturer's technical data	**Grand total, annual yield MWh, P50**		**10,348.7 MWh**
t h/a: Frequency distribution acc. Rayleigh Model	**Capacity factor**	**39.4%**	**3,450 h/a**

Excedance probability

Wind Farm: Base case		Unit	Value
Number of WTs		-	40
Gross electricity production wind farm		MWh/a	**413,947**
Losses wind farm		-	12.00%
Net electricity production windfarm P50	μ =	MWh/a	**364,273**

Uncertainties		U %	U^2
Wind data		8.0	64.0
Power curve		6.5	42.3
Wake effect		2.0	4.0
Standard deviation σ SQRT(ΣU^2)	σ =	-	10.5%
Energy yield base case exeedance probability =		**P50**	**364,273 MWh**
Energy yield exeedance probability =		**P90**	**315,256 MWh**

Input techno-economic parameters

Item	Unit	Zahlen-wert
Electrical capacity of eac WT	kW	3,000
Number of WT	-	40
Total installed capacity wind farm	kW	120,000
Operational & Economic parameters		
Life time	a	20
Construction time	a	1.5
Inflation rate	%	2.0%
Discount rate in real terms (WACC)	%	4.6%
Maintenance contract	Cent / kWh	1.00
Management/technical surveillance	% CAPEX / a	1.3%
Insurance	% CAPEX / a	0.5%
Reserves for decommissioning	% CAPEX / a	0.8%
Operating staff	Pers/a	4
Costs of personnel	th. €/(Pers. a)	70
Leasing costs for site	Cent / kWh	0.35

Weighted average cost of capital

Item		Unit	Equity	Debt
Asset shares		**%**	**20**	**80.0**
Expected returns after tax				
Risk-free returns/interest		% /a	5.0%	5.0%
Venture risk premium		% /a	5.0%	0.0%
Cost of capital in nominal terms after tax		**% /a**	**10.0%**	**5.0%**
Corporate tax	25%	% /a	3.3%	0.0%
Cost of capital in nominal terms before tax		% /a	13.3%	5.0%
Weighted average cost of capital before tax		**% /a**	**6.7%**	
./. Inflation		% /a	2.0%	
WACC inflation adjusted, before tax		**% /a**	**4.58%**	

CAPEX estimate

Item		Unit	Value
Power			
Capacity each WT		MW	3.00
Number of WTs		-	40
Capacity wind farm			120
Specific cost WT		€/MW	1,250
CAPEX estimate, 2013 prices, ±20%			
WT delivery and installation	100.0%	1,000 €	150,000
Civil works	3.0%	1,000 €	4,500
Fundaments	4.00%	1,000 €	6,000
Internal electrical wiring	5.00%	1,000 €	7,500
Grid connection, 110 kV	4.80%	1,000 €	7,200
Project development, Engineering	4.00%	1,000 €	6,000
Contingencies	5.00%	1,000 €	7,500
Total		**1,000 €**	**188,700**
Specific CAPEX		**€ /kW**	**1,570**

Cashflow analysis

For calculation for P50 insert 1; for P90 insert 2	1

Electricity price is determined with the "goal seek" function: of Excel:

Goal IRR on equity after tax for P50 = 10.0% (taken from spreadsheet IRR Analysis)

Item				Unit	Year of operation 1	2	3	4		15	16	17	18	19	20
Sales of electricity	**P50**	364.27 GWh		GWh	**364.3**	**364.3**	**364.3**	**364.3**	......	**364.3**	**364.3**	**364.3**	**364.3**	**364.3**	**364.3**
Electricity price (see note) *)		69.30 €/MWh	2.00% a	€/MWh	70.7	72.1	73.5	75.0	......	93.3	95.1	97.0	99.0	101.0	103.0
Revenues				**Mio. € / a**	**25.75**	**26.27**	**26.79**	**27.33**	......	**33.98**	**34.66**	**35.35**	**36.06**	**36.78**	**37.51**
Operating Expenses	CAPEX=	188.7 mln €		**mln €/a**	**9.12**	**9.28**	**9.45**	**9.62**	......	**11.85**	**12.08**	**12.32**	**12.56**	**12.82**	**13.08**
Maintenance contract	1.00 c/kWh	3.6 mln €	2.50% a	mln €/a	3.73	3.83	3.92	4.02	......	5.28	5.41	5.54	5.68	5.82	5.97
Management & supervision contract	1.30% a	2.5 mln €	2.50% a	mln €/a	2.51	2.58	2.64	2.71	......	3.55	3.64	3.73	3.83	3.92	4.02
Insurances	0.50% a	0.9 mln €	0.00% a	mln €/a	0.94	0.94	0.94	0.94	......	0.94	0.94	0.94	0.94	0.94	0.94
Reserves for decommissioning	0.80% a	1.5 mln €	0.00% a	mln €/a	1.51	1.51	1.51	1.51	......	1.51	1.51	1.51	1.51	1.51	1.51
Cost of personnel	280 th/a	0.3 mln €	3.00% a	mln €/a	0.29	0.30	0.31	0.32	......	0.44	0.45	0.46	0.48	0.49	0.51
Leasing cost for site	0.35 c/kWh	0.1 mln €	0.00% a	mln €/a	0.13	0.13	0.13	0.13	......	0.13	0.13	0.13	0.13	0.13	0.13
Operating Income				**mln €/a**	**16.63**	**16.98**	**17.34**	**17.70**	......	**22.13**	**22.58**	**23.03**	**23.49**	**23.96**	**24.44**
minus Depreciation	188.7 mln €		20 a	mln €/a	9.44	9.44	9.44	9.44	......	9.44	9.44	9.44	9.44	9.44	9.44
minus interest on loans	151.0 mln €	duration 15 a	5.00% a	mln €/a	7.55	7.04	6.54	6.04	......	0.50	-	-	-	-	-
Income before taxes				mln €/a	**(0.35)**	**0.50**	**1.36**	**2.23**	......	**12.19**	**13.14**	**13.60**	**14.06**	**14.53**	**15.00**
minus Corporate tax		25.00% rate		mln €/a	(0.09)	0.13	0.34	0.56	......	3.05	3.29	3.40	3.51	3.63	3.75
Income after tax				mln €/a	**(0.26)**	**0.38**	**1.02**	**1.67**	......	**9.14**	**9.86**	**10.20**	**10.54**	**10.89**	**11.25**
plus depreciation				mln €/a	9.44	9.44	9.44	9.44	......	9.44	9.44	9.44	9.44	9.44	9.44
Cashflow before amortization				mln €/a	**9.17**	**9.81**	**10.46**	**11.11**	......	**18.58**	**19.29**	**19.63**	**19.98**	**20.33**	**20.69**
Loan repayment	151.0 mln €	duration 15 a	Grace 0 a	mln €/a	10.06	10.06	10.06	10.06	......	10.06	-	-	-	-	-
Free cashflow				mln €/a	**(0.9)**	**(0.3)**	**0.4**	**1.0**	......	**8.5**	**19.3**	**19.6**	**20.0**	**20.3**	**20.7**
Debt Service Ratio (DCR)			**average**	**1.42**	**0.9**	**1.0**	**1.0**	**1.1**	......	**2.1**	-	-	-	-	-

*) **Note:** Electricity price is determined with the "goal seek" function of Excel: Goal IRR on equity after **10.0%**

Internal rate of return analysis

Item	Year of operation 0	1	2	3	4		15	16	17	18	19	20
		mln Euro /a										
IRR on investment	**8.1%**	**Function: IRR(F9:Z9)**										
Revenues	0.00	25.75	26.27	26.79	27.33	....	33.98	34.66	35.35	36.06	36.78	37.51
CAPEX	-188.70	0.00	0.00	0.00	0.00	....	0.00	0.00	0.00	0.00	0.00	0.00
Operating Expenses	0.00	-9.12	-9.28	-9.45	-9.62	....	-11.85	-12.08	-12.32	-12.56	-12.82	-13.08
Total	**-188.70**	**16.63**	**16.98**	**17.34**	**17.70**	....	**22.13**	**22.58**	**23.03**	**23.49**	**23.96**	**24.44**
IRR on equity before tax	**12.2%**	**Function: IRR(F18:Z18)**				....						
Revenues	0.00	25.75	26.27	26.79	27.33	....	33.98	34.66	35.35	36.06	36.78	37.51
Loan	150.96	0.00	0.00	0.00	0.00	....	0.00	0.00	0.00	0.00	0.00	0.00
CAPEX	-188.70	0.00	0.00	0.00	0.00	....	0.00	0.00	0.00	0.00	0.00	0.00
Operating Expenses	0.00	-9.12	-9.28	-9.45	-9.62	....	-11.85	-12.08	-12.32	-12.56	-12.82	-13.08
Loan repayment	0.00	-10.06	-10.06	-10.06	-10.06	....	-10.06	0.00	0.00	0.00	0.00	0.00
Interest on loans	0.00	-7.55	-7.04	-6.54	-6.04	....	-0.50	0.00	0.00	0.00	0.00	0.00
Total cashflow before tax	**-37.74**	**-0.98**	**-0.13**	**0.73**	**1.60**	....	**11.56**	**22.58**	**23.03**	**23.49**	**23.96**	**24.44**
IRR on equity after tax	**10.0%**	**Function: IRR(F22:Z22)**				....						
Corporate tax	-	0.09	-0.13	-0.34	-0.56	....	-3.05	-3.29	-3.40	-3.51	-3.63	-3.75
Total cashflow after tax	**-37.74**	**-0.89**	**-0.25**	**0.39**	**1.04**	....	**8.52**	**19.29**	**19.63**	**19.98**	**20.33**	**20.69**

Note: Spreadsheet is linked with spreadsheet cashflow analysis

Goal seek IRR on equity after tax: Set cell 20 to value 10%

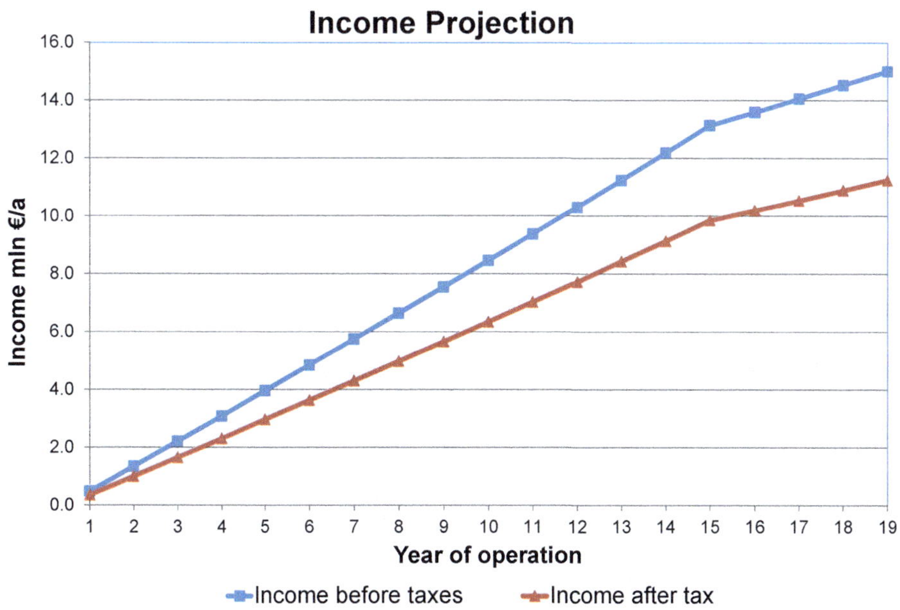
Income Projection
16.0
14.0
12.0
10.0
8.0
6.0
4.0
2.0
0.0
Income mln €/a
1
2
3
4
5
6
7
8
9
10
11
12
13
14
15
16
17
18
19
Year of operation
Income before taxes
Income after tax

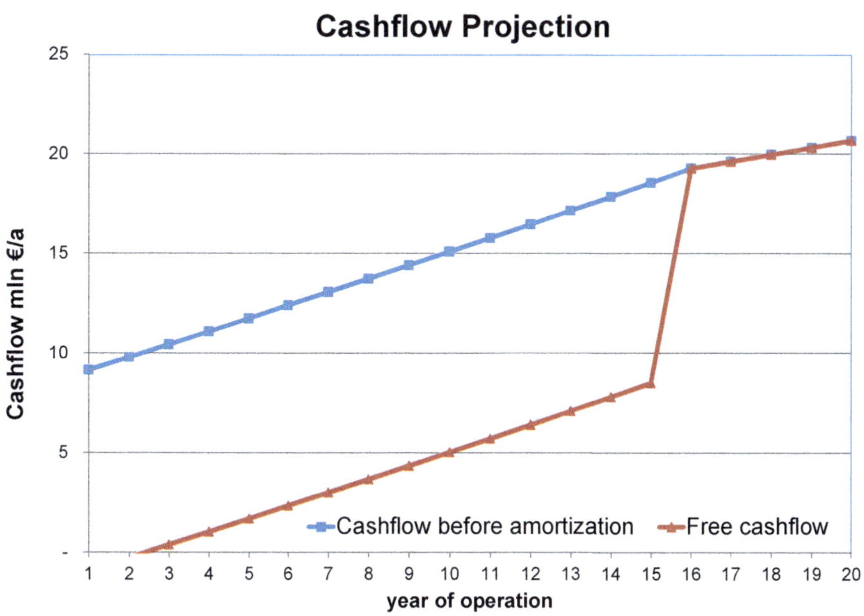
Cashflow Projection
25
20
15
10
5
-
Cashflow mln €/a
Cashflow before amortization
Free cashflow
1
2
3
4
5
6
7
8
9
10
11
12
13
14
15
16
17
18
19
20
year of operation

11.7 Techno-Economic Model of Parabolic Trough PPs

PK Panos Konstantin

The Power Supply Industry

Case Study

Integrated Techno-Economic Model
for
Parabolic Trough Solar Power Plants

Notes:

1. Cells with black characters include inputs
2. Cells with red characters include formulas
4. Download of Add-Ins (Macros) from website required
5. Description of Case Study in Section 5.5.3 in the book

Last update Octoer 2017

Summary of results

Item	Unit	100 MW		
		TES 0 h	TES 8 h	TES 12 h
Power balance				
Power output, solar field operation	MW	100		
Power output, TES operation	MW	85		
Number of collectors	-	655	1,400	1,773
Solar field heat production	MJ / s	271	579	734
Thermal Storage	hours	-	8	12
Solar multiple	-	1.0	2.1	2.7
Energy balance				
Annual irradiation DNI	kWh /(m^2 a)	2,400		
Solar heat to power block	GWh_t / a	591	1,263	1,599
Net electricity production	GWh_e /a	213	456	578
Financial parameters				
Discount rate, in real terms (WACC)	-	4.6%		
Project lifetime	years	25		
Capital expenditures, US$2014, ± 20%	mln US$	402	766	937
Electricity generation costs, in real terms, 2014				
Annual generation cost	mln US$ / a	39.2	71.0	87.1
of which capital cost		70%	73%	73%
Levelized electricity cost	**US$ / MWh**	**214**	**181**	**175**

Input solar field

Item	Unit	100 MW		
		TES 0 h	TES 8 h	TES 12 h
Site parameters				
Latitude	degrees	28		
Irradiance, solar noon equinox	W/m^2	900		
Annual irradiation	kWh /m^2 a	2,400		
Collector net aperture area	m^2	817		
Thermal storage				
Capacity per unit	MWh	925		
Number of units	-	-	2	3
Charge time	h	-	6	6
Discharge hours	h	-	8	12

*) Cycle calculation with KPRO® see also heat flow diagram and table in text part

For LEC calculations with escalation rates ? Insert yes ➡	yes	

Solar field thermal capacity

Item		Unit	100 MW		
			TES 0 h	TES 8 h	TES 12 h
Thermal storage					
Solar heat for power plant, solar field operation		MJ / s	271		
Solar heat for power plant, storage operation			234		
Storage capacity, total		MWh	0	1,850	2,775
Charging time		h	0	6	6
Discharge time, storage operation		h	0.0	7.9	11.9
Solar field heat production, required		MJ / s	271	579	734
Solar multiple			1.0	2.1	2.7
Sola field					
Latitude		digress	28		
Irradiance DNI, solar noon, summer solstice		W / m^2	900		
Incident angle modifier IAM		-	0.88		
Effective DNI		W / m^2	795		
Collector thermal efficiency	80% nominal	-	0.64		
Collector net aperture area		m^2	817		
Collector heat capacity		KJ /s	414		
Number of collectors		-	655	1,400	1,773
Total aperture area solar field		1000 m^2	535	1,144	1,448
Total heat production of solar field		MJ / s	271	579	734

Solar field size

Item	Unit	100 MW		
		TES 0 h	TES 8 h	TES 12 h
Size of solar field				
Direction of center line of collector	-	N-S	N-S	N-S
Net aperture area, Solar field	1000 m^2	535	1,144	1,448
North South dimension of Solar Field	m	1,315	2,585	3,220
East West dimension of Solar Field	m	1,641	1,739	1,759
Land area of Solar Field	1000 m^2	2,159	4,495	5,663
Factor Land area / Collector area	-	4.03	3.93	3.91
Number of Collector and loops				
Number of subfields (N-S)	-	4	8	10
Number of collectors	-	655	1,400	1,773
Number of Collectors for each loop	-	4	4	4
Number of loops	-	164	350	443
Collector size				
Length of one collector	m	148.50	148.50	148.50
Width of one collector	m	5.77	5.77	5.77
Width for Header Piping in center of Solar field	m	10	10	10
Center line distance between adjacent collectors	m	17.3	17.3	17.3
Longitudinal distance between adjacent collectors	m	3	3	3
Width of roads (around solar field)	m	35	35	35

CAPEX estimate

Item	Unit	100 MW		
		TES 0 h	TES 8 h	TES 12 h
Exchange rate	**Euro/US$**	1.4		
Nominal plant size				
Rated electric power, gross	MW_e	100	100	100
Number of collectors	-	655	1,400	1,773
Aperture area of solar field	1000 m²	535	1,144	1,448
Thermal storage units	-	-	2	3
Thermal storage	MWh	-	1,850	2,775
EPC Contract Costs	mln US$	**325.9**	**638.1**	**794.0**
Solar Field	mln US$	156.6	295.2	366.0
HTF System	mln US$	33.0	61.8	77.9
Thermal Energy Storage	mln US$	-	126.0	189.0
Power Block	mln US$	107.8	107.8	107.8
Balance of Plant	mln US$	28.5	47.3	53.3
Engineering	mln US$	18.7	31.9	35.7
Contingencies	mln US$	39.1	63.8	71.5
Owners Costs	mln US$	17.9	31.9	35.7
Grand Total, 2014, ± 20%	mln US$	**401.7**	**765.7**	**937.0**
Specific CAPEX	US$ / kW	4,017	7,657	9,370

OPEX estimate

Item	Unit	100 MW		
		TES 0 h	TES 8 h	TES 12 h
Consumables				
Lease cost	US$ / m²	1.0	1.0	1.0
Number of operating staff	-	45	55	65
Manpower cost (average)	1000 US$ / a	65.0	65.0	65.0
Price diesel fuel	US$ / liter	0.8	0.8	0.8
Fuel consumption	1000 Liter / a	120	120	120
Raw water	US$ / m³	1.00	1.00	1.00
Annual raw water consumption	1000 m³ / a	73.9	157.9	199.9
HTF Consumption	t / a	60.8	54.3	64.1
HTF price	US$ / t	3,000	3,000	3,000
Annual OPEX				
Fixed O&M Costs:	**1000 US$/a**	**9,972**	**17,641**	**21,798**
Solar field & storage system	1000 US$/a	1,896	4,830	6,329
Power block	1000 US$/a	1,363	1,551	1,611
Personnel	1000 US$/a	2,925	3,575	4,225
Lease	1000 US$/a	2,159	4,495	5,663
Insurance	1000 US$/a	1,629	3,190	3,970
Variable O&M Costs:	**1000 US$/a**	**672**	**1,101**	**1,355**
Fuel	1000 US$/a	96	96	96
Water	1000 US$/a	74	158	200
HTF refill	1000 US$/a	182	163	192
Other consumables & residues *)	1000 US$/a	320	684	866
Total OPEX	**1000 US$/a**	**10,644**	**18,742**	**23,153**

*) Electricity import, nitrogen, chemicals

Techno-economic model for calculation of electricity generation cost

Item	Unit	100 MW		
		TES 0 h	TES 8 h	TES 12 h
Solar Field				
Aperture area, total	1000 m^2	535	1,144	1,448
Solar field heat output	MJ /s	271	579	734
Heat supply to power block, solar field direct	MJ / s	271	271	271
Heat supply to power block, TES operation	MJ / s	234	234	234
Solar Multiple	-	1.0	2.1	2.7
Thermal Storage	MWh	0	1,850	2,775
Charging time	h	-	6.0	6.0
Discharging time	h	-	7.9	11.9
Technical Parameter, Power Plant				
Rated power output, field operation	MW	100	100	100
Power output, gross, TES operation	MW	85	85	85
Electrical efficiency, gross	%	36.9%	36.9%	36.9%
Condenser	-	Air cooled		
Annual Production				
Annual irradiation DNI	kWh /m^2 a	2,400.0	2,400.0	2,400.0
Solar heat to power block	GWh_t / a	590.8	1,263.0	1,599.1
Electricity production	GWh_e / a	213.4	456.2	577.6
of which in TES operation	GWh_e / a	0.0	242.8	364.2
Auxiliary consumption	GWh_e / a	29.9	63.9	80.9
Net electricity production	GWh_e / a	183.5	392.3	496.7
Capacity factor	-	24.4%	52.1%	65.9%
Full load hours, ref. field rated output	h / a	2,134	4,562	5,776
Techno-Economic Parameters				
Basic year	-	2014	2014	2014
Project life time	a	25	25	25
Discount rate, in real terms	%	4.6%	4.6%	4.6%
CAPEX, US$2014; ±20%	**mln US$**	**401.7**	**765.7**	**937.0**
specific	US$ / kW	4,017	7,657	9,370
Levelized annual costs, in real terms				
Annualized Capex	1000 US$ /a	27,298	52,037	63,677
OPEX, fixed *)	1000 US$ /a	11,122	17,641	21,798
OPEX, variable *)	1001 US$ /a	793	1,299	1,598
Total	**1000 US$ /a**	**39,213**	**70,977**	**87,072**
Levelized electricity cost	**US$ / MWh**	**213.66**	**180.91**	**175.29**

*) Escalation rates

11.8 Techno-economic Model of Solar Tower Power Plant

PK **Panos Konstantin**

The Power Supply Industry

Case Study
Integrated Techno-Economic Model
for
Solar Tower Power Plants

Notes:

1. Cells with black characters include inputs
2. Cells with red characters include formulas
4. Download of Add-Ins (Macros) from website required

Last update October 2017

Summary of Results

Item	Einheit Unit	100 MW		
		TES 9 h	TES 12 h	TES 15 h
Technical parameters				
Rated power output	MW	100		
Site latitude	grd	28		
Number of heliostats	-	7,158	8,978	11,074
Solar irradiation	kWh / m^2 a	2,400		
Net electricity production	GWh /a	379.6	476.2	587.3
Financial parammeters				
Discount rate in real terms	-	4.6%		
Project lifetime	a	25		
CAPEX, US$ 2014, ±20	**mln US$**	**784**	**933**	**1,094**
Electricity generation costs in real terms				
Annual costs	mln US$ /a	66.9	79.4	92.8
of which capital cost		79.7%	79.9%	80.1%
Levelized electricity cost	**US$ / MWh**	**176.3**	**166.7**	**158.1**

Key Performance Parameters

Item		Unit	100 MW		
			TES 9 h	TES 12 h	TES 15 h
Solar Field					
Latitude		degrees	28		
Design Reference DNI		W /m^2	900		
Design Reference DNI			Solar noon, Equinox		
Solar efficiency (DNI to heat)			80.0%		
Number of Heliostats		-	7,158	8,978	11,074
Aperture area, total	121 m2	1000 m^2	866	1,086	1,340
Tower height		m	280	315	320
Solar field heat output, design point		MJ /s	624	782	965
Receiver thermal power	72.5%	MJ /s	452	567	699
Solar heat to power block		MJ / s	238	238	238
Solar Multiple		-	1.90	2.38	2.94
Thermal Storage		MWh_t	2,143	2,857	3,571
Technical Parameters, Power Block cycle					
Rated power output, field operation		MW	100	100	100
Live steam parameters		bar / °C	155 / 550		
Electrical efficiency, gross		%	42.0%	42.0%	42.0%
Condenser cooling		-	ACC, air 32°C		

CAPEX Estimate

Item	Item	100 MW		
		TES 9 h	TES 12 h	TES 15 h
Nominal plant size				
Rated electric power, gross	MW_e	100	100	100
Receiver capacity	MW_t	475	594	713
Aperture area of solar field	1000 m²	866	1,086	1,340
Thermal storage	MWh	2,143	2,857	3,571
Exchange rate	€ / US$	1.4	1.4	1.4
EPC Contract Costs	**mln US$**	**700.4**	**833.1**	**976.7**
Solar Field	mln US$	225.2	282.4	348.4
Reciever	mln US$	106.9	133.7	160.4
Tower	mln US$	14.0	14.7	15.4
Thermal Energy Storage	mln US$	64.3	85.7	107.1
Power Block	mln US$	150.0	150.0	150.0
Balance of Plant	mln US$	56.0	66.7	78.1
Engineering	mln US$	28.0	33.3	39.1
Contingencies	mln US$	56.0	66.7	78.1
Owners Costs	mln US$	84.1	100.0	117.2
CAPEX Grand Total ± 20%	**mln US$**	**784.5**	**933.1**	**1,093.9**
Specific CAPEX	US$ / kW	7,845	9,331	10,939

OPEX Estimate

Item	Unit	100 MW		
		TES 9 h	TES 12 h	TES 15 h
Technical-financial constraints				
Exchange rate	EURO / US$	1.40	1.40	1.40
Aperture Area	1000 m^2	866	1,086	1,340
Power generation, gross	GWh / a	417.2	523.3	645.4
EPC Price	**mln US$**	**602.4**	**718.5**	**844.1**
Solar Field	mln US$	225.2	282.4	348.4
Thermal Storage + HTF System	mln US$	171.2	219.4	267.6
Power block + BoB	mln US$	206.0	216.7	228.1
Number of operating staff	-	45	50	55
Manpower cost (average)	1000 US$ / a	85.0	85.0	85.0
Raw water	US$ / m^3	1.0	1.0	1.0
Annual raw water consumption	1000* m^3 / a	119.5	149.9	184.9
Annual OPEX				
Fixed O&M Costs:	**1000 US$/a**	**12,861**	**15,027**	**17,336**
Solar field & storage system	1000 US$ /a	3,964	5,018	6,160
Power block	1000 US$ /a	2,060	2,167	2,281
Personnel	1000 US$ /a	3,825	4,250	4,675
Lease	1000 US$ /a	-	-	-
Insurance	1000 US$ /a	3,012	3,592	4,220
Consumables:	1000 US$ /a	**745**	**935**	**1,153**
Water	1000 US$ /a	120	150	185
Other consumables & residues *)	1000 US$ /a	626	785	968
Total OPEX	**1000 US$ /a**	**13,606**	**15,962**	**18,489**

*) Electricity import, nitorgen, chemicals

Energy Balance and LEC calculation

Item		Unit	100 MW		
			TES 9 h	TES 12 h	TES 15 h
Annual energy balance					
Annual irradiation		$kWh/m^2 a$	2,400	2,400	2,400
Thermal heat from receiver, DNI to heat	50.3%	GWh_t / a	1,046	1,311	1,618
Electricity production	42.0%	GWh_e / a	417	523	645
Auxiliary consumption	9.0%	GWh_e / a	38	47	58
Net electricity production		GWh_e / a	380	476	587
Capacity factor		-	47.6%	59.7%	73.7%
Equivalent operating hours		h / a	4,172	5,233	6,454
Techno-Economic Parameters					
Project lifetime	25	a	25	25	25
Discount rate, in real terms	4.6%	%	4.6%	4.6%	4.6%
CAPEX US$ 2014, ±20%		**mln US$**	**784.5**	**933.1**	**1,093.9**
Levelized annual costs					
Annualized Capex		mln US$ /a	53.3	63.4	74.3
OPEX fixed		mln US$ /a	12.9	15.0	17.3
OPEX variable		1000 US$ /a	0.7	0.9	1.2
Total		**1000 US$ /a**	**67**	**79**	**93**
Levelized electricity cost		US$ / MWh	**176.28**	**166.70**	**158.06**
electricity-to-solar heat (DNI)			18.3%	18.3%	18.3%

WACC

Item		Einheit Unit	Eigen-kapital	
			Equity	Debt
Asset shares		**%**	**20**	**80.0**
Risk free rate of return / interest		% /a	5.0%	5.0%
Risk and venture premium		% /a	5.0%	0.0%
Subtotal after corporate tax		**% /a**	**10.0%**	**5.0%**
Corporate tax	25%	% /a	3.3%	0.0%
Cost of capital, before tax, nominal		% /a	13.3%	5.0%
Weighted average cost of capital (WACC)		**% /a**	**6.67%**	
./. Inflation		% /a	2.00%	
WACC inflation adjusted, before tax		**% /a**	**4.58%**	

11.9 Cost Allocation, Electrical Equivalent

PK **Panos Konstantin**

The Power Supply Industry

Case Study

Integrated Model

Cost Allocation _Electrical Equivalent

Extraction-Condensing CHP

Notes:

1. Cells with black characters include inputs
2. Cells with red characters include formulas

Last update October 2017

Item		Formulas	Unit	Values
Extraction-condensing cycle, rated output				
Rated electrical output, gross		P_{gross}	MW_e	110.0
Rated electrical output, net	9 kW	P_{net}		101.0
Heat extraction at 12 bar	100 t / h	Q_{12bar}	MW_t	70.7
Heat extraction at 6 bar	50 t / h	Q_{6bar}	MW_t	33.7
Firing power rate		Q_f	MW_t	359.7
Performance parameters				
Tota efficiency, cogeneration		η_{tot}	-	82.5%
Electricity-to-heat ratio, 12 bar steam		σ_{12bar}	kWh_e / kWh_t	0.335
Electrical equivalent, 12 bar steam		$\beta_{12\ bar}$	kWh_e / kWh_t	0.289
Electricity-to-heat ratio, 6 bar steam		σ_{6bar}	kWh_e / kWh_t	0.403
Electrical equivalent, 6 bar steam		$\beta_{6\ bar}$	kWh_e / kWh_t	0.255
Annual Energy Balance				
Electricity generation, gross, total	7,500 h/a	W_{gross}	MWh_e / a	825,161
of which cogenerated		$W_{cogen} = \Sigma\sigma_i x Q_i$	MWh_e / a	215,011
Electricity generation, net, total	7,500 h/a	W_{net}	MWh_e / a	757,661
Heat generation 12 bar	6,500 h/a	$Q_{t,\ 12\ bar}$	MWh_t / a	459,720
Heat generation 6 bar	4,500 h/a	$Q_{t,6\ bar}$	MWh_t / a	151,530
Fuel consumption		Q_f	MWh_t / a	2,579,647
Equivalent cond. cycle performance				
Cond. Equivalent power, gross		P_{cond_equ}	MW_e	139.1
Cond. equivalent elect. Efficiency		η_{cond_equ}	-	*38.66%*
Cond. equivalent annual electricity production		W_{cond_equ}	**MWh / a**	***996,690***
Annual electricity production, gross	7,500 h/a	$W_{e\_net} = P_{net} \times t_{FLH}$	MWh / a	825,161
Equivalent elec. production, 12 bar steam	459,720	$\beta_{12\ bar} \times Q_{t\_12\ bar}$	MWh / a	132,908
Equivalent elec. production, 6 bar steam	151,530	$\beta_{6\ bar} \times Q_{t\_6\ bar}$	MWh / a	*38,621*
Fuel consumption		Q_f	MWh_t / a	**2,578,079**

Electricity generation cost

Item			Unit	Value
Power and energy balance, cond. Equivalent				
Power output, gross			MW_e	139.1
Power output, net	9.0	*Pnet*	MW_e	130.1
Electricity generation, gross	7,500 h/a		GWh_e/a	1,042.9
Electricity generation, net	7,500 h/a	*Wnet*	GWh_e/a	975.4
Fuel consumption			GWh_t/a	2,578.1
Financial parameters				
Discount rate, in real terms			-	6.5%
Lifetime			a	30.0
Fuel price	85 €/tce	8.14 €/tce	€ / MWh_t	10.4
CAPEX	2001 €/kW		**mln €**	**278.2**
Annual generation costs, in real terms		***CT***	**mln € /a**	**59.5**
Annualized CAPEX		CC	mln € /a	21.3
OPEX fixed	3.5% x CAPEX	COF	mln € /a	9.7
Fuel cost		CF	mln € /a	26.9
Non-fuel variable costs	1.50 €/MWhe	CV	mln € /a	1.6
Specific electricity generation cost, in real terms				
Composite generation cost		$\mathbf{ce=CT/P_{e\_net}}$	**€ / MWhe**	**61.03**
Capacity cost		$\mathbf{cc_e=(CC+COF)/W_{e\_net}}$	€ /kW a	238.68
Energy cost		$\mathbf{cv_e=(CF+CV)/W_{e\_net}}$	€ / MWh_e	29.20

Item		Formulas	Unit	Value
Electricity generation cost				
Composite cost	*tFCH=7,500 h/a*	$ce=1000*cc_e/t_{FCH}+cv_e$	€ / MWh	61.03
Capacity cost		cc_e	€ / (kWa)	238.68
Energy cost		cv_e	€ / MWh	29.20
Heat generation cost, 12 bar steam				
Composite cost	*tFCH=6,500 h/a*	$c_{H12}=1000*cc_{12}/t_{FCH}+c_{v12}$	€ / MWh	19.06
Capacity cost	$\beta=0.289$	$cc_{12}=\beta_{12bar}\times cc_e$	€ / (kWa)	69.00
Energy cost		$cv12=\beta_{12bar}\times cv_e$	€ / MWh	8.44
Heat generation cost, 6 bar steam				
Composite cost	*tFCH=4,500 h/a*	$c_{H6}=1000*cc_6/t_{FCH}+cv_6$	€ / MWh	20.96
Capacity cost	$\beta=0.255$	$cc_6=\beta_{6bar}\times cc_e$	€ / (kWa)	60.83
Energy cost		$cv6=\beta_{6bar}\times cv_e$	€ / MWh	7.44

Control Calculation

Item	Output	Spec. Cost	Unit	Value
Electricity			**th. € /a**	***55,316***
Capacity	139.1	238.68	th. € /a	33,189
Energy	757,657	29.20	th. € /a	22,127
12 bar steam			**th. € /a**	***8,762***
Capacity	70.7	69.00	th. € /a	4,880
Energy	459,720	8.44	th. € /a	3,881
6 bar steam			**th. € /a**	***3,176***
Capacity	33.7	60.83	th. € /a	2,048
Energy	151,530	7.44	th. € /a	1,128
Total	-	-	th. € /a	**67,254**

Electricity	757,657	61.03	**th. € /a**	***46,239***
12 bar steam	459,720	19.06	th. € /a	8,762
6 bar steam	151,530	20.96	th. € /a	3,176
Total	-	-		**58,177**

Cycle Model and simulation

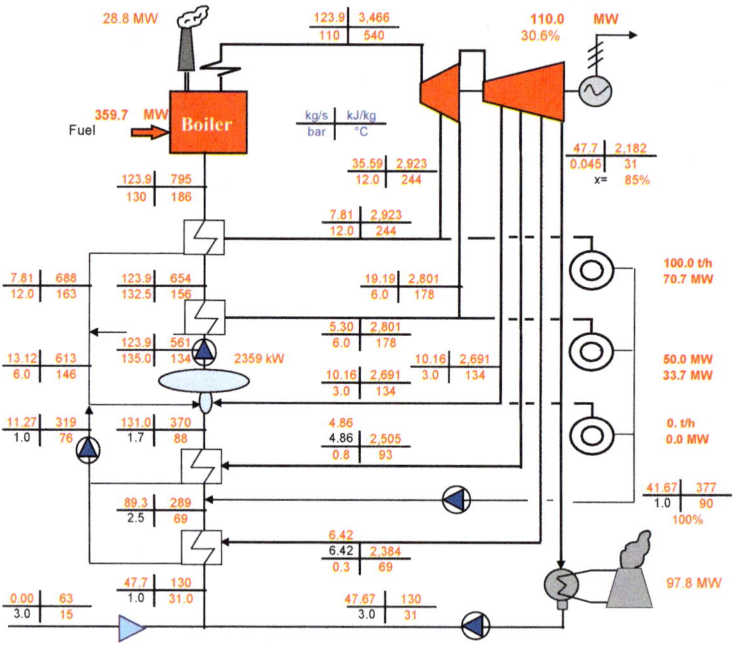

Generator power, gross	MW	110.0
Firing power rate:	MW	359.7
Live steam	MW	359.7
Reheat	MW	0
Electrical efficiency, gross	%	30.59%
Total efficiency	%	59.62%

11.10 Cost Allocation, Exergy, Backpressure CHP

PK **Panos Konstantin**

The Power Supply Industry

Case Study
Integrated Model
Cost Allocation_Exergy
Backpressure CHP

Notes:

1. Cells with black characters include inputs
2. Cells with red characters include formulas
3. Download of FluidEXL is required

Last update October 2017

Power and Energy Balance						
Item	Full load operating hours	Heat			Electricity	
		Output		Annual Production	Output	Annual production
	h/a	t / h	MW_t	MWh_t / a	MW_e	MWh_e / a
Steam 12 bar	5,000	25.0	19.1	95,410	2.6	13,249
Steam 6 bar	7,500	101.3	73.8	553,517	14.1	105,641
Steam 3 bar	7,500	22.6	15.9	118,898	-	-
Gas turbine	7,500	-	-	-	74.0	555,000
Total, gross	-	-	**108.7**	**767,826**	**90.7**	**673,890**
Total, net	-	-	**108.7**	**767,826**	**88.0**	**653,673**
					own consumption 3.0%	

Fuel consumption	-	η_{gt}	MW_t	η_{tot}	MWh_t /a
	-	37.2%	**243.8**	81.82%	**823,643**

Specific Exergy of the Extracted Heat						
Item	Pressure	Temp.	Enthalpy	Entropy	Exergy *)	
	bar	°C	kJ/kg	kJ/kg*K	kJ/kg	MWh_e / t
Steam 12 bar	12.0	284	3,012	6.97	564	0.157
Steam 6 bar	6.0	216	2,885	7.04	453	0.126
Steam 3 bar	3.0	164	2,790	7.15	356	0.099
Zero exergy *)	1.0	54	226	0.76	-	-
				*) Exergy conversion ratio		75.0%

Total Exergy of the Extracted Heat						
Item	Full load hours	Output	Specific exergy	Total exergy	Annual Production	Exergy
	h / a	t /h	MWh/t	MW	t /a	MWh /a
Steam 12 bar	5,000	25.0	0.157	3.9	125,000	19,598
Steam 6 bar	7,500	101.3	0.126	12.8	760,097	95,635
Steam 3 bar	7,500	22.6	0.099	2.2	169,431	16,733

Energy - Exergy Balance						
Item	Output			Energy		
	Output	Exergy	Exergy share	Annual production	Exergy	Exergy share
	MW	MW	%	MWh /a	MWh /a	%
Electricity, net	88.0	88.0	82.3%	653,673	653,673	83.2%
Steam 12 bar	19.1	3.9	3.7%	95,410	19,598	2.5%
Steam 6 bar	73.8	12.8	11.9%	553,517	95,635	12.2%
Steam 3 bar	15.9	2.2	2.1%	118,898	16,733	2.1%
Total	n.a.	**106.9**	**100.0%**	**1,421,499**	**785,639**	**100.0%**

Costs Estimation					
Item	**Unit**	**Fixed Costs**	**Item**	**Unit**	**Variable Costs**
Financial parameters					
CAPEX	Mio.€	120.00	Fuel NG	MWh /a	823,643
discount rate, real	-	6.5%	Price	€/MWh	25.00
Lifetime	a	25			
Annual costs					
Annualized CAPEX	th. €/a	9.84	Fuel Costs	th. €/a	20,591
O&M Cost	th. €/a	**1.80**	Non-fuel	th. €/a	3,369
Total	**T€/a**	**15,838**			**23,961**

Costs Allocation					
Item	**Capacity costs *)**		**Energy costs **)**		**Composite cost**
	Fixed costs th. € / a	**Specific Capacity Cost € / (kW*a)**	**Variable Costs th. € / a**	**Specific Energy Cost € / MWh**	**€ / MWh**
Electricity, net	13,038	143.7	19,936	29.58	48.93
Steam 12 bar	581	30.4	598	6.26	12.35
Steam 6 bar	1,889	25.6	2,917	5.27	8.68
Steam 3 bar	331	20.8	510	4.29	7.07
Total *)**	**15,838**	n.a.	**23,961**	n.a.	n.a.

*) Allocation based on the Exergy-power share

**) Allocation based on the exergy-energy share

Model and simulation of thermodynamic Cycle

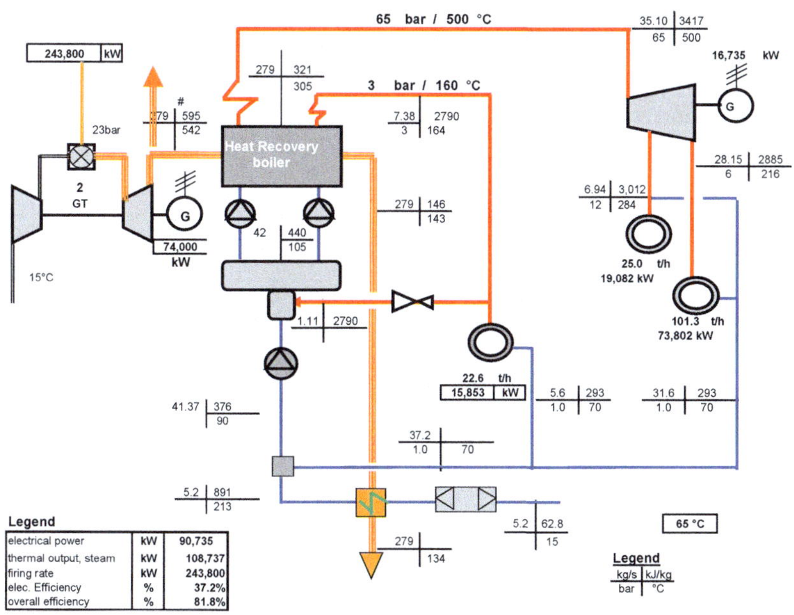

11.11 Modelling & Simulation Rankine Cycle, No-Reheat

PK Panos Konstantin

The Power Supply Industry

Case Study
Cycle Modelling & Simulation
Extraction-Condensing
No-Reheat

Notes:

1. Cells with black characters include inputs
2. Cells with red characters include formulas
3. Download of FluidEXL required for calculations
4. Read brief instruction for FluidEXL in the **Toolbox** in the book
5. Read introduction and notes in Case Study chapter of the book

The purpose of this Case Study is:

1. Training in calculation of thermodynamic cycles using FluidEXL
2. Calculation of performance parameters for cogeneration such as *s, b, hcond, htotal*
3. See also examples 6-3 and 6-4 in th book

Last update October 2017

Input Spreadsheet

Make inputs only here	Unit	Operation mode	
		Cond.	Cogen
Live steam	t/h	561.6	561.6
Pressure	bar	110	110
Temparature	°C	540	540
Saturation temperature	°C	318	318
Process Steam extraction	t/h	0	50
Pressure	bar	12	12
Terminal temperature difference	K	2.0	2.0
Process Steam extraction	t/h	0	45
Pressure	bar	6	6
Terminal temperature difference	K	3.0	3.0
Process Steam extraction	t / h	0	40
Pressure	bar	3	3
Terminal temperature difference	K	5	5
LP bleed	bar	0.8	0.8
Terminal temperature difference	K	5.0	5.0
LP bleed	bar	0.3	0.3
Terminal temperature difference	K	7	7
Condesing steam pressure	bar	0.045	0.045
Boiler efficiency	%	92.0%	92.0%
ST- internal efficiency	%	89.0%	89.0%
Generator efficiency	%	98.0%	98.0%
ST- mechanical efficiency	%	99.5%	99.5%
Condesater return rate	%	0%	100.0%
Condesater temperature	°C	0	90
make-up water	°C	15	15

Energy flow diagram – Condensing Mode of Operation, No-Reheat

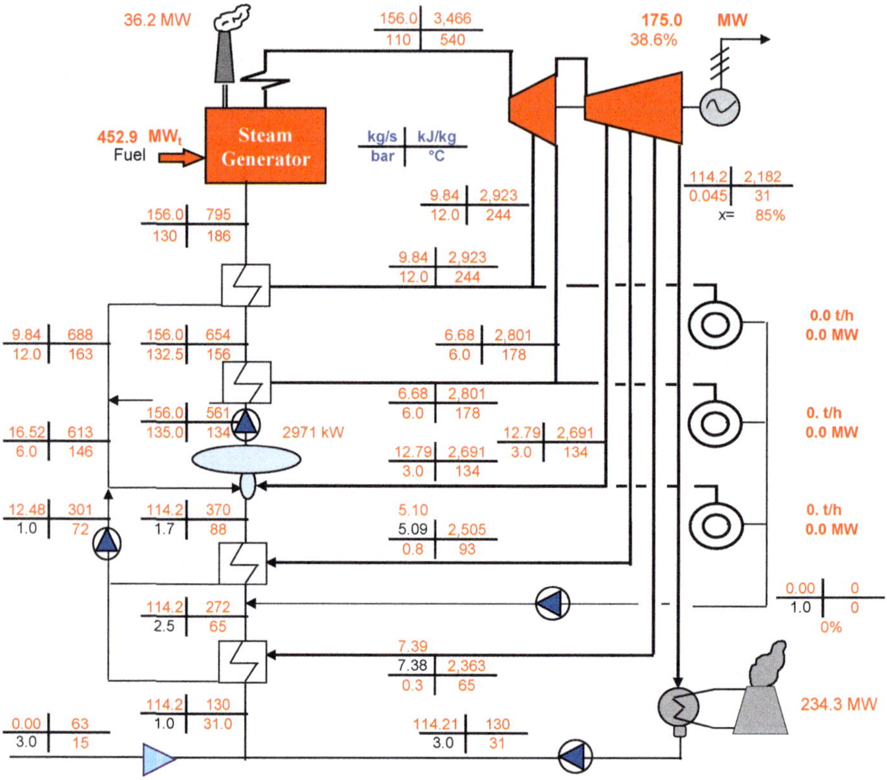

Panos Konstantin
Technologies & Economics

Power output, gross	MW	175.0
Fuel capacity	MW	452.9
Live steam	MW	452.9
Reheat steam	MW	0
Electrical efficiency	%	38.64%
Total efficiency	%	38.64%

Energy Flow Diagram – Cogeneration Mode of Operation, No-Reheat

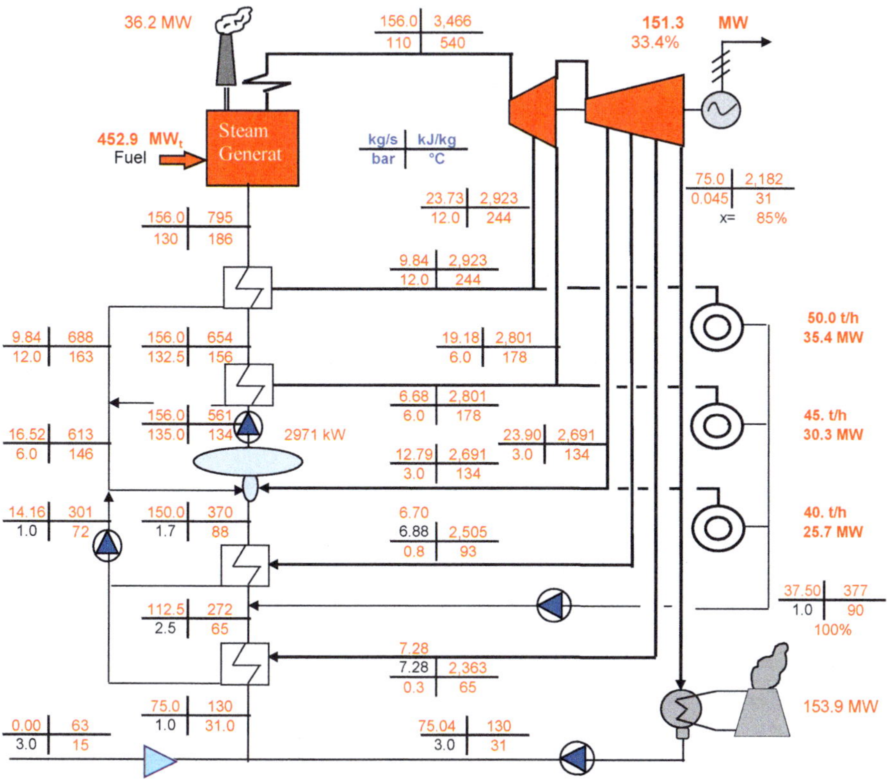

Panos Konstantin
Technologies & Economics

Generatorleistung, brutto	MW	151.3
Feuerungswärmeleistung:	MW	452.9
Frischdampf	MW	452.9
Zwischenüberhitzung	MW	0
Bruttowirkungsgrad	%	33.42%
Gesamtwirkungsgrad	%	53.59%

11.12 Modelling & Simulation Rankine Cycle, Reheat

PK Panos Konstantin

The Power Supply Industry

Case Study
Cycle Simulation
Extraction-Condensing-Reheat

Notes:

1. Cells with black characters include inputs
2. Cells with red characters include formulas
3. Download of FluidEXL required for calculations
4. Read brief instruction for FluidEXL in the **Toolbox** in the book
5. Read introduction and notes in Case Study chapter of the book

The purpose of this Case Study is:

1. Training in calculation of thermodynamic cycles using FluidEXL
2. Calculation of performance parameters for cogeneration such as σ, β, η_{cond}, η_{total}
3. See also exmpeles 6-3 ans 6-4 in the book

Last update October 2017

Input Spreadsheet

<table>
<tr><th rowspan="2">Item</th><th rowspan="2">Unit</th><th colspan="2">Operation Mode</th></tr>
<tr><th>Cond</th><th>Cogen</th></tr>
<tr><td>Live steam</td><td>t/h</td><td colspan="2">463.9</td></tr>
<tr><td>Pressure</td><td>bar</td><td colspan="2">185</td></tr>
<tr><td>Temperature</td><td>°C</td><td colspan="2">540</td></tr>
<tr><td>Saturation temperature</td><td>°C</td><td colspan="2">359</td></tr>
<tr><td>HP bleed points</td><td></td><td></td><td></td></tr>
<tr><td>Pressure HP bleed</td><td>bar</td><td colspan="2">40</td></tr>
<tr><td>Terminal temperature difference</td><td>K</td><td colspan="2">2.0</td></tr>
<tr><td>Pressure HP bleed</td><td>bar</td><td colspan="2">23.5</td></tr>
<tr><td>Terminal temperature difference</td><td>K</td><td colspan="2">1.0</td></tr>
<tr><td>Process Steam extraction</td><td>t/h</td><td>0</td><td>50</td></tr>
<tr><td>Pressure</td><td>bar</td><td colspan="2">12</td></tr>
<tr><td>Terminal temperature difference</td><td>K</td><td colspan="2">2.0</td></tr>
<tr><td>Process Steam extraction</td><td>t/h</td><td>0</td><td>45</td></tr>
<tr><td>Pressure</td><td>bar</td><td colspan="2">6</td></tr>
<tr><td>Terminal temperature difference</td><td>K</td><td colspan="2">3.0</td></tr>
<tr><td>Process Steam extraction</td><td>t / h</td><td>0</td><td>40</td></tr>
<tr><td>Pressure</td><td>bar</td><td colspan="2">3</td></tr>
<tr><td>Terminal temperature difference</td><td>K</td><td colspan="2">4</td></tr>
<tr><td>LP bleed</td><td>bar</td><td colspan="2">0.7</td></tr>
<tr><td>Terminal temperature difference</td><td>°C</td><td colspan="2">5</td></tr>
<tr><td>Condenser</td><td></td><td colspan="2"></td></tr>
<tr><td>Cooling water inlet</td><td>°C</td><td colspan="2">15</td></tr>
<tr><td>Cooling water temperature rise</td><td>K</td><td colspan="2">12</td></tr>
<tr><td>Terminal temperature difference</td><td>K</td><td colspan="2">4</td></tr>
<tr><td>Boiler efficiency</td><td>%</td><td colspan="2">92.0%</td></tr>
<tr><td>ST- internal efficiency</td><td>%</td><td colspan="2">91.0%</td></tr>
<tr><td>Generator efficiency</td><td>%</td><td colspan="2">98.6%</td></tr>
<tr><td>ST- mechanical efficiency</td><td>%</td><td colspan="2">99.5%</td></tr>
<tr><td>Condensate return rate</td><td>%</td><td colspan="2">100%</td></tr>
<tr><td>Temperature</td><td>°C</td><td colspan="2">90</td></tr>
<tr><td>make-up water</td><td>°C</td><td colspan="2">15</td></tr>
</table>

Energy flow diagram – Condensing Mode of Operation, Reheat

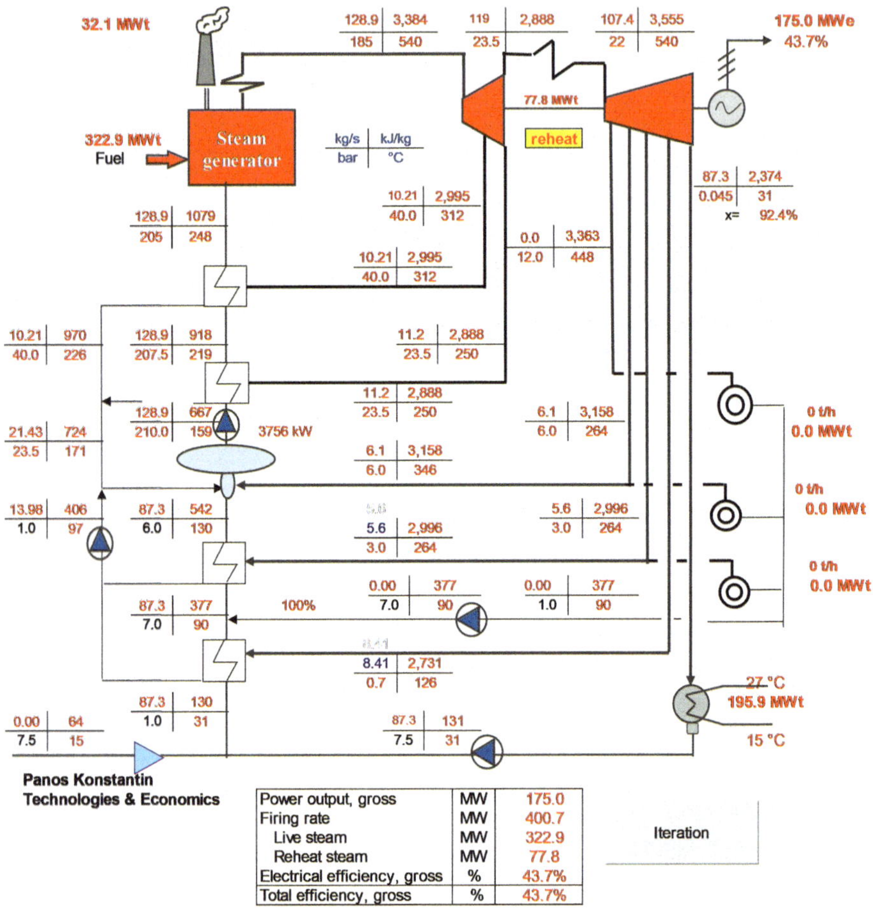

Power output, gross	MW	175.0
Firing rate	MW	400.7
Live steam	MW	322.9
Reheat steam	MW	77.8
Electrical efficiency, gross	%	43.7%
Total efficiency, gross	%	43.7%

Iteration

Energy Flow Diagram – Cogeneration Mode of Operation, Reheat

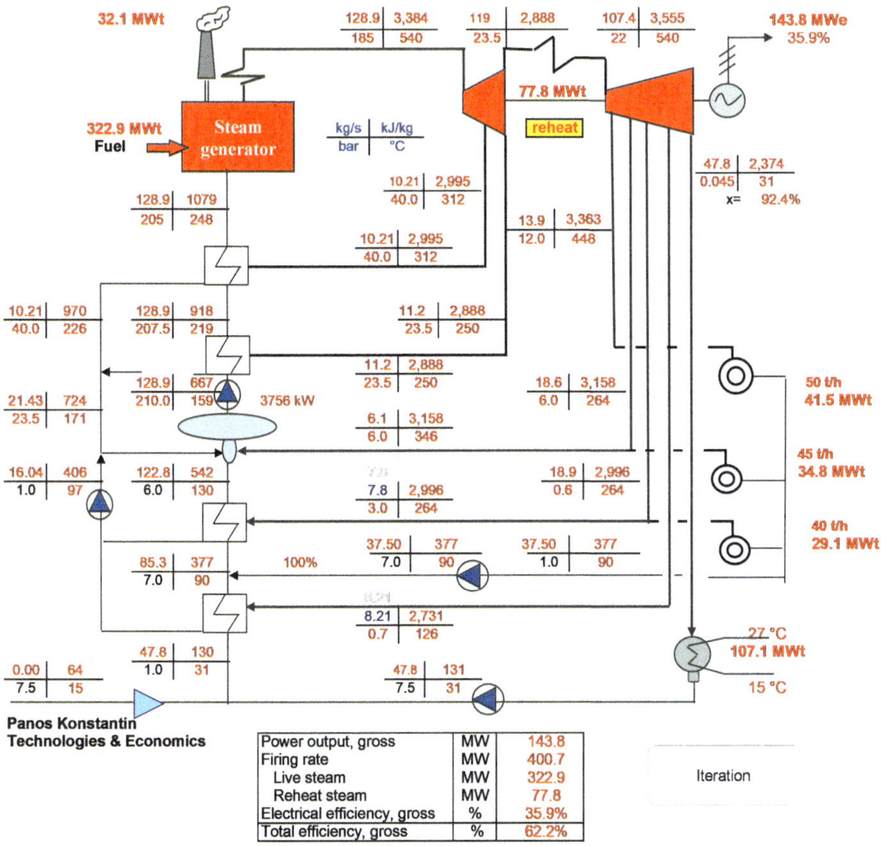

Power output, gross	MW	143.8
Firing rate	MW	400.7
Live steam	MW	322.9
Reheat steam	MW	77.8
Electrical efficiency, gross	%	35.9%
Total efficiency, gross	%	62.2%

11.13 Simulation of Extraction-condensing CCGT cycles

PK **Panos Konstantin**

The Power Supply Industry

Case Study
Cycle Simulation
Combined Cycle Gas Turbine

Notes:

1. Cells with black characters include inputs
2. Cells with red characters include formulas
3. Download of FluidEXL is required
4. Read notes in Chapter Case Studies of the book

The purpose of this Case Study is:

1. Training in calculation of thermodynamic cycles using FluidEXL
2. Calculation of performance parameters for cogeneration such as s, b, hcond, htotal
3. Read also examples 6-3 and 6-4 in th book

Last update October 2017

Input Gas Turbine Parameters

Item	Unit	Nominal Plant Size		Formula / Remark
		ISO	RSC	
Power plant				
Type of GT	-	Siemens SGT-700		
Frequency		50 Hz		
Fuel		NG		
Number of gas turbines		2		
Site Conditions				
Elevation	m	0	608	Elevation Corr. Factor applied
Elevation Correction Factor	-	1.00	0.93	Applied to Power, Mass flow
Ambient Temperature	°C	15	48	
	°F	59	118	
Unit capacity				
Rated Power output, gross (per GT)	MW_e	31.2		
Power output, gross	MW_e	31.2	23.6	y = -3E-05x2 - 0.0055x + 1.0889
Efficiency	-	36.4%	33.1%	y = -2E-05x2 - 0.0015x + 1.0263
Fuel consumption	MW_t	86	71	= power output / efficiency
	MMBTU / h	293	244	Million BTU/h
Rated mass flow, per turbine	lb / s	208		
Flue gas mass flow, total	lb / s	208	171	y = -3E-05x2 - 0.0035x + 1.0577
	kg / s	94	77	
Flue gas temperature	°C	528	570	y = 2E-05x2 + 0.0011x + 0.9794
	°F	983	1,057	

Source of ISO Data: Gas Turbine World, 2012 GTW Handbook

Input CCGT cycle Parameters

Item	Unit	Cond	Cogen
Number of gas turbines	-	2	
GT power output, each	MW	31.2	
Firing rate in LHV, each	MW	85.7	
Gas turbine make	-	Siemens SGT-700	
Flue gas HRG inlet, each	kg/s	94	
	°C	528	
HP live steam	bar	90	
	°C	520	
	t/h	80	
LP live steam (1 cond, 0 extraction)	-	1	0
	bar	4.5	
Pinch point, HRSG	K	14	
Delta feed - makeup water	K	20	
Extraction steam	t/h	0.0	20
	bar	16.0	
Backpressure steam extraction	t/h	0.0	18
	bar	8.0	
Cooling medium, condenser	water		
inlet Temperature	°C	12.0	
outlet temperature	°C	24.0	
Terminal Temperature Difference	°K	7.0	
condensing temperature	°C	31.0	
Boiler inlet	°C	105	
Condensate return from factory	%	100%	
	°C	90	
	bar	1.0	
Pinch point, preheater	K	25	
Makeup water	°C	15	

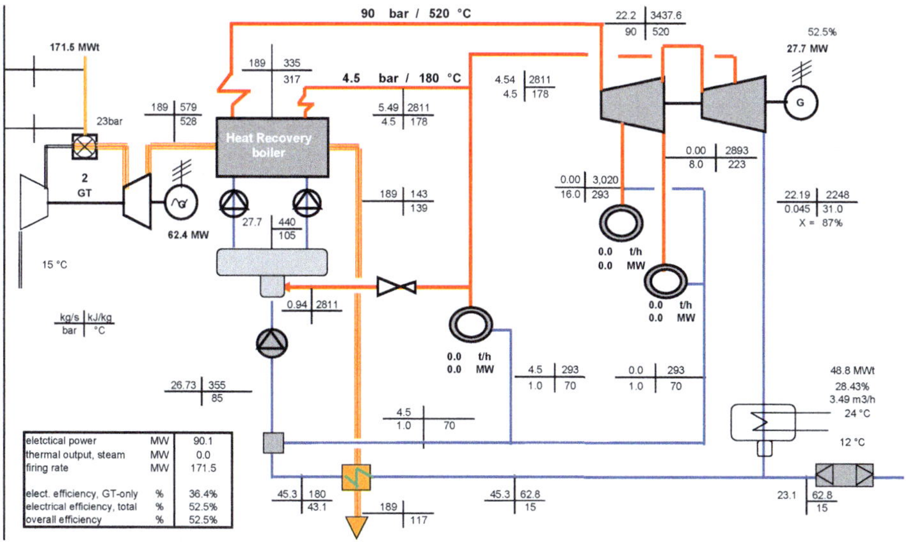
90 bar / 520 °C
4.5 bar / 180 °C
171.5 MWt
23bar
2
GT
62.4 MW
15 °C
Heat Recovery boiler
52.5%
27.7 MW
G
48.8 MWt
28.43%
3.49 m3/h
24 °C
12 °C
kg/s | kJ/kg
bar | °C
eletctical power MW 90.1
thermal output, steam MW 0.0
firing rate MW 171.5
elect. efficiency, GT-only % 36.4%
electrical efficiency, total % 52.5%
overall efficiency % 52.5%
0.0 t/h
0.0 MW
X = 87%

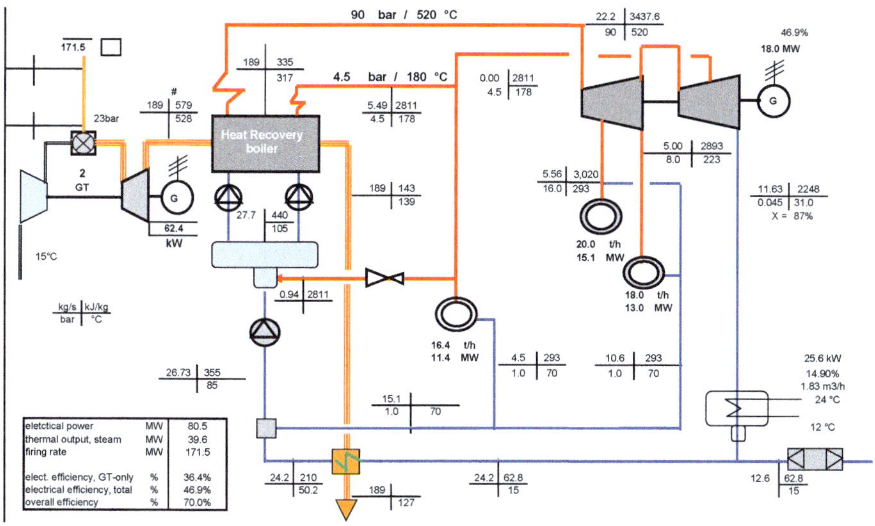
90 bar / 520 °C
4.5 bar / 180 °C
171.5
23bar
2
GT
G
62.4
kW
15°C
Heat Recovery boiler
46.9%
18.0 MW
25.6 kW
14.90%
1.83 m3/h
24 °C
12 °C
kg/s | kJ/kg
bar | °C
eletctical power MW 80.5
thermal output, steam MW 39.6
firing rate MW 171.5
elect. efficiency, GT-only % 36.4%
electrical efficiency, total % 46.9%
overall efficiency % 70.0%
20.0 t/h
15.1 MW
18.0 t/h
13.0 MW
16.4 t/h
11.4 MW
X = 87%

Bibliography and References

[1] H.-J. Kretzschmar and I. Kraft, Kleine Formelsammlung Technische Thermodynamik, 3. ed., Carl HANSER, Germany.

[2] D. K. Yadao, Thermal Engineering, Khanna Publishers, Dehli, India, 1999.

[3] W. Berties, Übungsbeispiele aus der Wärmelehre, 20 verbesserte Auflage ed., Fachbuchverlag Leipzig im Carl Hanser Verlag.

[4] Recknagel, Sprenger and Schramek, Taschenbuch für Heizung + Klima Technik, Oldenburg IndustrieVerlag München, 2002.

[5] IAPWS, International Association for the Properties of Water and Steam -, "Revised Release on the IAPWS Industrial Formulation 1997 for thermodynamic properties of water and steam," August 1997. [Online]. Available: http://www.iapws.org/relguide/IF97-Rev.pdf.

[6] University of Applied Sciences Zittau/Görlitz, Germany, Faculty of Mechanical Engineering, dept of Technical Thermodynamics, "FluidEXL graphics LibIF97," [Online]. Available: https://www.google.de/#q=fluidexlgraphics.

[7] Fichtner IT Consulting, "KPRO - Thermodynamic cycle modelling & simulation software _ KreisPROzess," Fichtner, Stuttgart.

[8] IEEE Power Engineering Society, "IEEE Standard Definitions for Use in Reporting Electric Generating Unit Reliability, Availability and Productivity," 2006.

[9] VGB PowerTech , "Availability of thermal power plants 2002 - 2011," Essen, Germany, 2012.

[10] P. Konstantin, Power and Energy Systems - Engineering Economics, SpringerVieveg, 2018, p. 200.

[11] U.S. Energy Information Administration, "Updated Capital Cost Estimates for Utility Scale Electricity Generating Plants," Washington, DC, April 2013.

[12] Fichtner IT Consulting AG, "KPRO - Thermodynamic Cycle Simulation Program," Fichtner, Stuttgart, Germany.

P. Konstantin and M. Konstantin, *The Power Supply Industry*,
https://doi.org/10.1007/978-3-319-72305-1

[13] GTW Handbook, USA: Gas Turbine World - A Pequot Publication, 2015.

[14] Gasturbinen Kenndaten und Referenzen, Germany: ASUE, 2015.

[15] Puyvelde, L. Pintelon and F. Van, Maintenance Decision Taking, 3000 Leuveb Belgium: Uitgevorij Acco, 2006.

[16] Babar and Fazal-ur-Rehman, Gas Turbines Maintenance Inspections and Calculations of Equivalent Operating Hours.

[17] SIEMENS, "Facts & Figures; Siemens H-Klasse Gasturbine," [Online]. [Accessed 02 2014].

[18] Energy Environmental Analysis - EEA/ICF, "Technology Characterization, Reciprorcating Engines," Arlington Virginia, 2008.

[19] Wärtsilä, "Reciprocating Engine vs. Gas Turbine, startup time," 29 01 2014. [Online].

[20] Wärtsilä Brochure, *Power Plants Solutions,* 2013.

[21] P. Konstantin, Power and Energy Systems - Engineering Economics, SpringerVieweg, 2018.

[22] eia, U.S. Energy Information Administration -, "Updated Cost Estimates for Utility Scale Electricity Generating Plants," US Department of Energy, Washington DC, April 2013.

[23] World Nuclear Power Association, "Nuclear fuel cycle, Nuclear power reactore," 5 1 2014. [Online].

[24] Word Nuclear Power Association, "Advanced Nuclear Power Reactors," 3 1 2014. [Online].

[25] Atomenergoproject - State Atomic Energy Corporation "Rosatom", "VVER-TOI design," Rosatom, 2015. [Online]. Available: www.Rosatom/rusia/en.

[26] World Nuclear Association, "Decommissioning Nuclear Power Plants," 04 2015. [Online]. Available: http://www.world-nuclear.org/info/Nuclear-Fuel-Cycle/Nuclear-Wastes/Decommissioning-Nuclear-Facilities/. [Accessed 20 09 2015].

[27] International Atomic Energy Agency, "IAEA Safety Standards Series - Decommissioning of Nuclear and Researche Reactors," [Online]. Available: http://www-pub.iaea.org/MTCD/publications/PDF/P079_scr.pdf. [Accessed 21 09 2015].

[28] P. Konstantin, Power and Energy Systems - Engineering Economics, SpringerVieweg, 2018, pp. Section 8.3.6 Nuclear fuel, Example 8-4.

[29] Worls Nuclear Association, "The Economics of Nuclear Power," October 2013. [Online].

[30] Organization for Economic Coordination and Development - Nuclear Energy Association OECD/NEA, "Decommissioning Nuclear Power Plants - Policies, Strategies and Costs," 2003. [Online]. Available: http://www-ns.iaea.org/downloads/rw/projects/r2d2/workshop6/references/others/oecd-nea-decom-npps-eng-2003.pdf.

[31] Hydro Power Association, Brussels, Belgium, "http://www.thehea.org/hydropower/special-focus/pump-storage-power-plants/," [Online]. [Accessed 02 2014].

[32] bdew - Verband der Energie Wasser und Abwasser, [Online]. Available: https://www.bdew.de/bdew.nsf/id/DE_Schaubilder/$file/turbinen%20prinzip.jpg. [Accessed 02 2014].

[33] Hydro Saar - Hydak International, "Wasserkraftwerk," [Online]. Available: http://www.hydrosaar.com/de/branchen/erneuerbare-energien/wasserenergie/wasserkraftwerk.html. [Accessed October 2015].

[34] (NVE), Norwegian Water Resources and Energy Directorate, "Cost Base for Hydropower plants with a generating capacity of more than 10 000 kW," 2012. [Online]. Available: http://webby.nve.no/publikasjoner/veileder/2012/veileder2012_03.pdf. [Accessed October 2015].

[35] M. Ragheb and A. M. Ragheb, "Wind Turbines Theory - The Betz Equation and Optimal Rotor Tip Speed Ratio," [Online]. Available: http://www.intechopen.com/download/get/type/pdfs/id/16242. [Accessed September 2015].

[36] J. Layton, "How Wind Power Works," [Online]. Available: http://science.howstuffworks.com/environmental/green-science/wind-power2.htm.

[37] "Danish Wind Energy Association - Wind Energy Manual - Wind Energy Concepts," [Online]. Available: http://drømstørre.dk/wp-content/wind/miller/windpower%20web/en/stat/unitsw.htm#roughness.

[38] Bundesverband WindEnergie e.V. BWE, "Status des Windenergie Ausbaus in Deutschland 2015".

[39] P. Konstantin, Power and Energy Systems - Engineering Economics, SpringerVieweg, 2018, pp. Section 7.4- Uncertainty anlysis of energ production.

[40] P. Konstantin, Praxisbuch Energiewirtschaft, 4th Edition, 4th 2017 ed., Germany: Springer Vieweg; VDI, 2017.

[41] Photovoltaic and solar energy design tools, "Solar radiation database," [Online]. Available: http://photovoltaic-software.com/solar-radiation-database.php.

[42] NASA - Atmosperic Science Data Center, "Surface meteorology and Solar Energy," November 2015. [Online]. Available: https://eosweb.larc.nasa.gov/sse/.

[43] Photovoltaik & Solar Radiation Design Tools, "Solar Radiation Databases," SolarGis, [Online]. Available: http://photovoltaic-software.com/solar-radiation-database.php. [Accessed November 2015].

[44] Wheeling Jesuit University, "Exporing Environment Eathe on Fire," 2016. [Online]. Available: http://www.wju.edu/.

[45] Solar Praxis AG,, "Solar-is-future.com," [Online]. [Accessed 02 2014].

[46] Lawrence Berkeley National Laboratory, "Photovoltaic System Pricing Trends," 2013 Edition.

[47] NREL - National Energy Laboratory, USA, "Photovoltaic System Pricing Trends, historical, recent and near term projections," 2014 Edition.

[48] Fichtner GmbH & CO. KG, Stuttgart, Germany, *SOLPRO, Software Tool for Design and Performance Calculations of Parabolic Trough Solar Power Plants.*

[49] Deutsches Zentrum für Luft- und Raumfahrt (DLR) e. V., "http://www.dlr.de/sf/Portaldata/73/Resources/dokumente/Soko/soko2009/Poster/Visual-HFLCAL.pdf," [Online].

[50] University of Applied Science, Zittau/Görlitz, Gemany, "FluidEXL Graphics," Fachhochschule Zitau/Görlitz, Fachgebiet Technische Thermodynamik, Prof. Dr. Ing.habil H.J. Kretzschmar, Dr.-Ing. I.Stöcker, Zittau Germany.

[51] P. Konstantin, "Engineering Economics," in *Power and Energy Systems - Engineering Economics*, epubli, Germany, 2015.

[52] M. Milton Gussow, Theory and Problems of Basic Electricity, McGraw-Hill.

[53] American National Standards Institute , *Ansi C84.1-1989 Voltage Rating for Electric Power Systems and Equipment.*

[54] D. D.-I. Klaus-Dieter. [Online]. Available: http://www.forum-netzintegration.de/uploads/media/Schlichting_120614_Dettmann.pdf.

[55] Björn Jacobson, ABB Power Systems - HVDC, "Developments in Multiterminal HVDC," October 2011. [Online]. Available: http://www.ieee.ca/epec11/admin/04-0800-hvdc_plenary_jacobson.pdf.

[56] ABB, Jochen Kreusel, "High Voltage Direct Current (HVDC) power transmission," ABB, [Online]. Available: http://www.iass-potsdam.de/sites/default/files/files/kreusel_presentation.pdf.

[57] A. Canelhas, "High Voltage Direct Current," 22 September 2010. [Online]. Available: http://www.alstom.com/Global/Group/Resources/Documents/Investors%20document/Investor%20events/Analysts%20presentation/Analyst%20Day%20-%20HVDC%20technology.pdf.

[58] Amprion GmbH, Germany, "Procurement of control power and energy in Germany," [Online]. Available: http://www.amprion.net/en/control-energy.

[59] European Network of Transmission System Operators for Electricity - ENTSOe, "Production data," [Online]. Available: https://www.entsoe.eu/data/data-portal/production/Pages/default.aspx. [Accessed 2013].

[60] H. Khatib, Economic Evaluation of Projects in the Electricity Industry, Herts : The Institution of Enineering and Technolgy, 2003.

[61] NPTEL - National Program on Technology Enhanced; India, Government of, "NPTEL >> Electrical Engineering >> Restructured PoweMarket models based on contractual arrangements," [Online]. Available: http://nptel.ac.in/courses/108101005/14.

[62] C. Harris, Electricity markets, Pricing, structures and Economics, Chichester, West Sussex, England: Wiley Finance, John Wiley & Sons Ltd.

[63] E. Onaiwu, "How does bilateral trading differ from electricity pooling," [Online]. Available: www.dundee.ac.uk/cepmlp/gateway/files.php?file=cepmlp_car13....

[64] M. Kraus, "Liberalized Energy Markets - Do We Need Regulations," [Online]. Available: http://userpage.fu-berlin.de/~jmueller/fhw/ss04/RegulierungKraus.pdf.

[65] EEX - European Energy Exchange, "EEX Product Brochure Power," August 2014. [Online]. Available: http://www.eex.com/blob/66452/40623f9f22d203a09155547bc84c89f2/konzept-strom-release-03a-englisch-pdf-data.pdf.

[66] NordPool, "The Nordic Electricity Exchange - The Nordic Model for a Liberalized Electricity Market," [Online]. Available: http://www.nordpoolspot.com/Global/Download%20Center/Rules-and-regulations/The-Nordic-Electricity-Exchange-and-the-Nordic-model-for-a-liberalized-electricity-market.pdf.

[67] Investopedia, "Futures Fundamentals- Introduction," [Online]. Available: http://www.investopedia.com/university/futures/.

[68] US Project Management Institute, A Guide to the Project Management Body of Knowledge (BMBOK-Guide), 5th Edition ed., A. 99-001-2013, Ed., Project Management Institute, Inc., 2013.

[69] International Federation of Consulting Engineers - FIDIC, "Contract and Agreements Collection," 22 February 2016. [Online]. Available: http://fidic.org/bookshop/about-publications.

[70] P. Konstantin, "Power and Energy Systems - Engineering Economics," 2018.

[71] P. Konstantin, "Power and Energy Systems - Engineering Economics," SpringerVieweg, 2018.

Toolbox

List of Tool Guides

P. Konstantin and M. Konstantin, *The Power Supply Industry*,
https://doi.org/10.1007/978-3-319-72305-1

Tool Guide 1: Brief instruction for software tool **FluidEXL**^{Graphics}

FluidEXL is a professional, powerful tool for calculating thermodynamic properties of water and steam. It includes a large number of functions and can be downloaded from the examples' section on the author's website. There are two versions available for 32 and 64 bits windows. The software is thoroughly described in the readme file "FluidEXL-Graphics-LibIF97-Doc-PK.pdf". After download it is automatically installed by the "insert function facility – f_x" of Excel, under category "Water IAPWS-IF9", where all the functions appear as a list.

FluidEXL is used in this book for calculation of water/steam properties in several examples and for simplified thermodynamic cycle simulations in MS-Excel®.

The calculations are directly conducted in the selected cells of Excel®. After selection of a cell and the appropriate function, a menu as the following will appear:

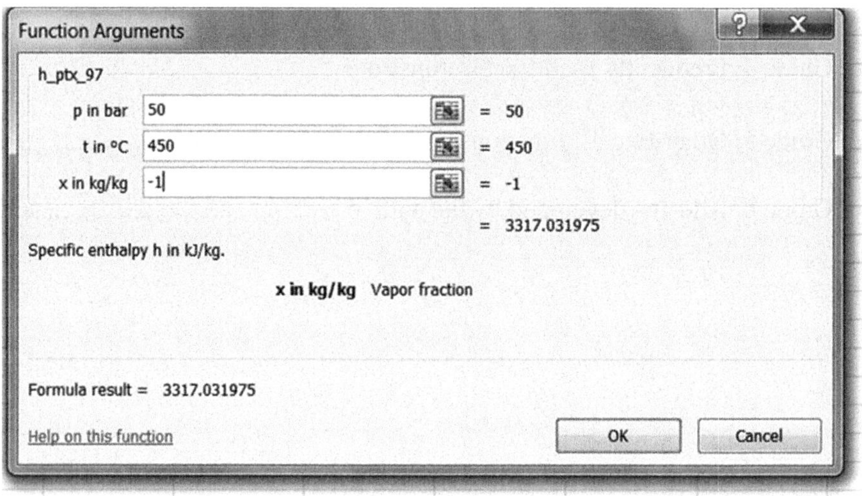

Important Note: For calculation of water/steam properties maximum two argument entries are required. If three arguments are shown, as in the menu above, one of them must be set as "-1" (minus 1).

As already mentioned, the FluidEXL is thoroughly described in the readme file. In the following table only some functions, very frequently used in the examples of the book, are explanatory presented for the beginning.

Users are, however, advised to study the readme file, as the tool provides many additional applications especially for students and engineers.

Property of function	Syntax	Function name in the list *)	Arguments for entry			Unit of result
Saturation temperature	ts=f(p)	ts_p_97	bar	-	-	°C
Saturation pressure	ps=f(t)	ps_t_97	-	°C	-	bar
Specific enthalpy:						
superheated steam	h=f(p,t,x)	h_ptx_97	bar	°C	-1	kJ/kg
wet steam			bar	-1	x	
saturated steam			bar	-1	1	
saturated steam			-1	°C	1	
water (not saturated)			bar	°C	-1	
saturated water			-1	°C	0	
saturated water			bar	-1	0	
Specific entropy:						
superheated steam	s=f(p,t,x)	s_ptx_97	bar	°C	-1	kJ/(kg K)
saturated steam			bar	-1	1	
Specific exergy **)	e=f(p,t,x,tu)	e_ptx_tu	bar	°C	-1	kJ/kg
Vapor fraction	x=f(h,p)	x_ph_97	bar	h	-	kg/kg

*) "**97**" stands for IAPWS Industrial Formulation **97** for the Thermodynamic Properties of Water and Steam

**) tu: ambient temperature °C

IAPWS: International Association for the Properties of Water and Steam

Note: FluidEXL is practiced in **Case Study 11.1** on the example of a Rankine cycle with supercritical steam parameters depicted in T-s diagram, and in a complete cycle calculation of a simple steam power plant in **Case Study 11.2**. Both Case Studies are presented in **Chapter 11, Case Studies** and are available as softcopies in Excel® on the author's website. The tool is moreover used in numerous application examples of the book.

Tool Guide 2: Download instruction[36] for Cycle Modelling & Simulation tool KPRO®

Fichtner's KPRO® (**K**reisprozess **R**echnungs **O**ptimierer[37]) is a powerful, highly professional software tool. It is being steadily developed and updated since 1980 based on the requirements and cumulative experience from real world projects and adapted in latest state-of–the-art information technology.

It enables modelling and performance simulation of all kind of power generation systems, also very complex ones such as power and steam supply systems of entire chemical factories or refineries.

The modelling of thermodynamic cycles for power generation with KPRO® is achieved by means of "graphic elements" (i.e., gas turbines, steam generators, steam turbines, heat exchangers, condensers etc.) and their connecting steam and water pipelines. The associated energy flow diagram is drawn directly on the screen with the aid of a graphic input system. The elements of the cycle, their connections with each other and the related thermodynamic parameters (pressure, temperature, enthalpy, mass flow) are depicted automatically for each graphic element on the flow diagram – Picture 1 below.

The modelling of the cycles is usually carried out for design flow rate at full load operation. In part load operation, the system configuration is kept the same, however, the tool considers changes and deviations of process characteristics against the full load operation such as effects of Stodola's cone law for turbines, changes of temperature differences (TTD) between steam and water in preheaters and condensers, which have an impact on the performance of the cycles.

The results of the KPRO® cycle simulation model are presented in a heat flow diagram, containing all the thermodynamic parameters of the process.

The results of KPRO® simulations are also used as reference in warranty tests for power plant performance in projects.

Users of the tool need a strong background in thermodynamics of cycles and must also be familiar in modelling with graphic elements. Hence, potential users have to spend some time in reading the instruction and practicing the tool.

Students and new users are advised to start practicing with the simple tool in Excel using the FluidEXL. A step-by-step instruction for beginners is introduced in the Tool Guide 3 and practiced in **Case Studies 11.1 and 11.2**.

[36]) **Note:** You find a brief download instruction for KPRO® at the end of this Guide (after the graph) and on the example section on the author's website.

[37]) Cycle Calculation and Optimization Program

Picture 1: Model and performance simulation of a 600 MW Rankine cycle

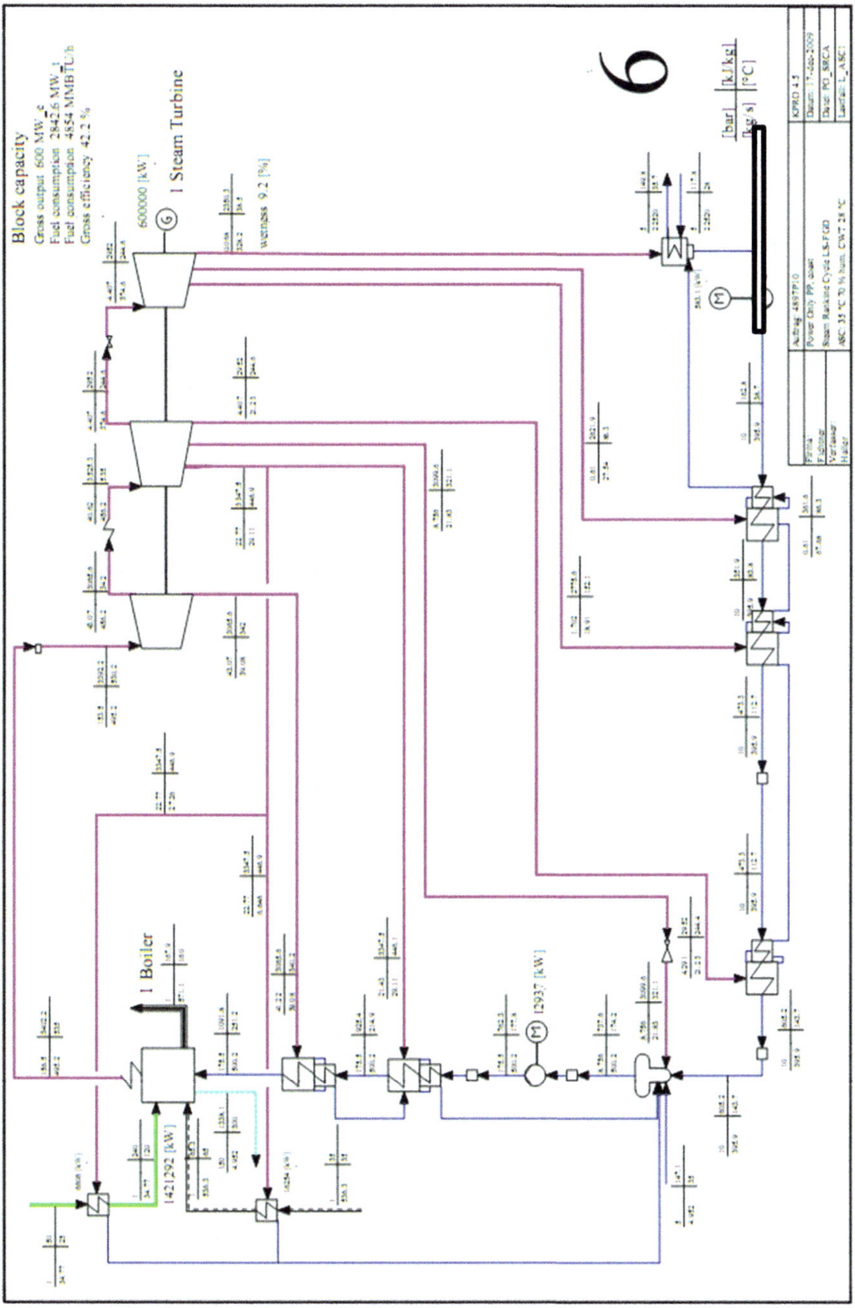

Download instruction for KPRO®

Fichtner IT Consulting AG has to grant a license for each user of KPRO®, which is linked to the MAC-address of the target computer. For readers of the book only and solely non-commercial use, the license is free of charge and is made available for six months.

For granting the license, Fichtner IT needs the contact data of the user (First and family name, Institution/company (or private), E-mail address) and the MAC-address of the target computer for installation. For determining the MAC-address, Fichtner IT provides a small software tool – "make-kpro-licence-info.bat" – that must be executed on the target computer (it is available in the examples section on the author's website). Potential users need to follow the steps below:

- Unzip the file on the desktop or in a hard disk of your preference
- Execute the file, the tool reads the computer's MAC-address and creates an text-file (.txt) in the same folder
- Send the text file to Fichtner IT to compile your license
- Afterwards, the user can download KPRO® from Fichtner IT FTP-server (www.fit.fichtner.de) and install his personal license on this computer.

Fichtner IT has also available mail instruction for download in English or German, including all the required information (Readme, KPRO-download-link).

Tool Guide 3: Methodology of Calculation of a simple Rankine cycle

Thermodynamic cycles are commonly calculated with professional software tools available on the market such as Fichtner's KPRO®. However, some background knowledge about the thermodynamics of cycle calculations is indispensable for users of such kind of tools, especially for those with little experience.

In the following, a stepwise guidance for calculation of a simple Rankine power cycle, as depicted in the figure below, is provided. In principle, the same procedure is also applied for more complex cycles. For calculations of water/steam properties FluidEXL is used. The figures depicted below are also part of the cycle calculation.

Picture 2: Simplified mass-energy flow diagram

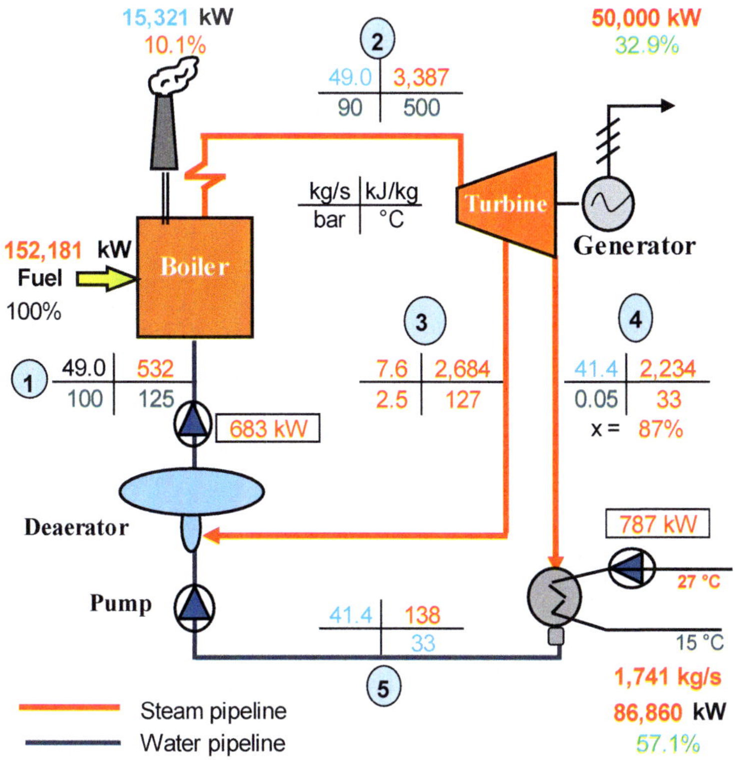

Note: The cycle calculation is presented in chapter 11 as **Case Study 11.2**. It is also available as softcopy on the author's website.

Picture 3: Enthalpy- Entropy (T-s) diagram

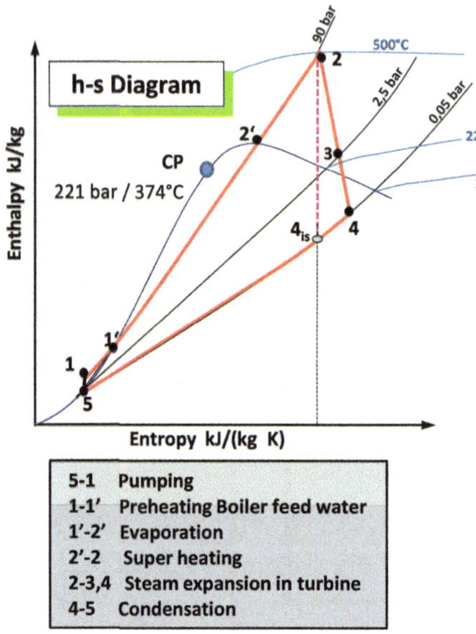

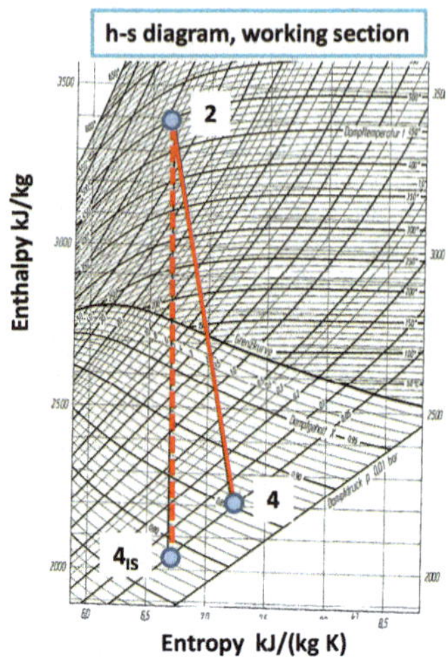

Source: Technologies & Economics, Author's own illustration

Stepwise guide (steps referred to energy flow diagram)

1. Define pressure and temperature for Point 2 (P2)
2. Define condensation temperature ts=t_{cw_out}+TTD (ts =condenser cooling water outlet temperature plus temperature difference steam-water, called Terminal Temperature Difference – TTD)
3. Calculate enthalpy *h_ptx* and entropy *s_ptx* for P2
4. Calculate saturation pressure $p_{s}\_t_{s}$
5. Drag a line in h-s diagram for isentropic expansion points 2 to 4_{is}
6. Calculate enthalpy for $P4_{is}$ (*h_ps*). Entropy is the same as of P2
7. Calculate isentropic enthalpy difference Δh_{is} between 2, 4_{is}
8. Calculate real enthalpy difference for points 2 to 4: $\Delta h_{real} = \eta_i \cdot \Delta h_{is}$ (η_i is the internal efficiency of the turbine)
9. Calculate real enthalpy for point for P4: $h_{4\_real} = h_2 - \cdot \Delta h_{real}$,
10. Drag a line 2 to 4
11. Calculate enthalpy for point 3 (repeat steps 6, 7, 8)
12. Calculate mass flow for deaerator (formula derived from mass and energy balance, m_{ls}: live steam)

$$\dot{m}_d = \frac{\dot{m}_{ls} \cdot (h_1 - h_5)}{(h_3 - h_5)} \quad \left[\frac{\text{kg}}{\text{s}}\right]$$

13. Calculate steam mass flow to condenser

$$\dot{m}_c = \dot{m}_{ls} - \dot{m}_d$$

14. Calculate gross power output (η: mechanical efficiencies, turbine, generator, gear)

$$P_{gross} = \eta_T \cdot \eta_{Gen} \cdot \eta_{gear} \cdot (\dot{m}_c \cdot (h_2 - h_4) + \dot{m}_d \cdot (h_2 - h_3) \quad [\mathrm{kW}]$$

15. Calculate thermal fuel input to the boiler (η_B: boiler efficiency)

$$\dot{Q} = \frac{\dot{m}_{ls} \cdot (h_2 - h_1)}{\eta_B} \quad [\mathrm{kW_t}]$$

16. Calculate gross cycle electrical efficiency and heat rate

Electrical efficiency: $\eta_e = \dfrac{P}{\dot{Q}} \quad [\mathrm{kW_t}]$

Heat rate: $\dot{q} = \dfrac{\dot{Q}}{P} \quad \left[\dfrac{\mathrm{kWh_t}}{\mathrm{kWh_e}}\right] = 3{,}6 \cdot \dfrac{\dot{Q}}{P} \quad \left[\dfrac{\mathrm{MJ}}{\mathrm{kWh_e}}\right]$

17. Calculate heat reject in condenser

$$Q_c = \dot{m}_c \cdot (h_4 - h_5) \quad [\mathrm{kW_t}]$$

18. Calculate cooling water flow

$$\dot{m}_{cw} = \frac{\dot{Q}\left[\frac{\mathrm{kJ}}{\mathrm{s}}\right]}{c\left[\frac{\mathrm{kJ}}{\mathrm{kg \times K}}\right] \bullet \Delta t\,[K]} \quad \left[\frac{\mathrm{kg}}{\mathrm{s}}\right] \qquad \dot{V} = 3.6 \cdot \dot{m}_{cw} \quad \left[\frac{\mathrm{m^3}}{\mathrm{h}}\right]$$

19. Electrical cooling water pump capacity (mechanical efficiencies: η_P, pump, η_e, motor)

$$P_{P\_el} = \frac{\dot{m}_{cw}\left[\frac{kg}{s}\right] \cdot g\left[\frac{m}{s^2}\right] \cdot H\,[m]}{1000 \cdot \eta_P \cdot \eta_e} \quad [\mathrm{kW}]$$

20. Feed water pump capacity

$$P_{P\_el} = \frac{m_{ls}\left[\frac{kg}{s}\right] \cdot 10^5 \cdot \Delta p\,[bar]}{1000.\rho\left[\frac{kg}{m^3}\right] \cdot \eta_P \cdot \eta_e} = \frac{100 \cdot m_{ls}\left[\frac{kg}{s}\right] \cdot \Delta p\,[bar]}{\rho\left[\frac{kg}{m^3}\right] \cdot \eta_P \cdot \eta_e} \quad [\mathrm{kW}]$$

Tool Guide 4: Frequently used Excel® functions

In the following some of frequently used Excel functions are listed and explained. The name of the function in the German Excel version is shown within parenthesis.

Future value function FV (ZW): Returns the future value of an investment based on periodic constant payments and a constant interest rate.
Syntax: *FV(Rate, Nper, Pv, Fv, Type)*

Present value function PV (BW): Returns the present value of a series of periodic future equal payments and a constant interest rate.
Syntax: *PV(Rate; Nper; Pmt; Fv, Type)*

Net present value function NPV (NBW): Calculates the net present value of future payments using a constant discount rate. In contrast to the PV function the payments can be equal or unequal, positive (cash inflows) or negative (cash outflows).
Syntax: *NPV(Rate, value 1, value 2, ...value n)*
Note: If the values 1 to n are arranged in series, just mark the range of series (e.g. A2:A20) instead inserting the values one by one.

Annual equivalent amounts (annuities) function PMT (RMZ): Calculates the annualized constant amounts of the present value of an initial payment (principal) at a constant interest rate.
Syntax: *PMT(Rate; Nper; Pv; Fv; Type)*

Where:
Rate: Interest rate in % per period
Nper: *Number of periods*
Fv: Future value (in the examples in this book it is usually zero)
Type: For payments at the year's end zero, at the year's beginning 1 (in the examples in this book it is usually zero)
Pmt: The constant future payments, annuities
Pv: The present value of CAPEX or principal

Note: Excel assumes payments to be cash-outflows and returns them as negative amounts. In Cost Models in this book all series are costs, hence it does not make sense to designate as negative values to them.

> All payment series have positive values therefore a minus sign must be inserted in front of the function to get positive values.

Tool Guide 5: Other Excel® functions

Internal Rate of Return function IRR (IKV): Returns the Internal rate of return of a periodic series of payments consisting of cash inflows (positive values) and cash- outflows (negative values). The payments must occur in regular intervals (e.g. yearly) but do not have to be equal.
Syntax: *IRR(range of values; guess)*
Where:
Range of values: just mark the range of payment series (e.g. A2:A20)
Guess: give an estimated IRR otherwise Excel will assume 10%.

Modified Internal Rate of Return function MIRR (QIKV): As IRR, however, the returns are reinvested. MIRR considers both the IRR of the initial investment and the interest rate of the reinvestment.
Syntax: *MIRR(Range of values; finance rate; reinvest rate)*
Where:

Range of values:	Mark the range of payment series (e.g. A2:A20)
Finance rate:	Interest rate used in the initial investment
Reinvest rate:	Interest rate expected for reinvestments

Exceedance Probability calculation PXX: Exceedance probability is calculated using the NORM.INV function as follows.
Syntax: *XP=(2 μ - Norm.Inv(Probability;mean μ; standard deviation σ)*
Where:
Probability: PXX, a number between 0 and 1 (or PXX/100); this is the probability corresponding to normal distribution.
Mean μ: This is the arithmetic mean μ of the distribution
Standard deviation σ: This is a positive number.

Example:

Exceedance Probability		
Mean value μ= 50 GWh/a		
PXX	σ=5	σ=10
	GWh / a	GWh / a
P95	41.8	33.6
P90	43.6	37.2
P75	46.6	43.3

Note: The cells PXX are customized: e.g. cell with 90; Format cells, custom, “P90”; insert Probability in the function P90/100 instead of 0.9.

“Goal seek” and **“Table”** are frequently used in examples. They can be found under “Data”, “what if analysis”.

Goal seek will find the right input to obtain a certain result (e.g. find rate to get NPV=0, this is the IRR).

Tables allow calculations of many different possible inputs at the same time (placed under “data” then ”table”).This is a very useful tool for sensitivity analysis.

Tool Guide 6: Add-Ins developed by the author

> **Note:** The Add-Ins are available for download as a Macro in the examples section of the author's website.

BWSesc: Calculates the present value (PV) of a series of payments escalating with a constant escalation rate.
Syntax: *BWSesc(escalation rate%; Discount rate %; Number of periods, starting value without escalation).*The Add-In is based on the equation shown below.

$$PV_{n\_esc} = P_0 \cdot \sum_{t=1}^{t=n} \frac{p^t}{q^t} = P_0 \cdot \frac{\left(q^n - p^n\right)}{(q-p)\cdot q^n}$$

Where:
P_0 : Constant payment each period before escalation
$q=1+i$: Discount factor, i : annual interest rate
$p = 1+j$: Escalation factor, j : escalation rate (may be also <0 !)
n: Number of years of the discounting period
Constraints: $q \neq p$; $q \neq 0$; $p \neq 0$
If $j > 0$ the PV of the series is increasing, if $j < 0$ it is decreasing!

ANesc: Converts a series of payments escalating with a constant escalation rate in constant annual equivalent amounts (annuities). The function calculates first the NPV of the series and uses the PMT function to convert the NPV in annuities. The function is used for levelizing escalating payment series.
Syntax: *ANesc(escalation rate%; Discount rate%; Number of periods, starting value without escalation).* The Add-In is based on the equation:

$$P_{AN\_esc} = LC = P_0 \cdot \frac{\left(q^n - p^n\right)\cdot p}{(q-p)} \times \frac{(q-1)}{q^n - 1} \qquad \left[\frac{\text{CU}}{\text{a}}\right]$$

Where:
P_{AN_esc}: Annuity of a series of escalating payments
P_0: Constant payment without escalation
$q= 1+i$: Discount factor, with "i" discount rate (interest rate)
$p= 1+j$: Escalation factor or geometric gradient, with "j" escalation rate
n : Number of periods (years)

IntCon: Calculates the interest during construction (IDC) assuming disbursement of loan in constant installments over the construction time. The Ann-In is developed based on the FV function of Excel.
Syntax: *IntCon(Capex; loan % of CAPEX; years of construction, interest for loans, periods per year).*

Annexes

Table of Annexes

P. Konstantin and M. Konstantin, *The Power Supply Industry*,
https://doi.org/10.1007/978-3-319-72305-1

Annex 1: Selected SI-units related to mechanical and thermal energy

Quantity	**Symbol** (italics)	**SI-Unit**	**In practice often used**	**Notes**
acceleration	a	m/s^2		
acceleration of free fall	g	m/s^2		
area	A	m^2		km
Celsius temperature	t, ϑ	°C		
density	ρ	kg/m^3	kg/m^3	
energy	E	J = W.s	kJ, MJ, GJ, TJ kWh, MWh, Wh	1 kWh = 3.6 MJ 1 GWh = 3.6 GJ
force	F	$N=kg.m/s^2$	kN	
frequency	f	Hz = 1/s (Hertz)		
heat flow rate	$\Phi_H, \dot{Q}$	W = J/s	kJ/s, GJ/h	
heat quantity	Q	J	kJ, MJ, GJ	kWh_t, MWh_t
heat transfer coefficient	U	$W/(m^2.K)$	$kW/(m^2.K)$	
length or path	l, s	m	cm, mm	also km
Mass	m	kg	t (for tonne)	1 t = 1000 kg
mass flow rate	$q_m, \dot{m}$	kg/s	t/h	
molar mass	M	kg/kmol		
molar volume	V_m	m^3/mol		ideal gas, air 22.4 $nm^3/kmol$
power	P	W= J/s	kW, MW, GW	
pressure (force divided by area)	p	$Pa = N/m^2$ (Pascal)	Pa for Pascal MPa, kPa, bar	1 bar=10^5 Pa 1 MPa = 10 bar
specific enthalpy	h	J/(kg.)	kJ/(kg)	
specific entropy	s	J/(kg.K)	kJ/(kg.K)	
specific heat capacity	c	J/(kg.K)	kJ/(kg.K)	
specific volume	υ	m^3/kg		
stress	σ	N/m^2		
thermal conductivity	λ	W/(m.K)	kW/(m K)	
thermodynamic temperature	T, θ	K	K for Kelvin	273.15 K=0°C
time, time interval	t	s	h, min, day, a	
velocity	υ, w, c	m/s		km/h
volume	V	m^3	1 l = 1 dm^3	l: litre
volume flow rate	$\Phi_V, \dot{V}$	m^3/s	m^3/h	
weight	G, W	N	kN	
work, energy	W	J = N.m	kJ, MJ, GJ kWh, MWh, GWh	

Annex 2: SI Unit System, Prefixes, Decimals and multiples

Prefix	Symbol	Factor	Example	
			Name	Unit
Micro	μ	10^{-6}	Micrometer	μm
Mil	m	10^{-3}	Millimeter	mm
Centi	c	10^{-2}	Centimeter	cm
Deci	d	10^{-1}	Decimeter	dm
Hecto	h	10^{2}	Hectoliter	hl
Kilo	k	10^{3}	Kilogram	kg
Mega	M	10^{6}	Megawatt	MW
Giga	G	10^{9}	Gigawatt	GW
Tera	T	10^{12}	Tera joule	TJ
Peta	P	10^{15}	Peta joule	PJ

Annex 3: Conversion tables for units

Imperial	multiply by		Metric	multiply by		Imperial
Length						
inches (in)	x	25.40	mm	x	0.039	inches
feet (ft)	x	30.48	cm	x	0.033	feet
feet (ft)	x	0.305	m	x	3.281	feet
yard (yd)	x	0.914	m	x	1.094	yard (yd)
miles (mi)	x	1.609	km	x	0.621	miles
Area						
square inch (si)	x	0.6452	cm^2	x	1.550	square inch (si)
square feet (sf)	x	0.0929	m^2	x	10.764	square feet (sf)
square yard (sy)	x	0.8361	m^2	x	1.196	square yard (sy)
Volume						
gallon (IG)	x	4.546	ltr	x	0.220	gallon (IG)
US gallon	x	3.785	ltr	x	0.264	US gallon
cubic feet (cf)	x	28.320	ltr	x	0.035	cubic feet (cf)
cubic feet (cf)	x	0.0283	m^3	x	35.311	cubic feet (cf)
scf	x	0.0268	nm^3	x	37.327	scf
Note: Standard cubic feet (scf): volume at 60°F/15.6°C and 14.7 psi/1.013bar; Normal cubic meter (nm^3) : Gasvolume at 1.013 bar, 0°C						
Mass						
pound (lb)	x	0.4536	kg	x	2.205	lb
ton	x	1.016	t	x	0.984	ton
Notes: 1 ton = 2,240 lb; 1 t = 1000 kg						
Force/weight						
pound-force (lbf)	x	4.448	N	x	0.225	pound-force (lbf)
Pressure						
psi	x	0.069	bar	x	14.503	psi
psi	x	6.895	kPa	x	0.145	psi
Note: 1 Pa=N/m^2; 1 bar = 10^5 Pa; 1 kPa=0.01bar						
Energy						
BTU (or Btu)	x	1.055	kJ	x	0.948	BTU
BTU	x	0.293	Wh	x	3.412	BTU
MMBTU	x	1.055	GJ	x	0.948	MMBTU
MMBTU	x	0.293	MWh	x	3.412	MMBTU
toe (ton oil equivalent)	x	41.850	GJ	x	0.024	toe
boe (barrel oil equivalent	x	1.697	MWh	x	0.589	boe
boe (bbl=159 l)	x	6.110	GJ	x	0.164	boe
Therms	x	0.106	GJ	x	9.479	Therms
Therms	x	29.310	kWh	x	0.034	Therms

Continuation Annex 3, conversion tables for unite

Imperial		multiply by	Metric		multiply by	Imperial
Power, Energy flow						
BTU / h	x	0.293	J/s = W	x	3.412	Btu / h
MMBTU/ h	x	0.293	MJ/s = MW	x	3.412	MMBtu / h
TR (tons refrigeration)	x	3.517	kW	x	0.284	TR
hp	x	0.746	kW	x	1.341	hp
ton of refrigeration	x	3.517	kW	x	0.284	ton of refrigeration
Calorific Value						
BTU / lb	x	2.326	kJ / kg	x	0.430	BTU / lb
BTU / lb	x	0.000646	kWh / kg	x	1547.8	BTU / lb
MMBTU / t	x		ce	x		
MMBTU / t	x		ce	x		
BTU / ft^3	x	37.260	kJ / m^3	x	0.027	BTU / ft^3
BTU / ft^3	x	10.349	kWh / m^3	x	0.097	BTU / ft^3
BTU /scf (gas)	x	39.382	kJ / nm^3	x	0.025	BTU / scf
BTU /scf (gas)	x	0.01094	kWh / nm^3	x	91.412	BTU /scf (gas)
Volume flow, output rates						
cfm	x	0.472	l / s	x	2.119	cfm
MIGD	x	4.546	1000 m^3 /d	x	0.220	MIGD
SCFD (gas)	x	0.027	1000 nm^3 /d	x	37.327	SCFD
MIGD: Million Imperial gallon per day (Water unit)						
SCFD: Standard cubic feet per day						
Viscosity						
1 St	x	$10^{\circ -4}$ m^2/s		x		
1 cSt	x	$10^{-6} m^2/s$		x		
Energy Prices						
US$ / MMBTU	x	0.948	US$ / GJ	x	1.055	US$ / MMBTU
US$ / MMBTU	x	3.412154	US$ / MWh	x	0.293	US$ / MMBTU
US$ / MMBTU	x	5.442	US$ / Bb	x	0.184	US$ / MMBTU
US$ / Bb	x	0.174	US$ / GJ	x	5.743	US$ / Bb
US$ / Bb	x	0.627	US$ / MWh	x	1.595	US$ / Bb

Annex 4: Properties of pure substances

Gas	Molar mass M kg /kmol	Gas constant R Nm / kg K	Density ρ_n 0°C /1.013 bar kg / m^3	Specific heat capacity c_{p0} kJ/kg K	Isentropic exponent κ $\kappa=c_p/c_v$
Atmospheric air, dry	28.96	287.1	1.29	1.00	1.40
Nitrogen N_2	28.01	296.8	1.25	1.04	1.40
Hydrogen H_2	2.02	4,124.8	0.09	14.30	1.41
Oxygen O_2	32.00	259.8	1.43	0.92	1.40
Carbon dioxide CO_2	44.01	188.9	1.98	0.84	1.31
Carbon on oxide CO	28.01	297.0	1.25	1.05	1.40
Methane CH_4	16.04	518.3	0.72	2.23	1.30
Butane C_4H_{10}	55.12	143.1	2.56	1.69	1.09
Propane C_3H_8	44.10	188.6	1.94	1.70	1.13
Water vapor H_2O	18.02	461.5	f(p,t)	1.86	f(p,t)

Important relations:	$c_p-c_v=R$	$\kappa=c_p/c_v$	$c_p=\kappa/(\kappa-1)\cdot R$	$c_v=1/(\kappa-1)\cdot R$	$\kappa=1/(1-R/c_p)$

Annex 5: Specific heat capacity of working fluids

Working fluid	c_p kJ / (kg K)	Note
Water		
1 bar/15 °C	4.189	Makeup water
15 bar/125 °C	4.252	Boiler feed water, industrial boilers
220 bar/180 °C	4.321	Boiler feed water, power plant boilers
Steam		
12 bar/250 °C	2.272	Industrial boilers
180 bar/540 °C	2.894	Live steam, power plants
Combustion air		
1 bar/25 °C	1.004	
1 bar/200 °C	1.011	
Flue gas		
1 bar/550 °C	1.027	Gas turbine exhaust gases
1 bar/120 °C	1.098	Chimney outlet

Annex 6: Chemical equation of combustion of pure fuels

Fuel		Oxygen		Combustion products		Reaction enthalpy	Unit
C	+	**O_2**	=	**CO_2**	+	**393.5**	**MJ**
1 kmol	+	1 kmol	=	1 kmol	+	393.5	MJ
12 kg	+	32 kg	=	44 kg	+	393.5	MJ
12 kg	+	22.4 mn3	=	22.4 mn3	+	393.5	MJ
1 kg	+	1.9 mn3	=	1.9 mn3	+	32.8	MJ/kg
S	+	**O_2**	=	**SO_2**	+	296.6	**MJ**
1 kmol	+	1 kmol	=	1 kmol	+	296.6	MJ
32 kg	+	32 kg	=	64 kg	+	296.6	MJ
32 kg	+	22.4 mn3	=	22.4 mn3	+	296.6	MJ
1 kg	+	0.70 mn3	=	0.70 mn3	+	9.3	MJ/kg
2 H_2	+	**O_2**	=	**2 H_2O**	+	241.8	**MJ**
2 kmol	+	1 kmol	=	2 kmol	+	241.8	MJ
44.8 m3 i. N.	+	22.4 mn3	=	44.8 mn3	+	241.8	MJ
1 m3 i. N.	+	0.5 mn3	=	1.0 mn3	+	5.4	MJ/nm3

Note: mn3 : Normal cubic meter

Annex 7. Heating values of selected fuel gases

Fuel gas	HHV MJ/m_n^3	LHV MJ/m_n^3
Hydrogen H_2	12.74	10.78
Carbon monoxide CO	12.63	12.63
Methane CH_4	39.82	35.88
Ethylene C_2H_4	63.41	59.46
Ethan C_2H_6	70.29	64.35
Azethylene C_2H_2	58.49	56.49
Propane C_3H_8	101.24	93.24
Butane C_4H_{10}	133.78	123.57

Annex 8: Properties of main combustion substances

Substamce	Symbol	Mol mass*) M kg / kmol	Normal density ρ_N kg / m_n^3	Normal spec heat c_{pN} kJ / (kg K) **)	Gas constant R Nm/ (kg K)
Solids					
Carbon, pure	C	12.00	-	-	-
Sulfur	S	32.00	-	-	-
Calcium	Ca	40.00	-	-	-
Combustion air					
Nitrogen	N_2	28.00	1.25	1.04	296.8
Oxygen	O_2	32.00	1.43	0.91	259.8
Hydrogen	H_2	2.00	0.09	14.05	4,116.0
Air, dry (21 % O_2; 79% N_2)	-	29.00	1.29	1.00	287.00
Flue gas					
Carbon dioxide	CO_2	44.00	1.98	0.83	188.9
Carbon monoxide	CO	28.00	1.25	1.04	296.80
Sulfur dioxide	SO_2	64.00	2.92	0.61	129.80
Nitrogen dioxide	NO_2	46.00	2.05		180.70
Ammonia	NH_3	17.00	0.77	2.18	488.2
Steam	H_2O	18.00	0.80	1.86	461.4

Note: Normal conditions($_N$) 1.013 bar; 0°C

Annex 9: Best practice air-to-fuel ratios and O_2 content of selected fuels

Combustion and Fuel Type	Air-to-Fuel ratio $\lambda = V_A/V_{A\_min}$ -	Corresponding O_2-content im flue gas %	V_{FGD}/V_A ratio min.
Solid fuels in boilers			
Hard coal	1.39	6.0	0.97
Lignite	1.39	6.0	0.97
Wood	1.40	6.0	0.99
Liquid fuels in boilers			
Light fuel oil	1.16	3.0	0.93
Heavy fuel oil	1.16	3.0	0.94
Gaseous fuels in boilers			
Natural gas H	1.15	3.0	0.91
Blast furnace gas	1.38	3.0	2.30
Converter gas	1.24	3.0	1.42
Coke oven gas	1.15	3.0	0.92
Mine gas	1.15	3.0	0.93
Sewage gas	1.16	3.0	0.95
Landfill gas	1.16	3.0	0.98
Gas turbines, natural gas	3.27	15.0	0.91
Engine, natural gas	1.28	5.0	0.91

V_A: Combustion air volume; V_{A_min}: minimum combustion air volume;
V_{FGD}: Dry flue gas volume

Annex 10: Fuel properties, HHV, LHV, V_A,V_{FG}, CO_2

Fuel	Trade	HHV		LHV		V_{A_min}		V_{FGW_min}	
	unit	MJ	kWh	MJ	kWh	m_n^3 pro		m_n^3 pro	
	3.6	per unit		per unit		unit	kWh_t	unit	kWh_t
Solid fuels									
carbon, pure	kg	32.80	9.11	32.80	9.11	8.89	0.98	8.89	0.98
Anthracitic coal	kg	31.53	8.76	30.95	8.60	8.40	0.98	8.57	1.00
Steam hard coal, 7 Mcal/kg	kg	30.32	8.42	29.27	8.13	7.97	0.98	8.31	1.02
Lignite, 30c, 50%w	kg	11.70	3.25	10.53	2.92	3.17	1.08	4.03	1.38
Wood, dry, 15% w	kg	15.49	4.30	14.21	3.95	3.83	0.97	4.56	1.15
Peat, dry	kg	19.44	5.40	18.09	5.03	5.03	1.00	5.79	1.15
Household carbage	kg	9.10	2.53	8.50	2.36	2.04	0.87	2.61	1.10
Liquid fuels									
crude oil	kg	44.5	12.4	41.9	11.6	11.5	0.99	12.2	1.05
light fuel oil	kg	45.4	12.6	42.2	11.7	11.1	0.95	11.8	1.01
light fuel oil	l	39.0	10.8	36.3	10.1	9.6	0.95	10.2	1.01
Heavy fuel oil	kg	42.3	11.8	40.2	11.2	10.6	0.95	11.2	1.00
Heavy fuel oil 380	kg	43.4	12.0	41.0	11.4	11.2	0.98	11.8	1.04
Propane, liquid C_3H_8	kg	50.1	13.9	46.2	12.8	11.8	0.9	12.8	1.0
Butane, liquid C_4H_{10}	kg	49.5	13.7	45.7	12.7	11.4	0.9	12.4	1.0
					-				
Gaseous fuels					-				
Hydrogen H_2	m_n^3	12.74	3.54	10.78	2.99	2.38	0.79	2.88	0.96
Methane CH_4	m_n^3	39.82	11.06	35.88	9.97	9.52	0.96	10.50	1.05
Natural gas L (Netherlands)	m_n^3	35.18	9.77	31.90	8.86	8.40	0.95	9.40	1.06
Natural gas H (Russia)	m_n^3	40.14	11.15	36.26	10.07	9.80	0.97	10.90	1.08
Propane, gas C_3H_8	m_n^3	101.24	28.12	93.24	25.90	23.80	0.92	25.80	1.00
Butane, gas C_4H_{10}	m_n^3	133.78	37.16	123.57	34.33	30.94	0.90	33.44	0.97
Blast furnace gas	m_n^3	4.04	1.12	3.99	1.11	0.64	0.58	1.51	1.36
Converter gas	m_n^3	8.15	2.26	8.10	2.25	1.54	0.68	2.21	0.98
Coke oven gas	m_n^3	20.59	5.72	18.36	5.10	4.58	0.90	5.32	1.04
Mine gas	m_n^3	29.87	8.30	26.91	7.48	6.90	0.92	7.90	1.06
Sewage gas	m_n^3	24.15	6.71	21.74	6.04	6.19	1.02	7.19	1.19
Landfill gas	m_n^3	21.90	6.08	19.73	5.48	5.24	0.96	6.24	1.14

Note: Data are calculated for the stated chemical composition, there can be significant deviation for several fuels

Annex 10 Fuel Properties, Continuation

Fuel	Trade unit	V_{FGD_min} m_n^3 pro		CO_{2_max} %	V_{Atr}/ V_{A_min}	CO_2-factors ref. to LHV*)	
	3.6	unit	kWh_t	in V_{FGD}	-	kg / GJ	kg / MWh_t
Solid fuels							
carbon, pure	kg	8.89	0.98	21.0%	1.00	108	390
Anthracitic coal	kg	8.28	0.96	19.6%	0.99	95	342
Steam hard coal, 7 Mcal/kg	kg	7.76	0.95	18.8%	0.97	95	342
Lignite, 30c, 50%w	kg	3.08	1.05	18.2%	0.97	114	410
Wood, dry, 15% w	kg	3.81	0.97	20.6%	0.99	0	0
Peat, dry	kg	4.86	0.97	16.9%	0.97	88	317
Household carbage	kg	2.20	0.93	19.6%	1.07	45	162
Liquid fuels							
crude oil	kg	10.9	0.93	16.1%	0.94	80	288
light fuel oil	kg	10.4	0.89	15.5%	0.93	74	266
light fuel oil	l	8.9	0.89	15.5%	0.93	74	266
Heavy fuel oil	kg	10.0	0.89	16.1%	0.94	78	281
Heavy fuel oil 380	kg	10.7	0.94	16.4%	0.95	78	281
Propane, liquid C_3H_8	kg	11.8	0.92	12.6%	1.00	64	230
Butane, liquid C_4H_{10}	kg	11.5	0.90	1.0%	1.00	64	230
Gaseous fuels							
Hydrogen H_2	m_n^3	1.88	0.63	-	0.79	0	0
Methane CH_4	m_n^3	8.52	0.85	0.12	0.89	55	198
Natural gas L (Netherlands)	m_n^3	7.70	0.87	0.12	0.92	56	202
Natural gas H (Russia)	m_n^3	8.90	0.88	0.12	0.91	56	202
Propane, gas C_3H_8	m_n^3	23.81	0.92	0.13	1.00	64	230
Butane, gas C_4H_{10}	m_n^3	30.95	0.90	0.01	1.00	64	230
Blast furnace gas	m_n^3	1.47	1.33	0.31	2.30	267	961
Converter gas	m_n^3	2.18	0.97	0.36	1.42	190	684
Coke oven gas	m_n^3	4.22	0.83	0.11	0.92	49	176
Mine gas	m_n^3	6.40	0.86	0.15	0.93	70	252
Sewage gas	m_n^3	5.89	0.98	0.17	0.95	0	0
Landfill gas	m_n^3	5.14	0.94	0.20	0.98	0	0

Annex 11: Image of a Modern Steam Power Plant – Walsum, Germany

Note: Cooling tower is also used as chimney

Annex 12: KPRO®ᵐ Model & simulation of 800 MW Rankine cycle

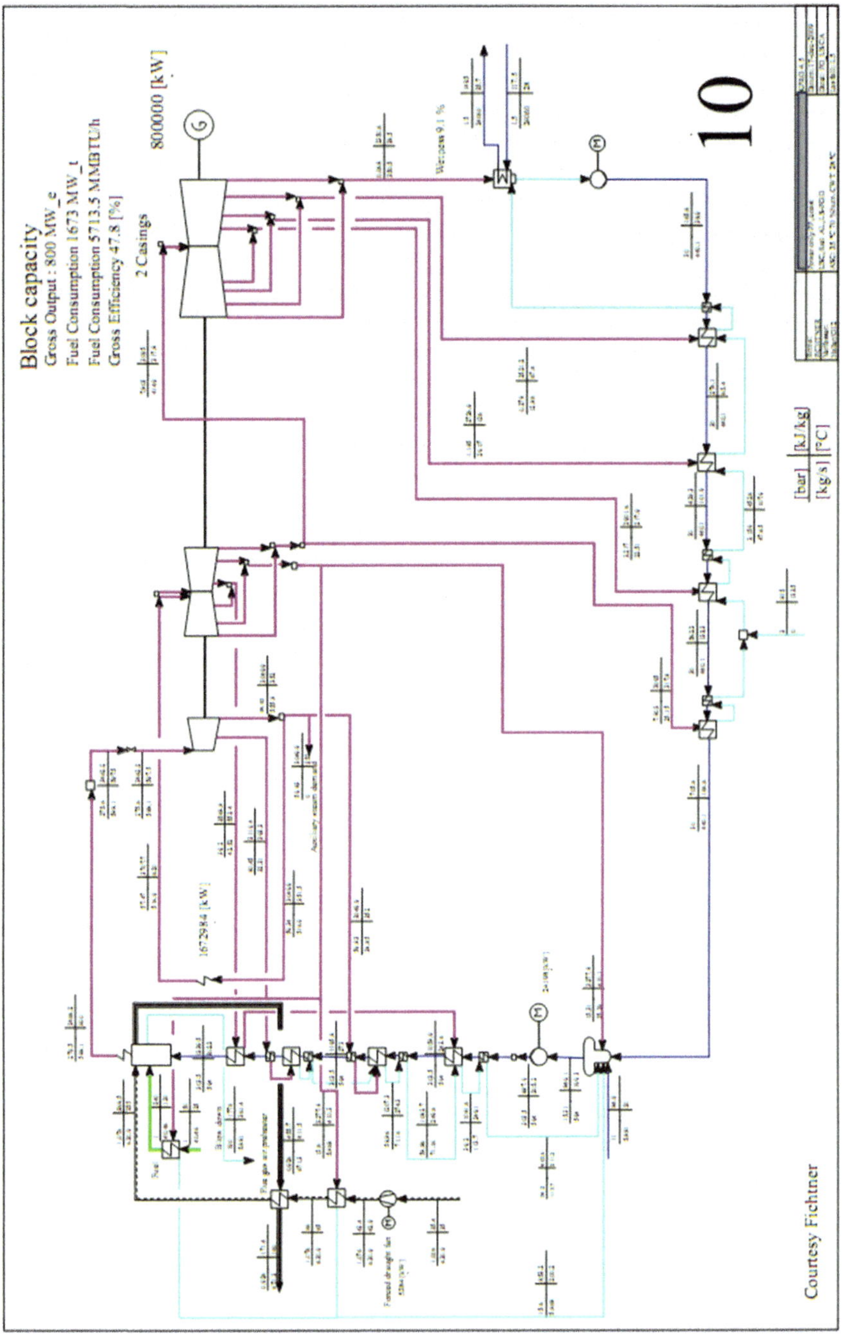

Annex 13: Schematic of the seawater FGD

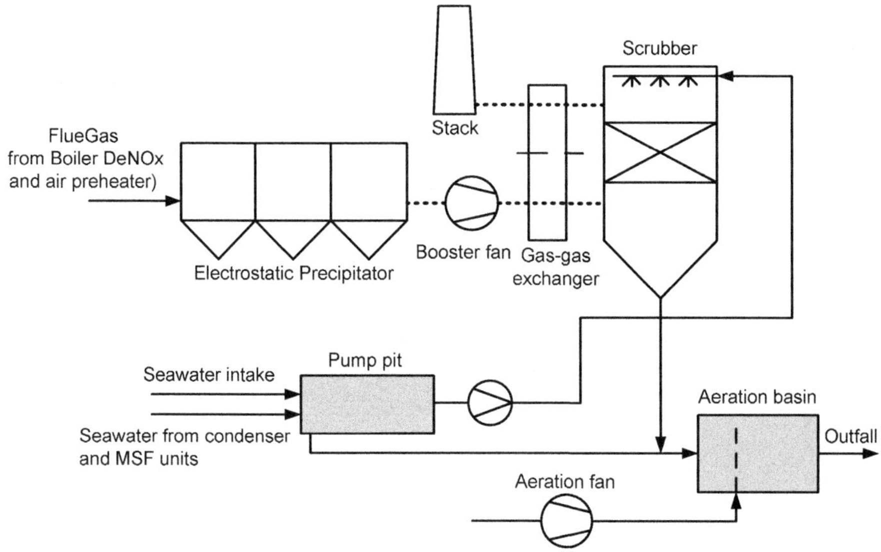

Annex 14: Schematic of the Limestone/Gypsum FGD with DeNox

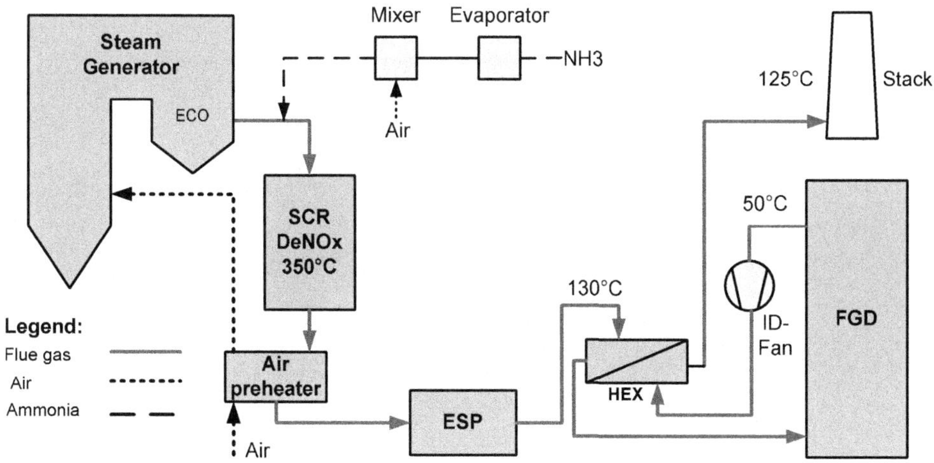

Annex 15: Schematic of the WSA/SWNOX technology

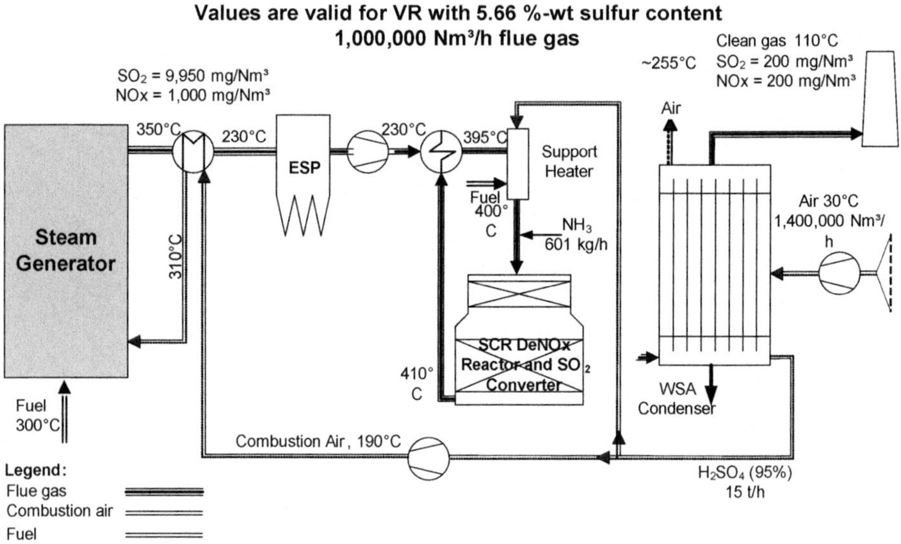

Annex 16: Combined Cycle Gas Turbine Power Plant

Source: Fichtner Archive

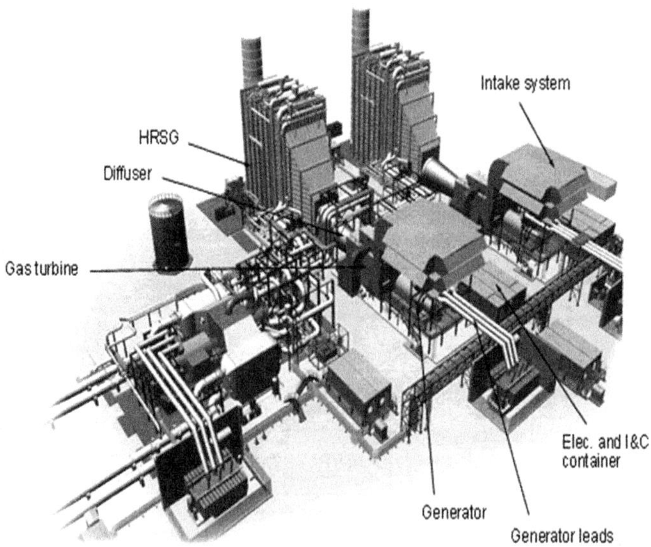

Annex 17: KPRO® m Model & simulation of 300 CCGT cycle

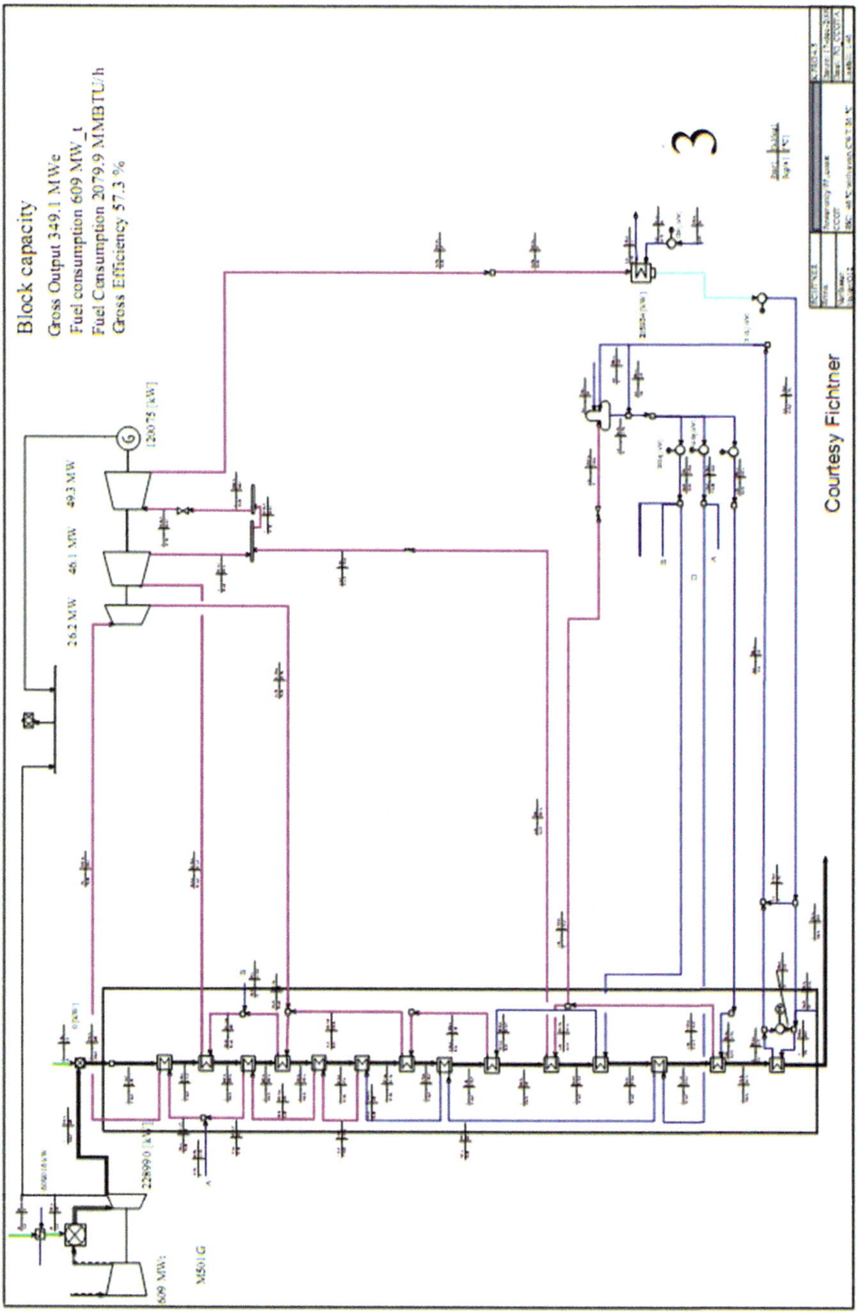

Annex 18: Guide values for capital expenditures for fossil fired PPs

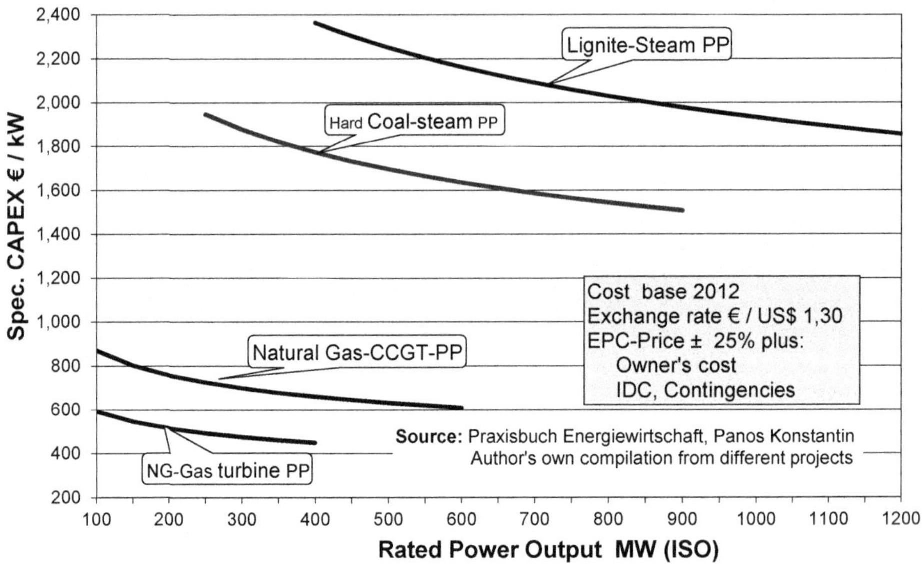

Annex 19: Guide values CAPEX for Gas turbines 1-60 MW

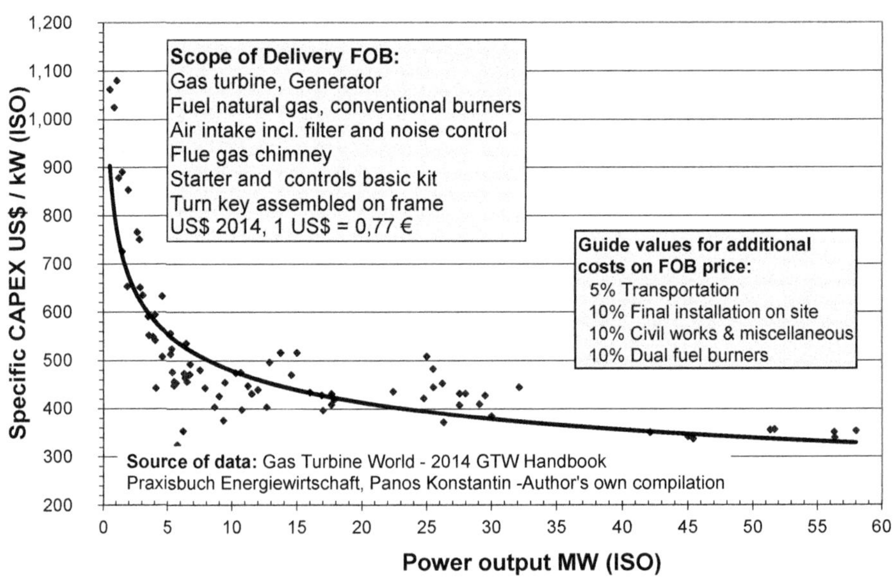

Annex 20: Guide values CAPEX for Gas turbine PPs

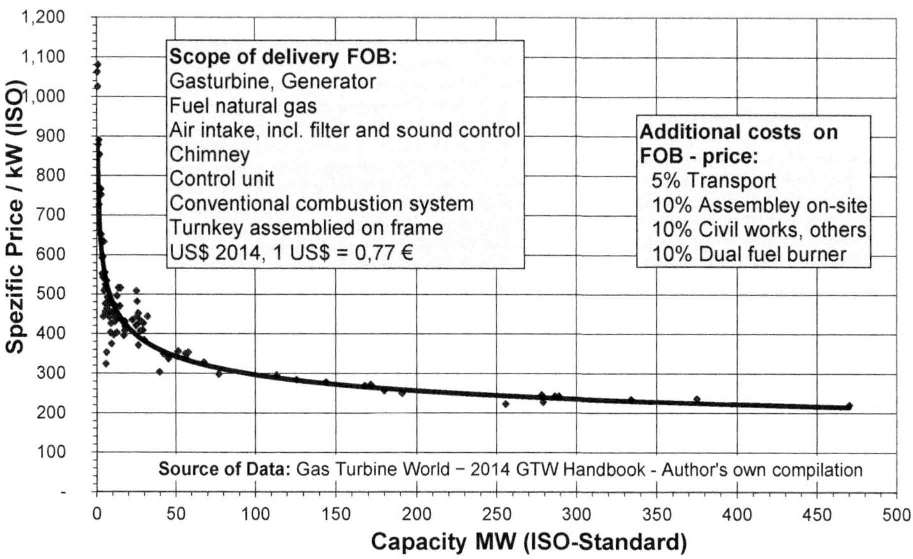

Annex 21: Guide values CAPEX of CCGT power plants

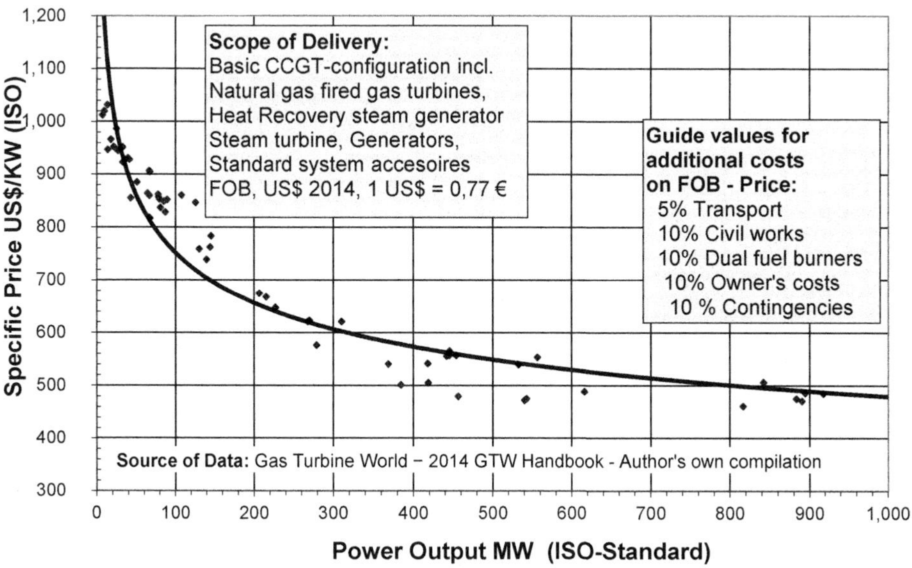

Annex 22: Roughness Classes and Roughness length

Roughness Class	Roughness Length z_0 m	Energy Index %	Landscape Type
0	0.0002	100	Water surface
0.5	0.0024	73	Completely open terrain with a smooth surface, e.g.concrete runways in airports, mowed grass, etc.
1	0.03	52	Open agricultural area without fences and hedgerows and very scattered buildings. Only softly rounded hills
1.5	0.055	45	Agricultural land with some houses and 8 metre tall sheltering hedgerows with a distance of approx. 1250 metres
2	0.1	39	Agricultural land with some houses and 8 metre tall sheltering hedgerows with a distance of approx. 500 metres
2.5	0.2	31	Agricultural land with many houses, shrubs and plants, or 8 metre tall sheltering hedgerows with a distance of approx. 250 metres
3	0.4	24	Villages, small towns, agricultural land with many or tall sheltering hedgerows, forests and very rough and uneven terrain
3.5	0.8	18	Larger cities with tall buildings
4	1.6	13	Very large cities with tall buildings and skycrapers

Note: Definitions according to the European Wind Atlas, **WAsP**.tion.

Source: Danish Wind Industry Association www.windpower.org/en/knowledge/windpower_wiki.html

Annex 23: Selected types of onshore wind turbines

Item		Unit	Technical Parameters					
Type			**GE 1.6-100**	**N117**	**V112 - 3075**	**SWT-3.3-130**	**3.4M114**	**E126EP4**
Manufacturer		-	GE	Nordex	Vestas	Siemens	Senvion	Enercon
Nominal Capacity		kW	1,600	2,400	3,075	3,300	3,400	4,200
Hub height		m	80/96	91/120/141	84/94/119	135	93/119/143	129/159
Wind class (IEC)		-	IB	IIIA	IIA	IA & IIA	IIA	IIA
Plant concept	variable spedd		variable					
	gear		yes	yes	yes	no	yes	no
Rotor								
Rotor diameter		m	100	116.8	112	130	114	127
Number of Blades		-	3					
Sweep area		m^2	7,850	10,715	9,852	13,300	10,207	12,661
Blade position			single blade pinch					
Generator		-						
Brake system	pitch control		yes					
Cut-in wind speed		m / s	3.5	3.0	3.0	3.5	3.0	3.0
Rated wind speed			11.0	7.5 - 13.2	12.0	12-13	12.0	-
Cut-out wind speed		m / s	25	20	25	25	22	25

Source: Manufacturers Datasheets

Annex 24: Selected types of offshore wind turbines

Item		Unit	Technical Parameters				
Type			**SWT-3.6-120**	**M5000-116**	**BARD 5.0**	**6M**	**V164-4C**
Manufacturer		-	Siemens	Areva	Bard	REpower	Vestas
Nominal Capacity		kW	3,600	5,000	5,000	6,150	8,000
Hub height		m	90	90/102	90	85 - 95	site specific
Wind class (IEC)		-	IA	IA	IC	IB	S
Plant concept	variable speed		yes	yes	-	yes	yes
	gear		yes				
Rotor							
Rotor diameter		m	120	116	122	126	164
Number of Blades		-	3				
Sweep area		m^2	11,300	10,568	11,690	12,469	21,124
Speed (variable)		U / min	5 - 13	5,9 - 14,8	not specified	12,1	4.8-12.1
Blade position							
Generator							
Brake system	pitch control		yes	redundant	not specified	yes	yes
Cut-in wind speed			3-5	4.0	3	-	4
Cut-out wind speed		m / s	25	25	25	30	25

Source: Manufacturer's Datasheet

Annex 25: Transportation of a wind rotor blade (60 m)

Annex 26: Shepherds Flat Wind Farm South

Project Name	Shepherds Flat Wind Farm
Facility	South
Country	Oregon, USA
Total Capacity	290 MW
Wind turbines	116 x GE 2.5 MW

Annex 27: Anholt Offshore Wind Power plant

Annex 28: Nor I Parabolic Trough Power Plant

Project name	NOR I
Country	Marocco
Technology	Prabolic trough
Power block capacity	160 MW
Solar field	537,000 Collectors
SCA Model	Sener trough
Solar radiation	2,635 kWh/(m^2 a)
Thermal storage	2-tank, molten salt
Commissioning	February 2016

Annex 29: Crescent Dunes Solar Power Project

Project name	Crecent Dunes
Country	Nevada, USA
Technology	Solar power tower
Power block capacity	110 MW
Steam Rankine cycle	115 bar live steam
Solar field area	1.197.148 m^2
Number of heliostats	10,347
Tower height	195 m (640 ft.)
Receiver temperature	288°C / 565°C
Solar radiation	2,685 kWh/(m^2 a)
Thermal storage	2-tank, molten salt, direct
Storage capacity	10 hours
Start of production	Nov-15

Source of data: NREL

Glossary

The terms and performance parameters used in this book are thoroughly described and explained in each chapter. In the following, the most important key-terms of each chapter are summarized, grouped in topics and listed according to their relevance, meaning and context. They are listed in non-alphabetic order and briefly explained.

P. Konstantin and M. Konstantin, *The Power Supply Industry*,
https://doi.org/10.1007/978-3-319-72305-1

Topic 11-1: Technical Thermodynamics at a Glance

Technical thermodynamics deals with processes of interconversion of energies into each other.

Energy E [J]**:** The ability of a system to produce technical work.

Forms of energy in physics: mechanical, thermal, potential (energy of height), chemical, nuclear.

According to the 1st law of thermodynamics, energy cannot be created or destroyed; it can only be converted into other forms of energy. The sum of energies of a closed system remains constant.

Forms of energy according to the stage of conversion:

Primary energy: The energy contained in natural resources as coal, oil, uranium, solar radiation, wind.

Final energy: The energy produced after conversion of primary energy in a technical process such as gasoline, diesel oil electricity. In simple words, the energy forms consumers buy on the market

Useful energy: The energy produced in end-use appliances from final energy such as space heating, electrical light, movement of tools.

Work W [J]**:** In a thermodynamic process or cycle, work is either generated by conversion of thermal energy into mechanical energy in a heat engine (e.g., internal ombustion engine) or consumed and converted into thermal energy in a driven engine (e.g., heat pump, or refrigerator). There are two forms:

Pressure-volume work used to expand the volume in a closed system

Shaft work transferred out or into a system (machine), also known as technical work

Power P [W]**:** is work divided by the time it has been performed (P=W/t)

Mass and weight: The amount of substance measured in kg. Weight is a force defined as mass kg x gravity m/s^2

Enthalpy H [J], specific h [J/kg]**:** Thermal energy contained in a mass flow.

Entropy S [J/K], specific s [J/kg K], is a property of state that never decreases in a closed system undergoing a process.

Exergy E [J] specific e [J/kg]**:** The maximum amount of mechanical work that an energy stream can produce in a reversible process against the environment. After reaching environmental level, exergy becomes zero. In contrast to energy, exergy is destructed in the course of a process.

Thermodynamic process is an operation in the course of which a thermodynamic system changes its properties of state (pressure, volume, temperature), from one equilibrium to another. They are distinguished upon the property of the system which remains constant (in brackets): isochoric (volume), isobaric (pressure), isothermal (temperature), adiabatic (heat transfer zero), polytropic, isenthalpic.

- A *reversible process* is a theoretical, idealized process that can be reversed without the expense of energy. Such a process is not feasible.
- All natural processes are *irreversible*.

Thermodynamic cycle: A thermodynamic cycle is a cycling succession of processes that involve changes of the properties of state of the *working fluid* (steam, gas, air). The cycle's result is generation or utilization of work.

An **Ideal gas** satisfies the equation of state $p \cdot V / T = \text{constant}$. *Real gases* satisfy the above equation only approximately. For technical applications, however, the accuracy margin is in most cases sufficient.

Topic 11-2: Techno-economics of Power Systems

Capital expenditures (CU): Acronym *CAPEX*, the initial capital outlay for an investment project to generate future returns.

Annualized CAPEX (CU/a): The initial capital outlay of a project converted into constant annual equivalent amounts (annuities) for a period equal to the lifetime of a project. Alternative term *capital costs*.

Operating expenses (CU/a): Acronym *OPEX*, Cash outflows during the operation phase of a project, e.g., for fuels, personnel, maintenance. Note: operating expenses are costs.

Costs CU/a: In general, regularly recurrent outlays such as operating expenses, corporate tax and also depreciation which is a non-cash item. Usually they are assumed to be due at the year's end. In general they are composed of *fixed costs* and *variable costs* (see definitions).

Specific cost (e.g. CU/unit): Annual costs divided by the production units (e.g. kWh) in a certain period (a). Alternative term used *per unit cost* (e.g. CU/kWh).

Fixed costs (CU/a): Costs that do not depend on the output level, e.g., cost of personnel, cost of maintenance, capital cost.

Variable costs (CU/a): Costs directly dependent on the output level such as fuel costs, costs for consumables and residues.

Capacity cost (CU/kWa): Fixed costs (CU/a) divided by the net power output (kW) in a certain period (a).

Energy cost alternative term **volume cost** (CU/kWh) or (CU/MWh) such as fuel costs plus non-fuel variable costs divided by the net electricity production (kWh) in the period (a).

Composite cost, specific (CU/kWh): Capacity cost (CU/kWa) plus energy cost (CU/kWh) converted to per production unit cost (CU/ kWh) – see item conversion functions below.

Revenues (CU/a): Price of product multiplied by the production amount.

Income, gross (CU/a): Revenues minus expenses.

Income, net (CU/a): Revenues minus operational expenses, interest on loans, and corporate tax.

Cashflow CU/a: The difference between revenues and costs; amount available for repayment of loans and dividends for equity investors (see also free cash flow).

Free cashflow (CU/a): Cash flow minus amortization of loans; amount available for dividend payments to equity investors and building of reserves to cover future costs.

Topic 11-3: Techno-economics of Power Systems – Cost Functions

Total cost(s) (CU/a): The sum (C_T) of fixed (C_F) and variable costs (C_V) for the production of given amount of the product (x); mathematically expressed it is:

$$C_T = C_F\left[\frac{\text{CU}}{\text{a}}\right] + C_V\left[\frac{\text{CU}}{\text{kWh}}\right]\times\left[\frac{\text{kWh}}{\text{a}}\right] \quad \left[\frac{\text{CU}}{\text{a}}\right]$$

Incremental cost(s) (CU/kWh): The difference in total costs (ΔC_T) caused by an increase or decrease of output (increment Δx) divided by associated number of units of output (Δx). The fixed costs remain thereby the same in absolute terms and influence the incremental costs (see equation below). Incremental costs are typically expressed on a per unit basis (see also total costs and marginal costs). Mathematically expressed it is:

$$\frac{\Delta C_T}{\Delta x} = \frac{C_F\left[\frac{\text{CU}}{\text{a}}\right] + C_V\left[\frac{\text{CU}}{\text{kWh}}\right]\cdot\Delta x\left[\frac{\text{kWh}}{\text{a}}\right]}{\Delta x\left[\frac{\text{kWh}}{\text{a}}\right]} \quad \left[\frac{\text{CU}}{\text{kWh}}\right]$$

Marginal cost(s) (CU/kWh): The change in (variable) costs for an increase of output (production) by one additional unit; the specific fixed costs do not influence the marginal cost. In mathematical terms marginal cost is the first derivative of the total cost:

$$\frac{dC_T}{dx} = C_V \quad \left[\frac{\text{CU}}{\text{kWh}}\right]$$

There is a distinction between *short run* marginal costs (SRMC) and *long run marginal costs* (LRMC). The definition above is referred to the former. LRMC include cost also for expansion of the production capacities to meet growing demand, e.g., in the course of expansion planning for a power system.

Composite electricity cost is the average cost in (CU/MWh) consisting of capacity cost and energy cost: Capacity cost is usually given in CU/kWa and Energy cost in CU/MWh. The conversion formula is:

$$Composite\ cost = \frac{Capacity\ cost\,[CU/(kWa)]}{Full\ load\ hours\,[h/a]} + Energy\ cost\,[CU/kWh]$$

Topic 11-4: Techno-economics of Power Systems – Operational terms

Operating time t_{OPH} (h/a): The time of the year where a power plant unit is performing generation function.

Full capacity hours t_{FCH} (h/a): Energy production of a power plant unit in a certain period of time, e.g., a year (kWh/a) divided by the rated capacity of the unit (kW).

Full load hours t_{FLH} (h/a): As above, however referred to the peak load of the grid: Often erroneously the term "full load hours" is also used instead of full capacity.

Equivalent operation hours EOH (h/a): The term is used in maintenance contracts. In addition to the operating hours it takes into account equivalent hours for start-ups, fuel changes and other operational parameters.

Capacity factor CF (-): Energy production of a power plant unit in a certain period of time, e.g., a year (kWh/a), divided by the hours (h/a) in this period. This corresponds to the average output (kW) during the period.

Load factor LF: As CF, however, referred to the peak load of the grid. Often the term "load factor" is erroneously used instead of CF.

Relation: $t_{FC} = CF \cdot (h/period)$.
e.g., period one year 8760 h/a CF=0.6→ t_{FC}= 0.6×8760= 5.256 h/a

Topic 11-5: Thermal power plants, fossil fueled

Thermodynamic cycles for power generation:

Steam Rankine Cycle: Consisting of a steam generator (boiler), a steam turbine as the prime mover of the generator.

Brayton or Joule Cycle: Using a gas turbine as the prime mover of the electric generator.

A *Combined Cycle* is a combination of a Brayton and a Rankine Cycle.

Steam generator (or boiler)**:** Main component of a Rankine Cycle consisting of economizer (feed water heater) evaporator, superheater and reheater.

Boiler feed water: High pressure water entering the steam generator (also known as boiler) consisting of condensate return and makeup water.

Makeup water: Chemically treated water for use in the boiler.

Deaerator: A power plant component removing air from condensate returned from the condenser before entering the boiler.

Simple Cycle Gas Turbine (SCGT) PP: A power plant using a gas turbine as the prime mover of the generator.

Combined Cycle Gas Turbine (CCGT): A power plant consisting of a gas turbine-generator set, heat recovery steam generator (HRSG) producing high pressure steam from the hot flue gases of the gas turbine, and a steam turbine generator.

Topic 11-6: Nuclear Power Plants

Pressurized Water Reactor (PWR): consisting of a primary high pressure water circuit placed within the containment and a secondary steam circuit with a steam turbine generator. The HP-water is heated up in the reactor vessel, where the nuclear fusion takes place, and gives up its heat in the steam generator heat exchanger.

Boiling water Reactor (BWR): Consists of only one steam circuit

Burnup: This is the equivalent to heating value of fossil fuels given in MWd/kg (instead of MWh/t in fossil fuels).

Topic 11-7: Power from Renewable Energy – Hydropower

Types of hydro power plants:

- Run-of-river power plants, base load duty
- Run-of-river pondage power plants, base load and peak
- Dam power plants, base load and peak
- Pump storage power plants, peak load

Hydro turbines:

- Kaplan turbines, used in run-of-river PPs, 0.1 MW to 50MW
- Francis turbines, for general use, 0.1 MW to 1000 MW
- Pelton turbines, used in pump-storage PPs, 0.1 MW to 400 MW

Topic 11-8: Power from Renewable Energy – Wind Power

Main components of wind turbines: Rotor, Rotor blades, Shaft, Nacelle casing, Hub, Tower, Electrical equipment

Performance curve of Wind Turbines (maximum 7.5 MW):

- Annual average wind speed (reference height 30 m)
- Wind speed at hub height
- Rated wind speed (power output constant)
- Cut-in wind speed, typically 3 m/s (start of power generation)
- Cut-off wind speed, typically 25 m/s (shut off of wind turbine)
- Performance coefficient c_p

Wind farms: Onshore, Offshore

Topic 11-9: Power from Renewable Energy – Solar Power

Solar Radiation or Insolation:
Radiance W/m^2, Irradiation $kWh/(m^2d)$ or $kWh/(m^2 a)$

Solar Irradiation:

- Direct Horizontal Irradiation DHI
- Diffuse Irradiation DIF
- Global Horizontal Irradiation GHI (DHI+DIF)

Direct Normal Irradiation DNI

Solar Power Plants:

- Photovoltaic PV, use both DHI and DIF
- Concentrated Solar Power CSP technologies, use DNI only
  - Parabolic Trough
  - Solar Tower
  - Fresnel

Topic 11-10: Cogeneration of Heat and Power

Combined Heat & Power plants - CHPs

- Internal combustion engines CHPs
- Gas turbine CHPs
- Steam turbine CHPs – Condensing-extraction, backpressure
- Combined Cycle Gas Turbine CCGT CHPs

Performance parameters of cogeneration

- Electricity-to-heat ratio σ kWh_e/kWh_t – Electricity generated by the extracted steam
- Electrical equivalent β kWh_e/kWh_t – Loss of electricity production caused by the steam extraction against condensation
- Total efficiency of cogeneration $\eta_{tot-cogen}$ - Electricity plus heat production divided by the fuel energy consumption for cogeneration

Acronyms, Abbreviations and Symbols

€	Euro
°F	Degrees Fahrenheit; F=9/5x°C+32
A	Annus (Latin), year
A	Acceleration, m/s^2
A	Area m^2
AC	Alternating current
acc.	According
AH	Arabian heavy crude oil
AL	Arabian light crude oil
AM	Relative air mass, Solar , length of sun's rays
a_n	Annuity factor
ANU	Annuity
API	Standard Coal Index
ARA	Amsterdam, Rotterdam Antwerp
BAFA	Bundesanstalt für Wirtschaft und Ausfuhrkontrolle
BAFA	German Federal Agency for Export Control
bbl	Barrel oil
bcm	Billion cubic meter
Bps	Basis point 1%=100 Bps, 0.01% = 1 Bps
BWR	Boiling water reactor
CAPEX	Capital expenditures
CCGT	Combined cycle gas turbine power plant
CF	Capacity factor (see also LF load factor)
CHP	Combined heat and power plant
CIF	Cost insurance freight (oversea shipping)
COGEN	Cogeneration of power and heat cycle
COP	Performance ratio of refrigeration
cosφ	Power factor $cos\varphi=P/S$

P. Konstantin and M. Konstantin, *The Power Supply Industry*,
https://doi.org/10.1007/978-3-319-72305-1

c_p	Specific heat capacity at constant pressure [J/kg K]
CP	Critical point of water 8221bar/374°C)
c_p	Performance coefficient of wind
c_{p_betz}	Betz performance coefficient
CPI	Consumer price index
CRA	Credit Rating Agency
CSP	Concentrated Solar Power
Ct	Cent (USD or Euro)
CU	Currency unit use in formulas as neutral term
c_v	Specific heat capacity at constant volume [J/kg K]
D	Diameter m
DC	Direct current
DCF	Discounted cash flow
DeNox	Denitrification plant (NOx reduction)
DHI	Direct horizontal irradiation
DIF	Diffuse irradiation
DisCo	Distribution System Company
DLR	German Aerospace Center
DLR	Deutsches Zentrum für Luft und Raumfahrt
DNI	Direct normal irradiation kWh/m^2 a
DSCR	Debt service coverage ratio
DWT	Deadweight tonnage (ship transport capacity)
E	Energy, J, kWh
E	East
e.g.	Latin “exempli gratia”; stands “for example” see also “i.e.”
E; e	Exergy kJ; kJ/kg
EBIDA	Earnings before interest depreciation and amortization
EBIT	Earnings before interest and taxes
EBITDA	Earnings before interest taxes depreciation amortization
EEX	European Power Exchange
EEX	European Energy Exchange
EHV	Extra High Voltage, 230 to 765 kV
EnBW	Energie Baden Württemberg (German Utility)
ENTSO	Transmission System Operator for Electricity (Europe)
EOH	Equivalent Operating Hours (incl. start-ups etc.)
EPC	Engineering Procurement Construction contract

EPC	Engineering Procurement Construction
ESP	Electrostatic Precipitator (Dust Filter)
E_t	Expenses at time t
etc.	Et cetera (Latin: and all other things, and so on)
Evap	Evaporation cooling process for Gas Turbine inlet air
EURO	Euro currency, symbol €
Excel®	Table calculation software (Microsoft)
F	Force, N
FIDIC	Féderation Internationale des Ingénieurs Conseils (French)
FIDIC	International Federation of Consulting Engineers (English)
FGC	Flue Gas Cleaning
FGD	Flue Gas Desulphurization
FluidEXL	Software tool for water/steam parameters calculation
FOB	Free on board, overseas shipping
FV	Future value
G	Acceleration of gravity (fall) [m/s^2]
Gcal	Giga calorie, see also kcal
GCV	Gross calorific value, see also NCV, LHV
GenCo	Generation System Company
GHI	Global horizontal irradiation
GIS	Gas Insulated Switch gear
GPV	Gross product value (refinery term)
GWh	Gigawatt hours
H	Head of water (Pressure) m
H, h	H Enthalpy [J], h specific [J/kg]
HEX	Heat Exchanger
HEX	Heat exchanger
HFO	Heavy Fuel Oil
H_G	Global irradiation on tilted plain kWh/m^2 a
Hi	Inferior heating value (scientific term), same LHV, NCV
Hl	Hectoliter (100 liter oil)
H_o	Oberer Heizwert (higher heating value)
HOB	Heat Only Boiler
HP	High pressure
HRSG	Heat Recovery Steam Generator
Hs	Superior heating value (scientific term); same HHV; GCV

HTF	Heat Transfer Fluid (solar)
Hu	Unterer Heizwert (lower heating value), same LHV, NCV
HV	High Voltage, 69 to 230 kV
HVAC	High Voltage Alternating Current
HVDC	High Voltage Direct Current
Hz	Hertz, frequency cycle /s
I	Electric current A
I	Interest rate %/a
I	Ampere, electrical current A
i.e.	Latin "it est"; stands for "that is", see also "e.g."
IAEA	International Atomic Energy Agency
IAM	Incident angle modifier
ICE	Internal combustion engine
IDC	Interest during construction
IEA	International Energy Agency
IEC	International Electro technical Commission
i_{eff}	Effective interest rate %/a
i_n; i_r	Nominal, real interest rate %/a
I_o	Capital expenditures for investment
IPP	Independent power producer
IPWP	Independent power and water producer
i_r; i_n	Real, nominal interest rate %/a
IRR	Internal rate of return
IRROE	Internal rate of return on equity
IRROI	Internal rate of return on investment
ISCC	Integrated Solar Combined Cycle
ISO	International Standard Organization
J	Escalation rate %/a, on top of inflation
Kcal	Kilo calorie, old thermal unit, 4.187 Joule
KPRO®	Cycle simulation software tool (Fichtner)
kVA	Kilo Volt Ampere, apparent power
kW	Kilowatt; multiples: MW, GW, TW
kWh	Kilowatt hour, Multiples: MWh, GWh, TWh
L	Length, distance, m
lb	Pound, English weight unit
LEC	Levelized electricity cost

LF	Load factor, see also CF capacity factor
LFO	Light fuel oil
LFR	Linear Fresnel Reflectors
LLCR	Loan life coverage ratio
LNG	Liquified natural gas
LOLP	Loss of Load Probability
Lot	Contract for delivery of good or works (e.g., civil works)
LP	Low pressure, see also HP high, IP medium pressure
LPG	Liquified petroleum gas
LRMC	Long Run Marginal Costs
LV	>0.6 kV
LWR	Light water reactor
M	Mass, kg
M	Molar mass, kg/Mol
Max	Maximum
MBD	Million barrels per day
MCP	Market Clearing Price
Min	Minimum
Mln	Million
MP	Medium pressure, see also HP, LP
MPP	Maximum Power Point (PV modules)
MV	Medium Voltage, 0.6 to 69 kV
MVA	Mega Volt Ampere, apparent power
MW	Megawatt
MWh	Megawatt hours
N	Number of years lifetime
N	Amount of substance, n=m/M
N	North
NAR	Net as received (coal term)
NCV	Net calorific value, see also LHV: lower heating value
NG	Natural gas
NPC	Net Present cost
NPV	Net present value
NREL	National Energy Laboratory, USA
NSCR	Non-Selective Catalytic Reduction (CO, NOx, CHm)
NYMEX	New York Mercantile Exchange

O&M	Operation and maintenance
OECD	Organization of Economic Cooperation and Development
OPEC	Organization of oil Exporting Countries
OPEX	Operation expenses
OTC	Over the Counter
P	Active power $P=U \cdot I\ cos\varphi$
P	Power kW, MW, GW
P	Pressure, Pa, bar
p Escalation	
P&L	Profit & Loss statement
p_g	Gage pressure
PLCR	Project life coverage ratio
PMBOK	Project Management Body of Knowledge
PN	Pressure normed
p_o	Over pressure; $p_o=p_b+p_g$
PP	Power plant
p_p	Barometric pressure
PPI	Producer price index
PPP	Pool Purchase Price
PR	Annual performance ratio % (PV plants)
PSP	Pool Selling Price
PV	Present value
PV	Photovoltaics
PWR	Pressurized water reactor
PXX	Probability ,e.g., P50, P90
Q	Reactive power Var
Q	Discount factor (1+ discount rate *i* %)
R	Electric resistance Ω, kΩ
R	Gas constant, specific [J/kg K]
$\bar{R}$	Universal gas constant [J/mol K]
rmp	Revolutions per minute
rms	Root mean square value, e.g., electric current I_{rms} voltage V_{rms}
ROI	Return on investment
RSC	Reference site Conditions
R_t	Revenues at time t
S	Apparent power VA, kVA, MVA

S, s	S entropy J/K, s specific J/kg K
SC	Super Critical steam parameters
SCR	Selective Catalytic Reduction (removes NOx only)
SG	Steam generator (boiler)
SMP	System Marginal Price
SNOX	Flue Gas Cleaning process (combined FGD & DeNox)
SRMC	Short Run Marginal Costs
SSG	Solar Steam Generator
SSGT	Single Cycle Gas Turbine power plant
STC	Standard Test Conditions (1000 W/m^2, AM 1.5)
SubC	Sub Critical steam parameter
SWOT	Strengths, Weaknesses, Opportunities, Threats analysis
T	Period s, a complete sinus wave of current or voltage
T	Thermodynamic temperature, K (Kelvin); T=°C+273
T	Time (s, h, a)
T&C	Transfer and Convertibility (Risk)
tce	Tons of coal equivalent
TES	Thermal Energy Storage
t_{FLH}	Full load hours, equivalent hours referred to the grid's peak
t_o	Reference time for discounting
t_{oe}	Tons of oil equivalent, one metric ton oil
TOP	Take or Pay, term commonly used for imported natural gas
TOR	Terms of Reference (Definition of scope in inquiries)
TP	Triple point of water (0.006bar/0°C)
t_{pb}	Payback period
Trafo	Electric transformer
TransCo	Transmission System Company
TSO	Transmission System Operator
U	Voltage V, kV
U, u	U Internal energy J, u specific J/kg
U_3O_8	Uranium Oxide, trade name yellow cake
UCTE	Union for Coordination of the Transmission of Electricity
UF_6	Uranium Hexafluorid
UHV	Ultra High Voltage, 765 to 1100 kV
UK	United Kingdom
UO_2	Uranium Oxide

US$	US Dollar
USA	United States of America
USC	Use of System Cost (fees for using electrical networks)
USC	Ultra Super Critical steam parameters
UST	Use of System Tariff
V_A	Combustion air volume
V_{FG}	Flue gas volume (V_{FGD}:dry, V_{FGW}:wet)
W	Speed of wind m/s
W	Work, J, kWh
WACC	Weighted Average Cost of Capital
WACC	Weighted average cost of capital
WB	World Bank
W_e	Electricity production (kWh/a)
WT	Wind turbine
WTI	West Texas Intermediate (crude oil quality)
Yr	Year
z_0	Roughness length terrain

Greek Characters

α,β,γ	Angles of triangle
a_s°	Sun's altitude or elevation
β	Electrical equivalent of extracted steam [kWh_e/kWh_t]
$\gamma_{s°}$	Solar azimuth
δ°	Sun's declination
Δ	Difference
ΔE	Difference of Expenses
ΔI	Difference of capital expenditures
η	Efficiency %
θ	Temperature °C
κ	Isentropic exponent, process of ideal gas
λ°	Longitude, e.g., 30° E (E for East)
μ	mi, Symbol for mean value (Gauß distribution)
π	Value 3.14 for circle calculations
ρ	Density kg/m^3
ρ	Density kg/m^3
σ	Electricity-to-heat ration of extracted steam [kWh_e/kWh_t]

σ	Standard deviation, statistics
Σ	Symbol for Sum
υ	Velocity m/s
υ	Specific volume m^3/kg
$\overline{\upsilon}$	Volume of ideal gas, 22.41 $m^3/kmol$
$\varphi°$	Phase difference (electricity)
$\varphi°$	Latitude, e.g., 30° N (N for north)
Φ	Flow rate e.g.: Φ_m:kg/s; Φ_Q:J/s; Φ_V:m^3/s

Index

P. Konstantin and M. Konstantin, *The Power Supply Industry*,
https://doi.org/10.1007/978-3-319-72305-1

The manufacturer's authorised representative in the EU is Springer Nature Customer Service Centre GmbH, Europaplatz 3, 69115 Heidelberg, Germany. If you have any concerns regarding our products, please contact ProductSafety@springernature.com

Printed and bound by CPI Group (UK) Ltd, Croydon, CR0 4YY
01/10/2026
02225871-0010